Vector Optimization

Springer
*Berlin
Heidelberg
New York
Hong Kong
London
Milan
Paris
Tokyo*

Johannes Jahn

Vector Optimization

Theory,
Applications,
and Extensions

With 62 Figures
and 14 Tables

 Springer

Professor Dr. Johannes Jahn
Universität Erlangen-Nürnberg
Naturwissenschaftliche Fakultät I
Institut für Angewandte Mathematik
Martensstraße 3
91058 Erlangen, Germany
jahn@am.uni-erlangen.de

ISBN 3-540-20615-9 Springer-Verlag Berlin Heidelberg New York

Cataloging-in-Publication Data applied for
A catalog record for this book is available from the Library of Congress.

Bibliographic information published by Die Deutsche Bibliothek
Die Deutsche Bibliothek lists this publication in the Deutsche Nationalbibliografie;
detailed bibliographic data available in the internet at *http.//dnb.ddb.de*

This work is subject to copyright. All rights are reserved, whether the whole or part of the material is concerned, specifically the rights of translation, reprinting, reuse of illustrations, recitation, broadcasting, reproduction on microfilm or in any other way, and storage in data banks. Duplication of this publication or parts thereof is permitted only under the provisions of the German Copyright Law of September 9, 1965, in its current version, and permission for use must always be obtained from Springer-Verlag. Violations are liable for prosecution under the German Copyright Law.

Springer-Verlag is a part of Springer Science+Business Media
springeronline.com

© Springer-Verlag Berlin Heidelberg 2004
Printed in Germany

The use of general descriptive names, registered names, trademarks, etc. in this publication does not imply, even in the absence of a specific statement, that such names are exempt from the relevant protective laws and regulations and therefore free for general use.

Cover design: Erich Kirchner, Heidelberg

SPIN 10973967   42/3130 – 5 4 3 2 1 0 – Printed on acid-free paper

Preface

In vector optimization one investigates optimal elements such as minimal, strongly minimal, properly minimal or weakly minimal elements of a nonempty subset of a partially ordered linear space. The problem of determining at least one of these optimal elements, if they exist at all, is also called a vector optimization problem. Problems of this type can be found not only in mathematics but also in engineering and economics. Vector optimization problems arise, for example, in functional analysis (the Hahn-Banach theorem, the lemma of Bishop-Phelps, Ekeland's variational principle), multiobjective programming, multi-criteria decision making, statistics (Bayes solutions, theory of tests, minimal covariance matrices), approximation theory (location theory, simultaneous approximation, solution of boundary value problems) and cooperative game theory (cooperative n player differential games and, as a special case, optimal control problems). In the last decade vector optimization has been extended to problems with set-valued maps. This new field of research, called set optimization, seems to have important applications to variational inequalities and optimization problems with multivalued data.

The roots of vector optimization go back to F.Y. Edgeworth (1881) and V. Pareto (1896) who has already given the definition of the standard optimality concept in multiobjective optimization. But in mathematics this branch of optimization has started with the legendary paper of H.W. Kuhn and A.W. Tucker (1951). Since about the end of the 60's research is intensively made in vector optimization.

It is the aim of this book to present various basic and important results of vector optimization in a general mathematical setting and to demonstrate its usefulness in mathematics and engineering. An extension to set optimization is also given. The first three parts are a revised edition of the former book [145] of the author. The forth part on engineering applications and the fifth part entitled extensions to set optimization are new material.

The theoretical vector optimization results are contained in the second part of this book. For a better understanding of the proofs several theorems of convex analysis are recalled in the first part. This part concisely summarizes the necessary background material and may be viewed as an appendix.

The main part of this book begins on page 102 with a discussion of several optimality notions together with some simple relations. Necessary and sufficient conditions for optimal elements are obtained by scalarization, i.e. the original vector optimization problem is replaced by an optimization problem with a real-valued objective map. The scalarizing functionals being used are certain linear functionals and norms. Existence theorems for optimal elements are proved using Zorn's lemma and the scalarization theory. For vector optimization problems with inequality and equality constraints a generalized Lagrange multiplier rule is given. Moreover, a duality theory is developed for convex maps. These results are also specialized to abstract linear optimization problems. The third part of this book is devoted to the application of the preceding general theory. For vector approximation problems the connections to simultaneous approximation problems are shown and a generalized Kolmogorov condition is formulated. Furthermore, nonlinear and linear Chebyshev problems are considered in detail. The last section is entitled cooperative n player differential games. These include optimal control problems. For these games a maximum principle is proved.

In the part on engineering applications the developed theoretical results are applied to multiobjective optimization problems arising in engineering. After a presentation of the theoretical basics of multiobjective optimization numerical methods are discussed. Some of these methods are applied to concrete nonlinear multiobjective optimization problems from electrical engineering, computer science and chemical engineering. The last part extends the second part of this book to set optimization. After an introduction to this field of research including basic concepts the notion of the contingent epiderivative is discussed in detail. Subdifferentials are the topic together with a comprehensive chapter on optimality conditions in set optimization.

This book should be readable for students in mathematics whose background includes a basic knowledge in optimization and linear functional analysis. Mathematically oriented engineers may be interested in the forth part on engineering applications.

The bibliography contains only a selection of references. A reader who is interested in further topics of vector optimization is requested to consult the extensive bibliographies of Achilles-Elster-Nehse [79], Nehse [82] and Stadler [84].

I am very grateful to Professors W. Krabs, R.H. Martin and B. Brosowski for their support and valuable suggestions. Moreover, I am indebted to A. Garhammer, S. Gmeiner, Dr. J. Klose, A. Merkel, B. Pfeiffer and H. Winkler for their assistance.

Erlangen, November 2003   Johannes Jahn

Contents

Preface   v

I   Convex Analysis   1

1 Linear Spaces   3
   1.1   Linear Spaces and Convex Sets 3
   1.2   Partially Ordered Linear Spaces 12
   1.3   Topological Linear Spaces 21
   1.4   Some Examples 32
   Notes 35

2 Maps on Linear Spaces   37
   2.1   Convex Maps 37
   2.2   Differentiable Maps 45
   Notes 59

3 Some Fundamental Theorems   61
   3.1   Zorn's Lemma and the Hahn-Banach Theorem 61
   3.2   Separation Theorems 71
   3.3   A James Theorem 81
   3.4   Two Krein-Rutman Theorems 87
   3.5   Contingent Cones and a Lyusternik Theorem 90
   Notes 99

II Theory of Vector Optimization — 101

4 Optimality Notions — 103
Notes 113

5 Scalarization — 115
5.1 Necessary Conditions for Optimal Elements of a Set . . 115
5.2 Sufficient Conditions for Optimal Elements of a Set . . 129
5.3 Parametric Approximation Problems 139
Notes 147

6 Existence Theorems — 149
Notes 159

7 Generalized Lagrange Multiplier Rule — 161
7.1 Necessary Conditions for Minimal and Weakly Minimal Elements 161
7.2 Sufficient Conditions for Minimal and Weakly Minimal Elements 174
    7.2.1 Generalized Quasiconvex Maps 174
    7.2.2 Sufficiency of the Generalized Multiplier Rule . 181
Notes 187

8 Duality — 189
8.1 A General Duality Principle 189
8.2 Duality Theorems for Abstract Optimization Problems 192
8.3 Specialization to Abstract Linear Optimization Problems 200
Notes 207

III Mathematical Applications — 209

9 Vector Approximation — 211
9.1 Introduction 211
9.2 Simultaneous Approximation 213
9.3 Generalized Kolmogorov Condition 216
9.4 Nonlinear Chebyshev Vector Approximation 218
9.5 Linear Chebyshev Vector Approximation 226

| | 9.5.1 Duality Results . . . . 227 |
| | 9.5.2 An Alternation Theorem . . . . 233 |
| Notes . . . . 241 |

10 Cooperative n Player Differential Games   243
10.1 Basic Remarks on the Cooperation Concept 243
10.2 A Maximum Principle 245
    10.2.1 Necessary Conditions for Optimal and Weakly Optimal Controls 247
    10.2.2 Sufficient Conditions for Optimal and Weakly Optimal Controls 259
10.3 A Special Cooperative n Player Differential Game . . . 270
Notes 277

IV   Engineering Applications   279

11 Theoretical Basics of Multiobjective Optimization   281
11.1 Basic Concepts 281
11.2 Special Scalarization Results 291
    11.2.1 Weighted Sum Approach 292
    11.2.2 Weighted Chebyshev Norm Approach 304
    11.2.3 Special Scalar Problems 307
Notes 311

12 Numerical Methods   313
12.1 Modified Polak Method 313
12.2 Interactive Methods 319
    12.2.1 Modified STEM Method 319
    12.2.2 Method of Reference Point Approximation . . . 323
               The Linear Case 325
               The Bicriterial Nonlinear Case 331
12.3 Method for Discrete Problems 336
Notes 339

13 Multiobjective Design Problems   341
13.1 Design of Antennas 342
13.2 Design of FDDI Computer Networks 349

13.2.1 A Cooperative Game	350
13.2.2 Minimization of Mean Waiting Times	352
13.2.3 Numerical Results	355
13.3 Fluidized Reactor-Heater System	357
13.3.1 Simplification of the Constraints	359
13.3.2 Numerical Results	361
13.4 A Cross-Current Multistage Extraction Process	363
Notes	366

V Extensions to Set Optimization 369

14 Basic Concepts and Results of Set Optimization 371
Notes 377

15 Contingent Epiderivatives 379
15.1 Contingent Derivatives and Contingent Epiderivatives . 379
15.2 Properties of Contingent Epiderivatives 383
15.3 Contingent Epiderivatives of Real-Valued Functions . . 387
15.4 Generalized Contingent Epiderivatives 391
Notes 395

16 Subdifferential 397
16.1 Concept of Subdifferential 397
16.2 Properties of the Subdifferential 399
16.3 Weak Subgradients 403
Notes 407

17 Optimality Conditions 409
17.1 Optimality Conditions with Contingent Epiderivatives . 409
17.2 Optimality Conditions with Subgradients 414
17.3 Optimality Conditions with Weak Subgradients 415
17.4 Generalized Lagrange Multiplier Rule 417
   17.4.1 A Necessary Optimality Condition 418
   17.4.2 A Sufficient Optimality Condition 427
Notes 432

Bibliography 435

List of Symbols	**459**
Index	**461**

Part I
Convex Analysis

Convex analysis turns out to be a powerful tool for the investigation of vector optimization problems in a partially ordered linear space for two main reasons. A partial ordering in a real linear space can be characterized by a convex cone and therefore theorems concerning convex cones are very useful. Furthermore, separation theorems are especially helpful for the development of a Lagrangian theory. In this first part which consists of three chapters we present all these results on convex analysis which are necessary for the following theory on vector optimization. The most important theorems are separation theorems, a James theorem and a Krein-Rutman theorem.

Chapter 1

Linear Spaces

Although several results of the theory described in the second part of this book are also valid in a rather abstract setting we restrict our attention to real linear spaces. For convenience, we summarize in this chapter the well-known definitions of linear spaces and convex sets as well as the definition of (locally convex) topological linear spaces and we consider a partial ordering in such a linear setting. Finally, we investigate some special partially ordered linear spaces and list various known properties.

1.1 Linear Spaces and Convex Sets

We recall the definition of a real linear space and present some other notations.

Definition 1.1. Let X be a given set. Assume that an addition on X, i.e. a mapping from $X \times X$ to X, and a scalar multiplication on X, i.e. a mapping from $\mathbb{R} \times X$ to X, is defined. The set X is called a *real linear space*, if the following axioms are satisfied (for arbitrary $x, y, z \in X$ and $\lambda, \mu \in \mathbb{R}$):

(a) $(x + y) + z = x + (y + z)$,

(b) $x + y = y + x$,

(c) there is an element $0_X \in X$ with $x + 0_X = x$ for all $x \in X$,

(d) for every $x \in X$ there is a $y \in X$ with $x + y = 0_X$,

(e) $\lambda(x + y) = \lambda x + \lambda y$,

(f) $(\lambda + \mu)x = \lambda x + \mu x$,

(g) $\lambda(\mu x) = (\lambda\mu)x$,

(h) $1x = x$.

The element 0_X given under (c) is called the *zero element* of X.

Definition 1.2. Let S and T be nonempty subsets of a real linear space X. Then we define the *algebraic sum* of S and T as

$$S + T := \{x + y \mid x \in S \text{ and } y \in T\}$$

and the *algebraic difference* of S and T as

$$S - T := \{x - y \mid x \in S \text{ and } y \in T\}.$$

For an arbitrary $\lambda \in \mathbb{R}$ the notation λS will be used as

$$\lambda S := \{\lambda x \mid x \in S\}.$$

It is important to note that the set equation $S + S = 2S$ does not hold in general for a nonempty subset S of a real linear space.

Definition 1.3. Let X be a real linear space. The set X' is defined to be the set of all linear mappings from X to $\mathbb{R}$. If we define for all $\varphi, \psi \in X'$ and all $\lambda \in \mathbb{R}$

$$(\varphi + \psi)(x) = \varphi(x) + \psi(x) \quad \text{for all } x \in X$$

and

$$(\lambda\varphi)(x) = \lambda\varphi(x) \quad \text{for all } x \in X,$$

then X' is a real linear space itself and it is called the *algebraic dual space* of X. The algebraic dual space of X' is denoted by X'' and it is called the *second algebraic dual space* of X.

1.1. Linear Spaces and Convex Sets

The most important class of subsets in a real linear space are convex sets.

Definition 1.4. Let S be a subset of a real linear space X.

(a) Let some $x \in S$ be given. The set S is called *starshaped* at $\bar{x}$, if for every $x \in S$
$$\lambda x + (1-\lambda)\bar{x} \in S \quad \text{for all } \lambda \in [0,1]$$
(see Fig. 1.1).

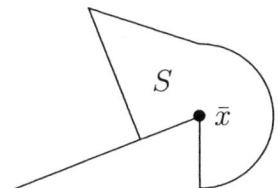

Figure 1.1: A set S being starshaped at $\bar{x}$.

(b) The set S is called *convex*, if for every $x, y \in S$
$$\lambda x + (1-\lambda)y \in S \quad \text{for all } \lambda \in [0,1]$$
(see Fig. 1.2 and 1.3).

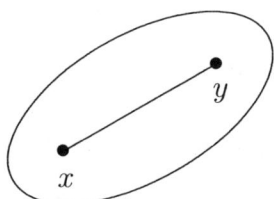 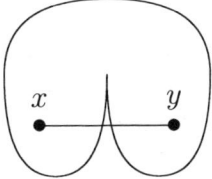

Figure 1.2: Convex set.   Figure 1.3: Non-convex set.

(c) The set S is called *balanced*, if it is nonempty and
$$\alpha S \subset S \quad \text{for all } \alpha \in [-1, 1].$$

(d) The set S is called *absolutely convex*, if it is convex and balanced.

Obviously, the empty set is convex and a set which is starshaped at every point is convex as well.

Remark 1.5. Let S and T be convex subsets of a real linear space X.

(a) The intersection of arbitrarily many convex sets is convex.

(b) If S and T are nonempty, then the algebraic sum $\alpha S + \beta T$ is convex for all $\alpha, \beta \in \mathbb{R}$. Consequently, for every $\bar{x} \in X$ the translated set $S + \{\bar{x}\}$ is convex as well.

Definition 1.6. Let S be a nonempty subset of a real linear space X. The intersection of all convex subsets of X that contain S is called the *convex hull* of S and is denoted $\mathrm{co}(S)$.

Remark 1.7. For two nonempty subsets S and T of a real linear space we obtain for all $\alpha, \beta \in \mathbb{R}$

$$\mathrm{co}(\alpha S + \beta T) = \alpha \mathrm{co}(S) + \beta \mathrm{co}(T).$$

Next, we consider sets which are algebraically open or closed.

Definition 1.8. Let S be a nonempty subset of a real linear space X.

(a) The set

$$\mathrm{cor}(S) := \{\bar{x} \in S \mid \text{for every } x \in X \text{ there is a } \bar{\lambda} > 0 \text{ with } \bar{x} + \lambda x \in S \text{ for all } \lambda \in [0, \bar{\lambda}]\}$$

is called the *algebraic interior* of S (or the *core* of S, see Fig. 1.4).

1.1. Linear Spaces and Convex Sets

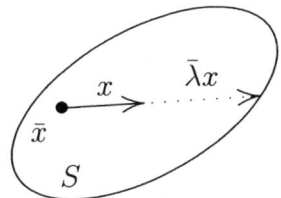

Figure 1.4: $\bar{x} \in \mathrm{cor}(S)$.

(b) The set S with $S = \mathrm{cor}(S)$ is called *algebraically open*.

(c) The set of all elements of X which do not belong to $\mathrm{cor}(S)$ and $\mathrm{cor}(X \setminus S)$ is called the *algebraic boundary* of S.

(d) An element $\bar{x} \in X$ is called *linearly accessible* from S, if there is an $x \in S$, $x \neq \bar{x}$, with the property

$$\lambda x + (1 - \lambda)\bar{x} \in S \quad \text{for all } \lambda \in (0, 1].$$

The union of S and the set of all linearly accessible elements from S is called the *algebraic closure* of S and it is denoted by

$$\mathrm{lin}(S) := S \cup \{x \in X \mid x \text{ is linearly accessible from } S\}.$$

In the case of $S = \mathrm{lin}(S)$ the set S is called *algebraically closed*.

(e) The set S is called *algebraically bounded*, if for every $\bar{x} \in S$ and every $x \in X$ there is a $\bar{\lambda} > 0$ such that

$$\bar{x} + \lambda x \notin S \quad \text{for all } \lambda \geq \bar{\lambda}.$$

These algebraic notions have a special geometric meaning. Take the intersections of the set S with each straight line in the real linear space X and consider these intersections as subsets of the real line $\mathbb{R}$. Then the set S is algebraically open, if these subsets are open; S is algebraically closed, if these subsets are closed; and S is algebraically bounded, if these subsets are bounded.

Lemma 1.9. *For a nonempty convex subset S of a real linear space we have:*

(a) $\bar{x} \in cor(S), \tilde{x} \in lin(S) \implies \{\lambda\tilde{x}+(1-\lambda)\bar{x}|\lambda \in [0,1)\} \subset cor(S)$,

(b) $cor(cor(S)) = cor(S)$,

(c) $cor(S)$ and $lin(S)$ are convex,

(d) $cor(S) \neq 0 \implies lin(cor(S)) = lin(S)$ and $cor(lin(S)) = cor(S)$.

A proof of Lemma 9 which is rather technical may be found in Kirsch-Warth-Werner [171, p. 9].

Another important class of subsets in a real linear space is introduced in

Definition 1.10. Let C be a nonempty subset of a real linear space X.

(a) The set C is called a *cone*, if

$$x \in C, \; \lambda \geq 0 \implies \lambda x \in C$$

(see Fig. 1.5)

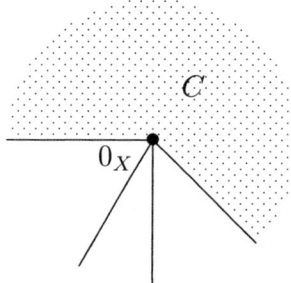

Figure 1.5: Cone.

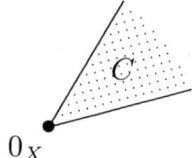

Figure 1.6: Pointed cone.

(b) A cone C is called *pointed*, if

$$C \cap (-C) = \{0_X\}$$

(see Fig. 1.6).

1.1. Linear Spaces and Convex Sets

(c) A cone C is called *reproducing*, if
$$C - C = X.$$
In this case one also says that C generates X.

(d) A nonempty convex subset B of a convex cone $C \neq \{0_X\}$ is called a *base* for C, if each $x \in C \setminus \{0_X\}$ has a unique representation of the form
$$x = \lambda b \text{ for some } \lambda > 0 \text{ and some } b \in B$$
(see Fig. 1.7).

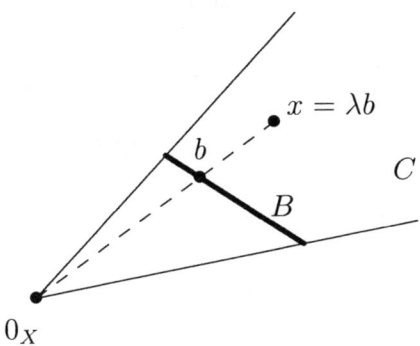

Figure 1.7: Base B for C.

Sometimes a cone is also called a wedge and a pointed wedge is called a cone. But in this book we use the terms in Definition 1.10.

By definition each cone contains the zero element of the real linear space. The simplest cones in a real linear space X are $\{0_X\}$ and X itself. $\{0_X\}$ is also called the *trivial cone*. From a geometric point of view a nontrivial cone is a set of rays emanating from the origin. Consequently, each cone is starshaped at 0_X.

For the investigation of partial orderings convex cones are very important. They are characterized by

Lemma 1.11. *A cone C in a real linear space is convex if and only if*
$$C + C \subset C.$$

Proof.

(a) Assume that C is a convex cone. Then for every $x, y \in C$ we have
$$\frac{1}{2}x + \frac{1}{2}y = \frac{1}{2}(x+y) \in C$$
implying $x + y \in C$. So, the inclusion $C + C \subset C$ is true.

(b) For arbitrary $x, y \in C$ and $\lambda \in [0, l]$ we obtain
$$\lambda x \in C \text{ and } (1-\lambda)y \in C.$$
With the inclusion $C + C \subset C$ we then get
$$\lambda x + (1-\lambda)y \in C,$$
i.e. the cone C is convex.

□

The algebraic interior of a convex cone has interesting properties listed below.

Lemma 1.12. *Let C be a convex cone in a real linear space X with a nonempty algebraic interior. Then:*

(a) $\text{cor}(C) \cup \{0_X\}$ is a convex cone,

(b) $\text{cor}(C) = C + \text{cor}(C)$.

Proof.

(a) Take arbitrary $\bar{x} \in \text{cor}(C)$ and $\mu > 0$. For every $x \in X$ there is a $\bar{\lambda} > 0$ with
$$\bar{x} + \frac{\lambda}{\mu}x \in C \text{ for all } \lambda \in [0, \bar{\lambda}].$$

1.1. Linear Spaces and Convex Sets

Since C is a cone, we get

$$\mu\left(\bar{x} + \frac{\lambda}{\mu}x\right) = \mu\bar{x} + \lambda x \in C \text{ for all } \lambda \in [0, \bar{\lambda}].$$

So, we obtain $\mu\bar{x} \in \text{cor}(C)$ and with Lemma 9, c) the assertion is obvious.

(b) The inclusion

$$\text{cor}(C) = \{0_X\} + \text{cor}(C) \subset C + \text{cor}(C)$$

is clear. For the proof of the converse inclusion we take arbitrary $\tilde{x} \in C$, $\bar{x} \in \text{cor}(C)$ and $x \in X$. Then there is a $\lambda > 0$ with

$$\bar{x} + \lambda x \in C \text{ for all } \lambda \in [0, \bar{\lambda}].$$

Since C is assumed to be convex, we conclude with Lemma 11

$$\tilde{x} + \bar{x} + \lambda x \in C \text{ for all } \lambda \in [0, \bar{\lambda}]$$

implying $\tilde{x} + \bar{x} \in \text{cor}(C)$. So, we conclude $C + \text{cor}(C) \subset \text{cor}(C)$.

$\square$

The following lemma gives a sufficient condition for a cone to be reproducing.

Lemma 1.13. *A cone C in a real linear space X is reproducing, if $\text{cor}(C) \neq \emptyset$.*

Proof. If $\text{cor}(C)$ is nonempty, take some $\bar{x} \in \text{cor}(C)$ and any $x \in X$. Then there is a $\bar{\lambda} > 0$ with $\bar{x} + \bar{\lambda}x \in C$ implying

$$x \in \frac{1}{\bar{\lambda}}C - \left\{\frac{1}{\bar{\lambda}}\bar{x}\right\} \subset C - C.$$

So, we get $X \subset C - C$ and together with the trivial inclusion $C - C \subset X$ we obtain the assertion. $\square$

Next, we turn our attention to the notion of a base B of a convex cone. Because of the convexity of B and the uniqueness of λ we have $0_X \notin B$.

Lemma 1.14. *Each nontrivial convex cone with a base in a real linear space is pointed.*

Proof. Let C be a nontrivial convex cone with base B. Take any $x \in C \cap (-C)$ and assume that $x \neq 0_X$. Then there are $b_1, b_2 \in B$ and $\lambda_1, \lambda_2 > 0$ with $x = \lambda_1 b_1 = -\lambda_2 b_2$ implying $\frac{\lambda_1}{\lambda_1 + \lambda_2} b_1 + \frac{\lambda_2}{\lambda_1 + \lambda_2} b_2 = 0_X \in B$. But this is a contradiction to the afore-mentioned remark. □

Definition 1.15. Let S be a nonempty subset of a real linear space. The cone

$$\text{cone}(S) := \{x \in X \mid x = \lambda s \text{ for some } \lambda \geq 0 \text{ and some } s \in S\}$$

is called the cone *generated* by S (see Fig. 1.8).

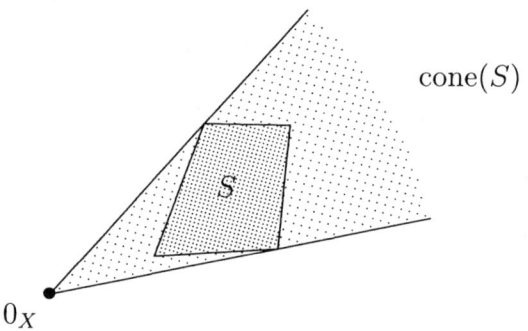

Figure 1.8: Cone generated by S.

It is an important property of a base B of a cone C that $\text{cone}(B) = C$. If $0_X \in \text{cor}(S)$ for a nonempty subset S of a real linear space X, then $\text{cone}(S) = X$.

1.2 Partially Ordered Linear Spaces

In addition to the linear structure of a space we consider a partial ordering which is given in many real linear spaces which are of practical interest.

1.2. Partially Ordered Linear Spaces

Definition 1.16. Let X be a real linear space.

(a) Each nonempty subset R of the product space $X \times X$ is called a *binary relation* R on X (we write xRy for $(x,y) \in R$).

(b) Every binary relation $\leq$ on X is called a *partial ordering* on X, if the following axioms are satisfied (for arbitrary $w, x, y, z \in X$):

   (i) $x \leq x$;
   (ii) $x \leq y$, $y \leq z$ $\implies$ $x \leq z$;
   (iii) $x \leq y$, $w \leq z$ $\implies$ $x + w \leq y + z$;
   (iv) $x \leq y$, $\alpha \in \mathbb{R}_+$ $\implies$ $\alpha x \leq \alpha y$.

(c) A partial ordering $\leq$ on X is called *antisymmetric*, if the following implication holds for arbitrary $x, y \in X$:

$$x \leq y, \ y \leq x \implies x = y.$$

In Definition 1.16, (b) with axiom (i) the partial ordering is reflexive and with (ii) it is transitive. The axioms (iii) and (iv) guarantee the compatibility of the partial ordering with the linear structure of the space.

Definition 1.17. A real linear space equipped with a partial ordering is called a *partially ordered linear space*.

It is important to note that in a partially ordered linear space two arbitrary elements cannot be compared, in general, in terms of the partial ordering. A significant characterization of a partial ordering in a real linear space is given by

Theorem 1.18. *Let X be a real linear space.*

(a) *If $\leq$ is a partial ordering on X, then the set*

$$C := \{x \in X \mid 0_X \leq x\}$$

is a convex cone. If, in addition, $\leq$ is antisymmetric, then C is pointed.

(b) If C is a convex cone in X, then the binary relation

$$\leq_C := \{(x,y) \in X \times X \mid y - x \in C\}$$

is a partial ordering on X. If, in addition, C is pointed, then $\leq_C$ is antisymmetric.

This theorem is easy to prove and is of great importance because a partial ordering can be investigated using convex analysis.

The next definition is based on the result of Theorem 1.18.

Definition 1.19. A convex cone characterizing a partial ordering in a real linear space is called an *ordering cone*.

Several authors also call an ordering cone a *positive cone*. We denote $\leq_C$ as a partial ordering induced by a convex cone C.

Example 1.20. For $X = \mathbb{R}^n$ the ordering cone of the componentwise partial ordering on $\mathbb{R}^n$ is given by

$$C := \{x \in \mathbb{R}^n \mid x_i \geq 0 \text{ for all } i \in \{l, \ldots, n\}\} = \mathbb{R}^n_+.$$

It is also called the natural ordering cone. Other ordering cones in $\mathbb{R}^n$ are for instance

$$\{x \in \mathbb{R}^n \mid x_i \geq 0 \text{ for all } i \in \{l, \ldots, m\} \text{ and}$$
$$x_i = 0 \text{ for all } i \in \{m+1, \ldots, n\}\} \quad \text{for some } 1 \leq m < n$$

or $0_{\mathbb{R}_n}$ and $\mathbb{R}^n$ itself. $\mathbb{R}_+$, $\mathbb{R}_-$, $\{0\}$ and $\mathbb{R}$ are the only ordering cones in $\mathbb{R}$. Ordering cones of special infinite dimensional linear spaces will be presented in subsection 1.4.

Definition 1.21. Let X be a partially ordered linear space. For arbitrary elements $x, y \in X$ with $x \leq y$ the set

$$[x, y] := \{z \in X \mid x \leq z \leq y\}$$

is called the *order interval* between x and y (see Fig. 1.9)

1.2. Partially Ordered Linear Spaces

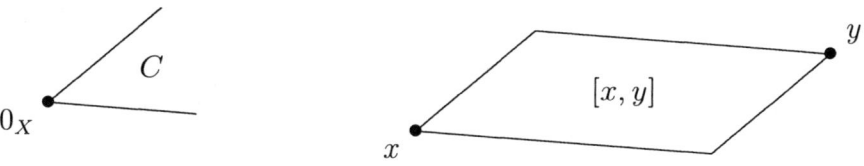

Figure 1.9: Order interval $[x, y]$.

If C is the ordering cone in a partially ordered linear space, then the order interval between x and y can be written as

$$[x, y] = (\{x\} + C) \cap (\{y\} - C).$$

Lemma 1.22. *Let X be a partially ordered linear space with the ordering cone C. Let $x, y \in X$ with $x \in \{y\} - C$ (i.e. $x \leq_C y$) be given arbitrarily. Then we have for $z := \frac{1}{2}(x + y)$:*

(a) *The order interval $[x - z, y - z]$ is absolutely convex.*

(b) *If $cor(C) \neq 0$ and $x \in \{y\} - cor(C)$, then $z \in cor([x, y])$.*

(c) *If C is algebraically closed, then $[x, y]$ is algebraically closed.*

(d) *If C is algebraically closed and pointed, then $[x, y]$ is algebraically bounded.*

Proof.

(a) With the equality

$$[x - z, y - z] = \left[-\frac{1}{2}(y - x), \frac{1}{2}(y - x)\right]$$

the assertion is obvious.

(b) Since

$$z = x + \frac{1}{2}(y - x) \in \{x\} + cor(C)$$

and
$$z = y - \frac{1}{2}(y - x) \in \{y\} - \operatorname{cor}(C),$$
we conclude $z \in \operatorname{cor}([x, y])$.

(c) Because of the equality $[x, y] = (\{x\} + C) \cap (\{y\} - C)$ this assertion is evident.

(d) First, if the pointed convex cone C is algebraically closed, then the complement set $X \backslash C$ is algebraically open. For if we assume that $X \backslash C$ is not algebraically open, then there is an $\bar{x} \in X \backslash C$ and an $h \in X$ so that for all $\bar{\lambda} > 0$
$$\bar{x} + \lambda h \in C \text{ for some } \lambda \in (0, \bar{\lambda}].$$
Since C is convex, we conclude for some $x := \bar{x} + \lambda h \in C$
$$\mu x + (l - \mu)\bar{x} \in C \text{ for all } \in (0, 1]$$
which implies $\bar{x} \in \operatorname{lin}(C) = C$. But this contradicts the assumption $\bar{x} \notin C$. So, the complement set $X \backslash C$ is algebraically open.

In order to prove that $[x, y]$ is algebraically bounded we take any $v \in [x, y]$ and any $w \in X \backslash \{0_X\}$. Then we consider the two cases $w \notin C$ and $w \in C$. Assume that $w \notin C$. Since $X \backslash C$ is algebraically open, there is a $\bar{\lambda} > 0$ with
$$w + \lambda(v - x) \in X \backslash C \text{ for all } \lambda \in [0, \bar{\lambda}].$$
The set $(X \backslash C) \cup \{0_X\}$ is a cone and, therefore, we obtain
$$\frac{1}{\lambda}(w + \lambda(v - x)) \in X \backslash C \text{ for all } \lambda \in (0, \bar{\lambda}]$$
or alternatively
$$\lambda(w + \frac{1}{\lambda}(v - x)) \in X \backslash C \text{ for all } \lambda \in \left[\frac{1}{\bar{\lambda}}, \infty\right).$$
But then we have
$$v - x + \lambda w \in X \backslash C \text{ for all } \lambda \in \left[\frac{1}{\bar{\lambda}}, \infty\right)$$

1.2. Partially Ordered Linear Spaces

and
$$v + \lambda w \notin \{x\} + C \text{ for all } \lambda \in \left[\frac{1}{\bar\lambda}, \infty\right)$$

which implies
$$v + \lambda w \notin [x, y] \text{ for all } \lambda \in \left[\frac{1}{\bar\lambda}, \infty\right).$$

Next, assume that $w \in C$. Since the ordering cone C is assumed to be pointed and $w \neq 0_X$, we conclude $w \notin -C$. With the same arguments as before there is a $\bar\lambda > 0$ with
$$v + \lambda w \notin [x, y] \text{ for all } \lambda \in \left[\frac{1}{\bar\lambda}, \infty\right).$$

Hence, the order interval $[x, y]$ is algebraically bounded.

$\square$

With a partial ordering on a real linear space it is also possible to introduce a partial ordering on the algebraic dual space.

Definition 1.23. Let X be a real linear space with a convex cone C_X.

(a) The cone
$$C_{X'} := \{x' \in X' \mid x'(x) \geq 0 \text{ for all } x \in C_X\}$$

is called the *dual cone* for C_X. The partial ordering in X' which is induced by $C_{X'}$ is called the *dual partial ordering*.

(b) The set
$$C_{X'}^{\#} := \{x' \in X' \mid x'(x) > 0 \text{ for all } x \in C_X \setminus \{0_X\}\}$$

is called the *quasi-interior* of the dual cone for C_X.

Notice that $C_{X'}$ is a convex cone so that Definition 1.23, (a) makes sense. For $C_X = \{0_X\}$ we obtain $C_{X'} = X'$, and for $C_X = X$ we have

$C_{X'} = \{0_{X'}\}$. If the quasi-interior $C_{X'}^{\#}$ of the dual cone for C_X is nonempty, then $C_{X'}^{\#} \cup \{0_{X'}\}$ is a nontrivial convex cone. With the following lemma we list some useful properties of dual cones without proof.

Lemma 1.24. *Let C_X and D_X be two convex cones in a real linear space X with the dual cone $C_{X'}$ and $D_{X'}$, respectively. Then:*

(a) $C_X \subset D_X \implies D_{X'} \subset C_{X'}$;

(b) $C_{X'} \cap D_{X'}$ *is the dual cone for* $C_X + D_X$;

(c) $C_X \cup D_X$ *and* $C_X + D_X$ *have the same dual cone;*

(d) $C_{X'} + D_{X'}$ *is a subset of the dual cone for* $C_X \cap D_X$.

In general, the quasi-interior of the dual cone does not coincide with the algebraic interior of the dual cone but the following inclusion holds.

Lemma 1.25. *If C_X is a convex cone in a real linear space X and X' separates elements in X (i.e., two different elements in X may be separated by an hyperplane), then*

$$cor(C_{X'}) \subset C_{X'}^{\#}.$$

Proof. The assertion is trivial for $C_X = \{0_X\}$ and for $cor(C_{X'}) = \emptyset$. If $C_X \neq \{0_X\}$ and $cor(C_{X'}) \neq \emptyset$, then take any $\bar{x} \in cor(C_{X'})$ and assume that $\bar{x} \notin C_{X'}^{\#}$. Consequently, there is an $x \in C_X \setminus \{0_X\}$ with $\bar{x}(x) = 0$. Since X' separates elements in X, there is a linear functional $x' \in X'$ with the property $x'(x) < 0$. Then we conclude

$$(\lambda x' + (1-\lambda)\bar{x})(x) < 0 \text{ for all } \lambda > 0$$

which contradicts the assumption that $\bar{x} \in cor(C_{X'})$. □

Conditions under which the quasi-interior of the dual cone is nonempty will be given in subsection 3.4. The following result is very similar to that of Lemma 1.25.

1.2. Partially Ordered Linear Spaces

Lemma 1.26. *If C_X is a convex cone in a real linear space X, then*

$$cor(C_X) \subset \{x \in X \mid x'(x) > 0 \text{ for all } x' \in C_{X'}\backslash\{0_{X'}\}\}.$$

Proof. Take any $\bar{x} \in cor(C_X)$ and any $x' \in C_{X'}\backslash\{0_{X'}\}$. Consequently, there are an $x \in X$ with $x'(x) < 0$ and a $\bar{\lambda} > 0$ with $\bar{x} + \bar{\lambda}x \in C_X$. Hence, we obtain $x'(\bar{x} + \bar{\lambda}x) \geq 0$ and

$$x'(\bar{x}) \geq -\bar{\lambda}x'(x) > 0$$

which leads to the assertion. □

A consequence of Lemma 1.26 is given by

Lemma 1.27. *Let C_X be a convex cone in a real linear space X.*

(a) *If $cor(C_X)$ is nonempty, then $C_{X'}$ is pointed.*

(b) *If $C_{X'}^{\#}$ is nonempty, then C_X is pointed.*

Proof.

(a) For every $x' \in C_{X'} \cap (-C_{X'})$ we have

$$x'(x) = 0 \text{ for all } x \in C_X$$

and especially for some $\bar{x} \in cor(C_X)$ we get $x'(\bar{x}) = 0$. With Lemma 1.26 we obtain $x' = 0_{X'}$, and this implies

$$C_{X'} \cap (-C_{X'}) = \{0_{X'}\}.$$

(b) Take any $x \in C_X \cap (-C_X)$. If we assume that $x \neq 0_{X'}$ we obtain for every $x' \in C_{X'}^{\#}$

$$x'(x) > 0 \quad \text{and} \quad x'(x) < 0$$

which is a contradiction. □

An important property of the quasi-interior of a dual cone is that it can be used to characterize the base of the original cone.

Lemma 1.28. *Let C_X be a nontrivial convex cone in a real linear space X.*

(a) For every $x' \in C_{X'}^{\#}$, the set $B := \{x \in C_X \mid x'(x) = 1\}$ is a base for C_X.

(b) In addition, let C_X be reproducing and let C_X have a base. Then there is an $x' \in C_{X'}^{\#}$ with
$$B = \{x \in C_X \mid x'(x) = 1\}.$$

Proof.

(a) Choose any $x' \in C_{X'}^{\#}$. Then we obtain for every $x \in C_X \setminus \{0_X\}$ $x'(x) > 0$ and, therefore, x can be uniquely represented as
$$x = x'(x) \frac{1}{x'(x)} x \quad \text{for} \quad \frac{1}{x'(x)} x \in B.$$
Hence, the assertion is evident.

(b) We define the functional $x' : C_X \setminus \{0_X\} \to \mathbb{R}_+$ with
$$x'(x) = \lambda(x) \text{ for all } x \in C_X \setminus \{0_X\}$$
where $\lambda(x)$ is the positive number in the representation formula for x. It is obvious that x' is positively homogeneous. In order to see that it is additive pick some elements $x, y \in C_X \setminus \{0_X\}$. Then we obtain
$$\frac{1}{x'(x) + x'(y)}(x + y) = \frac{x'(x)}{x'(x) + x'(y)} \frac{1}{x'(x)} x$$
$$+ \frac{x'(y)}{x'(x) + x'(y)} \frac{1}{x'(y)} y \in B$$
because $\frac{1}{x'(x)} x \in B$, $\frac{1}{x'(y)} y \in B$ and B is convex. Consequently, we get
$$x'(x + y) = x'(x) + x'(y) \text{ for all } x, y \in C_X \setminus \{0_X\}.$$

Hence, x' is a positively homogeneous and additive functional on $C_X \setminus \{0_X\}$. Next, we define $x'(0_X) := 0$ and we see that this extension is positively homogeneous and additive on C_X as well. Finally we extend x' to $X = C_X - C_X$ by defining

$$x'(x - y) := x'(x) - x'(y) \text{ for all } x, y \in C_X.$$

It is obvious that x' is positively homogeneous and additive on X, and since

$$x'(x - y) = x'(x) - x'(y) = -x'(y - x) \text{ for all } x, y \in C_X,$$

x' is also linear on X. With

$$x'(x) > 0 \text{ for all } x \in C_X \setminus \{0_X\}$$

we obtain $x' \in C_{X'}^{\#}$. The set equation

$$B = \{x \in C_X \mid x'(x) = 1\}$$

is evident, if we use the definition of x'.

$\square$

It is important to note that with Zorn's lemma one does not need the assumption $X = C_X - C_X$ (i.e., C_X is reproducing) in Lemma 1.28, (b). This assumption can be dropped as one may see in Lemma 3.3.

1.3 Topological Linear Spaces

In this section we investigate partially ordered linear spaces which are equipped with a topology. The important spaces as locally convex spaces and normed spaces are considered, and the connections between topology and partial ordering are examined.

Definition 1.29. Let X be a nonempty set.

(a) A *topology* $\mathcal{T}$ on X is defined to be a set of subsets of X which satisfy the following axioms:

(i) every union of sets of $\mathcal{T}$ belongs to $\mathcal{T}$;

(ii) every finite intersection of sets of $\mathcal{T}$ belongs to $\mathcal{T}$;

(iii) $\emptyset \in \mathcal{T}$ and $X \in \mathcal{T}$.

In this case $(X, \mathcal{T})$ is called a *topological space* and the elements of $\mathcal{T}$ are called *open* sets.

(b) Let $\mathcal{S}$ and $\mathcal{T}$ be two topologies on X. $\mathcal{S}$ is called *finer* than $\mathcal{T}$ (or $\mathcal{T}$ is called *coarser* than $\mathcal{S}$), if every $\mathcal{S}$-open set is $\mathcal{T}$-open.

(c) Let $(X, \mathcal{T})$ be a topological space, let S be a subset of X and let some $x \in X$ be a given element. The set S is called a *neighborhood* of x, if there is an open set T with $x \in T \subset S$. x is called an *interior* element of S, if there is a neighborhood T of x contained in S. The set of all interior elements of S is called the *interior* of S and it is denoted int(S). The set S is called *closed*, if $X \backslash S$ is open. The set of all elements of X for which every neighborhood meets the set S is called the *closure* of S and it is denoted cl(S). The set S is called *dense* in X, if $X \subset$ cl(S).

(d) A topological space $(X, \mathcal{T})$ is called *separable*, if X contains a countable dense subset.

(e) (i) A nonempty partially ordered set I is called *directed*, if two arbitrary elements in I are majorized in I.

(ii) A map from a directed set I to a nonempty set X is called a *net* and is denoted $(x_i)_{i \in I}$.

(iii) Let $(X, \mathcal{T})$ be a topological space. A net $(x_i)_{i \in I}$ is called to *converge* to some $x \in X$, if for every neighborhood U of x there is an $n \in I$ so that

$x_i \in U$ for all $i \geq n$

($\leq i$ denotes the partial ordering in I).

In this case we write $x = \lim_{i \in I} x_i$.

1.3. Topological Linear Spaces

(iv) Let $(X, \mathcal{T})$ be a topological space. An element $x \in X$ is called a *cluster point* of a net $(x_i)_{i \in I}$, if for every neighborhood U of x and every $n \in I$ there is an $i \in I$ with $i \geq n$ so that $x_i \in U$.

(f) A nonempty subset S of a topological space $(X, \mathcal{T})$ is called *compact*, if every net in S has a cluster point in S.

(g) Let $(X, \mathcal{S})$ and $(Y, \mathcal{T})$ be two topological spaces. A map $f : X \to Y$ is called *continuous* at some $x \in X$, if to every neighborhood V of $f(x)$ there is a neighborhood U of x with $f(U) \subset V$. $f : X \to Y$ is called *continuous* on X, if f is continuous at every $x \in X$.

(h) A topological space $(X, \mathcal{T})$ is called *separated* (or a Hausdorff space), if any two different elements have disjoint neighborhoods.

An important class of topological spaces are so-called metric spaces.

Definition 1.30.

(a) Let X be a nonempty set. A map $d : X \times X \to \mathbb{R}_+$ is called a *metric*, if (for all $x, y, z \in X$):

   (i) $d(x, y) = 0 \iff x = y$;
   (ii) $d(x, y) = d(y, x)$;
   (iii) $d(x, z) \leq d(x, y) + d(y, z)$.

   In this case (X, d) is called a *metric space*.

(b) A topological space $(X, \mathcal{T})$ is called *metrizable*, if its topology can be defined by a metric.

If (X, d) is a metric space, then for any $x \in X$ a set $S(x)$ is called a *neighborhood* of x, if there is an $\varepsilon > 0$ so that

$$\{y \in X \mid d(x, y) < \varepsilon\} \subset S(x).$$

The set of all neighborhoods of x defines a topology on X.

Next, we consider a topological space $(X, \mathcal{T})$ where X is now a real linear space. In this case we require that the topological and the linear structure of the space are compatible.

Definition 1.31. Let X be a real linear space and let $\mathcal{T}$ be a topology on X.

(a) $(X, \mathcal{T})$ is called a real *topological linear space*, if addition and multiplication with reals are continuous, i.e. the maps

$$(x, y) \mapsto x + y \text{ with } x, y \in X,$$
$$(\alpha, x) \mapsto \alpha x \text{ with } \alpha \in \mathbb{R} \text{ and } x \in X$$

are continuous on $X \times X$ and $\mathbb{R} \times X$, respectively. In many situations we use, for simplicity, the notation X instead of $(X, \mathcal{T})$ for a real topological linear space.

(b) A subset S of a real topological linear space X is called *bounded*, if for each 0_X-neighborhood U there is a $\lambda \in \mathbb{R}$ with the property $S \subset \lambda U$.

(c) A nonempty subset S of a real topological linear space X is called *complete*, if each Cauchy net in S converges to some $x \in S$ (i.e. for every net $(x_i)_{i \in I}$ in S with $\lim\limits_{(i,j) \in I \times I} (x_i - x_j) = 0$ there is an $x \in S$ with $x = \lim\limits_{i \in I} x_i$).

(d) A real topological linear space X is called *quasi-complete*, if each nonempty, closed and bounded set in X is complete.

In Lemma 1.9 we listed some results on the algebraic interior and closure of a set. Now we consider the relationships between these notions and the corresponding topological notions. For a proof of these results see Holmes [125, p. 59].

Lemma 1.32. *Let S be a convex set of a real topological linear space X. Then the closure $cl(S)$ is convex. For $int(S) \neq \emptyset$ we have:*

1.3. Topological Linear Spaces

(a) $int(S) = cor(S)$;

(b) $cl(S) = cl(int(S))$ and $int(S) = int(cl(S))$;

(c) $cl(S) = lin(S)$.

Definition 1.33. Let X be a real topological linear space.

(a) A subset $\mathcal{B}$ of the set $\mathcal{S}$ of neighborhoods of 0_X is called a *base of neighborhoods of* $0_{X'}$ if for every $a \in \mathcal{S}$ there is a set $T \in \mathcal{B}$ with $T \subset S$.

(b) If X has a base of convex neighborhoods of $0_{X'}$ it is called a real *locally convex topological linear space* or a real *locally convex space*.

It can be shown that every topological linear space has a base of balanced neighborhoods of the origin. But in many practical situations one needs convex neighborhoods of the origin and, therefore, locally convex spaces are very useful in practice.

For certain results in vector optimization we will assume that the algebraic sum of two sets is closed. A sufficient condition for the property of being closed is given by

Lemma 1.34. *In a real locally convex space X the algebraic sum of a nonempty compact set and a nonempty closed set is closed.*

For a proof see Robertson-Robertson [259, p. 53/54].

Next, we consider some other types of spaces which are important for the vector optimization theory.

Definition 1.35. Let X and Y be real linear spaces, and let C_Y be a convex cone in Y. A map $\|\cdot\| : X \to C_Y$ is called a *vectorial norm*, if the following conditions are satisfied (for all $x, z \in X$ and all $\lambda \in \mathbb{R}$):

(a) $\|x\| = 0_Y \iff x = 0_X$;

(b) $\|\lambda x\| = |\lambda| \, \|x\|$;

(c) $\|x + z\| \leq_{C_Y} \|x\| + \|z\|$.

If, in addition, $Y = \mathbb{R}$ and $C_Y = \mathbb{R}_+$ the map $\|\cdot\|$ is called a *norm* and it is denoted $\|\cdot\|$. If the condition (a) is not fulfilled, the map $\|\cdot\|$ is called a *seminorm*.

Definition 1.36. Let X be a real linear space equipped with a norm $\|\cdot\|$.

(a) The pair $(X, \|\cdot\|)$ is called a real *normed space* (a real normed space is a real topological linear space, if the topology is generated by the metric $(x, y) \mapsto \|x - y\|$).

(b) A complete real normed space is called a real *Banach space*.

A significant class of normed spaces are Hilbert spaces.

Definition 1.37. Let X be a real linear space.

(a) A map $\langle .,. \rangle : X \times X \to \mathbb{R}$ is called an *inner product*, if the following conditions are satisfied (for all $x, y, z \in X$ and all $\lambda \in \mathbb{R}$):

  (i) $\langle x, x \rangle > 0$ for $x \neq 0_X$;
  (ii) $\langle x, y \rangle = \langle y, x \rangle$;
  (iii) $\langle \lambda x, y \rangle = \lambda \langle x, y \rangle$;
  (iv) $\langle x + y, z \rangle = \langle x, z \rangle + \langle y, z \rangle$.

(b) If the real linear space X equipped with an inner product $\langle .,. \rangle$ is complete, the pair $(X, \langle .,. \rangle)$ is called a *Hilbert space* (it is a real normed space with the norm $\|\cdot\|$ defined by $\|x\| = \sqrt{\langle x, x \rangle}$ for all $x \in X$).

Next, we turn our attention to dual spaces and we list some important definitions.

1.3. Topological Linear Spaces

Definition 1.38. Let X be a real linear space and let Y be a nonempty subset of the algebraic dual space X'.

(a) For every $x' \in Y$ there is a seminorm $p : X \to \mathbb{R}$ given by

$$p(x) = |x'(x)| \text{ for all } x \in X.$$

The coarsest topology on X making all these seminorms continuous is called the *weak topology* on X generated by Y and it is denoted $\sigma(X, Y)$ (it is the weakest topology on X in which all linear functionals which belong to Y are continuous).

(b) If X is equipped with a topology, then the subspace X^* of all continuous linear functionals which belong to X' is called the *topological dual space* of X. For $Y = X^*$ $\sigma(X, X^*)$ is simply called the *weak topology* on X.

(c) If X is equipped with a topology, then the topology $\sigma(X^*, X)$ defined by the functionals

$$\varphi \mapsto \varphi(x) \text{ for all } x \in X \text{ and all } \varphi \in X^*$$

is called the *weak* topology*.

A characterization of a separable normed space is given by

Lemma 1.39. *A real normed space* $(X, \|\cdot\|)$ *is separable if and only if every ball in* X^* *is weak*-metrizable.*

For a proof of this lemma see, for instance, Holmes [125, p. 72].

Definition 1.40. A real normed space $(X, \|\cdot\|)$ is called *reflexive*, if the canonical embedding $J_X : X \to X^{**}$ defined by

$$J_X(x)(\varphi) = \varphi(x) \text{ for all } x \in X \text{ and all } \varphi \in X^*$$

is surjective.

Every reflexive real normed space is complete and, therefore, it is a real Banach space. For the applications the following assertion is important (see Holmes [125, p. 126/127]).

Lemma 1.41. *A real Banach space* $(X, \|\cdot\|)$ *is reflexive if and only if the closed unit ball* $\{x \in X \mid \|\cdot\| \leq 1\}$ *is weakly compact.*

If in a topological linear space a partial ordering is given additionally, it is important to know the relationships between the topology and the ordering. First, we present the notion of a normal cone.

Definition 1.42.

(a) Let X be a real linear space with a partial ordering. The finest locally convex topology on X for which every order interval is bounded is called the *order topology*.

(b) Let $(X, \mathcal{T})$ be a real topological linear space equipped with an ordering cone C. The convex cone C is called *normal* for the topology $\mathcal{T}$, if there is a base of neighborhoods of 0_X consisting of sets S with the property

$$S = (S + C) \cap (S - C).$$

For a normed space a normal ordering cone can be characterized by

Lemma 1.43. *Let* $(X, \|\cdot\|)$ *be a real normed space with an ordering cone* C. *The convex cone* C *is normal for the norm topology if and only if there is* $\lambda > 0$ *so that for all* $y \in C$

$$x \in [0_X, y] \implies \lambda \|x\| \leq \|y\|.$$

A proof of this lemma may be found in Peressini [249, p. 64].

Several results on normality are listed in the following lemma (see Peressini [249] and Borwein [35]).

Lemma 1.44.

(a) *In a real Banach space* X *an ordering cone* C_X *is normal for the norm topology if and only if the dual cone* C_{X^*} *is reproducing.*

1.3. Topological Linear Spaces

(b) In a real locally convex space X an ordering cone C_X is normal for the weak topology $\sigma(X, X^*)$ if and only if the dual cone C_{X^*} is reproducing.

(c) In a real locally convex space X a normal ordering cone is also normal for the weak topology $\sigma(X, X^*)$.

(d) In a real locally convex space an ordering cone with a bounded base is normal.

Order intervals play an important role for the definition of a norm in a real linear space. We list some properties.

Lemma 1.45. *Let X be a real linear space with an ordering cone C.*

(a) *If $C \neq X$ and $\mathrm{cor}(C) \neq \emptyset$, then there is a seminorm $\|\cdot\|$ on X with the property that for all $y \in \mathrm{cor}(C)$*

$$x \in \mathrm{cor}([0_X, y]) \implies \|x\| < \|y\|.$$

(b) *If $\mathrm{cor}(C) \neq \emptyset$ and C is algebraically closed and pointed, then there is a norm $\|\cdot\|$ on X with the property that for all $y \in C$*

$$x \in [0_X, y] \implies \|x\| \leq \|y\|.$$

(c) *If $\mathrm{cor}(C) \neq \emptyset$ and C has a weakly compact base, then there is a norm $\|\cdot\|$ on X with the property that for all $y \in \mathrm{cor}(C)$*

$$x \in [0_X, y] \implies \|x\| \leq \|y\|.$$

The real normed space $(X, \|\cdot\|)$ is even reflexive.

Proof.

(a) For an arbitrary $z \in \mathrm{cor}(C)$ we define a seminorm $\|\cdot\|$ on X using the Minkowski functional

$$\|x\| := \inf_{\lambda > 0} \left\{ \lambda \;\middle|\; \frac{1}{\lambda} x \in [-z, z] \right\} \text{ for all } x \in X.$$

With Lemma 1.22, (a) and (b) the order interval $[-z, z]$ is absolutely convex with $0_X \in \operatorname{cor}([-z, z])$ and, therefore, the Minkowski functional is indeed a seminorm (compare Dunford-Schwartz [84, p. 411]). Next, for an arbitrary $y \in \operatorname{cor}(C)$ we obtain

$$\frac{1}{\|y\|} \operatorname{cor}([0_X, y]) = \operatorname{cor}\left(\left[0_X, \frac{1}{\|y\|} y\right]\right)$$
$$\subset \operatorname{cor}([-z, z])$$
$$= \{x \in X \mid \|x\| < 1\}$$

resulting in

$$\operatorname{cor}([0_X, y]) \subset \{x \in X \mid \|x\| < \|y\|\}.$$

Then the assertion is obvious.

(b) The proof of this part is similar to that under (a). For an arbitrary $z \in \operatorname{cor}(C)$ we know that the Minkowski functional $\|\cdot\|$ on X given by

$$\|x\| := \inf_{\lambda > 0} \left\{ \lambda \mid \frac{1}{\lambda} x \in [-z, z] \right\} \text{ for all } x \in X$$

is a seminorm. Since, by Lemma 1.22, (d), the order interval is even algebraically bounded, $\|\cdot\|$ is indeed a norm. With Lemma 1.22, (c) the order interval $[-z, z]$ is also algebraically closed so that this order interval can be written as

$$[-z, z] = \{x \in X \mid \|x\| \leq 1\}.$$

Then we get for an arbitrary $y \in C \setminus \{0_X\}$

$$\frac{1}{\|y\|} [0_X, y] = \left[0_X, \frac{1}{\|y\|} y\right] \subset \{x \in X \mid \|x\| \leq 1\}$$

implying

$$[0_X, y] \subset \{x \in X \mid \|x\| \leq \|y\|\}.$$

This last inclusion is even true for $y = 0_X$. Then the assertion is evident.

(c) By Lemma 3.3 (see also Lemma 1.28, (b) under the additional assumption that C is reproducing) there is a linear functional x' which belongs to the quasi-interior of the dual cone of C so that the base B of C can be written as

$$B = \{x \in C \mid x'(x) = 1\}.$$

Since B is weakly compact, the set

$$S := \{x \in C \mid x'(x) \leq 1\}$$

is weakly compact as well, and $\operatorname{cor}(C) \neq \emptyset$ implies $\operatorname{cor}(S) \neq \emptyset$. For an arbitrary $z \in \operatorname{cor}(S)$ we define a norm using the order interval $[-z, z]$ which is weakly compact. Hence, with the same arguments as in part (b) we see that $\|\cdot\|$ given by the Minkowski functional is a norm and has the asserted monotonicity property. By construction the unit ball $[-z, z]$ is weakly compact and, therefore, the real normed space $(X, \|\cdot\|)$ is even reflexive. □

Another essential property of an ordering cone is the Daniell property.

Definition 1.46.

(a) Let X be a real topological linear space with an ordering cone C. The convex cone C is called *Daniell*, if every decreasing net (i.e. $i \leq j \Rightarrow x_j \leq x_i$) which has a lower bound converges to its infimum.

(b) Let X be a real topological linear space with an ordering cone C. X is called *boundedly order complete*, if every bounded decreasing net has an infimum.

Conditions ensuring the Daniell property are given by

Lemma 1.47.

(a) *Let X be a real topological linear space with an ordering cone C. If X has compact intervals and C is closed and pointed, then C is Daniell.*

(b) Let X be a real locally convex linear space with an ordering cone C. If X is reflexive and C is normal for the weak topology $\sigma(X, X^*)$, then C is weakly Daniell.

(c) If X is a real locally convex linear space and C is a complete ordering cone which has a bounded base, then C is Daniell.

For these results (and even some more) see Borwein [35].

1.4 Some Examples

In this section we discuss some important linear spaces with respect to their topology and their partial ordering. We restrict our attention only to some special classes and we do not present these spaces in the most general form.

Example 1.48.

(a) First, we consider for every $p \in [1, \infty)$ the sequence space

$$l_p := \left\{ (x_i)_{i \in \mathbb{N}} \mid x_i \in \mathbb{R} \text{ for all } i \in \mathbb{N} \text{ and } \sum_{i=1}^{\infty} |x_i|^p < \infty \right\}.$$

The real linear spaces l_p are separable Banach spaces with respect to the norm $\|\cdot\|_{l_p}$ given by

$$\|x\|_{l_p} := \left(\sum_{i=1}^{\infty} |x_i|^p \right)^{\frac{1}{p}} \text{ for all } x \in l_p.$$

For every $p \in [1, \infty)$ the so-called natural ordering cone is given by

$$C_{l_p} := \{ x \in l_p \mid x_i \geq 0 \text{ for all } i \in \mathbb{N} \}.$$

This ordering cone has no topological interior; every element of C_{l_p} belongs to the boundary. If we take another ordering cone D_{l_1} where

$$D_{l_1} := \{ x \in l_1 \mid \text{all partial sums of } x \text{ are non-negative} \},$$

1.4. Some Examples

then $C_{l_1} \subset D_{l_1}$ and for this ordering cone we have $\text{int}(D_{l_1}) \neq \emptyset$ (e.g. $(1, 0, 0, \ldots) \in \text{int}(D_{l_1})$).

The ordering cone C_{l_1} is Daniell and normal for the norm topology and it has weakly compact order intervals and a bounded base. The quasi-interior $C_{l_1^*}^{\#}$ of the dual cone is nonempty (for instance, the functional φ given by $\varphi(x) = \sum_{i=1}^{\infty} x_i$ belongs to $C_{l_1^*}^{\#}$).

(b) Another well-known sequence space is

$$l_\infty := \Big\{(x_i)_{i \in \mathbb{N}} \;\Big|\; x_i \in \mathbb{R} \text{ for all } i \in \mathbb{N} \text{ and } \sup_{i \in \mathbb{N}}\{|x_i|\} < \infty \Big\}.$$

This is a real (non-separable) Banach space with the norm $\|\cdot\|_{l_\infty}$ given by

$$\|x\|_{l_\infty} := \sup_{i \in \mathbb{N}}\{|x_i|\} \text{ for all } x \in l_\infty.$$

The natural ordering cone

$$C_{l_\infty} := \{x \in l_\infty \mid x_i \geq 0 \text{ for all } i \in \mathbb{N}\}$$

has interior elements (e.g. $(1, 1, \ldots) \in \text{int}(C_{l_\infty})$). The unit ball equals the order interval $[-(1, 1, \ldots), (1, 1, \ldots)]$. C_{l_∞} has also a base but this base is not bounded. The quasi-interior $C_{l_\infty^*}^{\#}$ of the dual cone is nonempty; for instance, the linear functional φ given by $\varphi(x) = \sum_{i=1}^{\infty} \dfrac{x_i}{2^i}$ is an element of $C_{l_\infty^*}^{\#}$.

Example 1.49. Let Ω be any compact Hausdorff space. The real linear space of all real-valued functions which are continuous on Ω is denoted $C(\Omega)$. It is a real normed space with

$$\|f\|_{C(\Omega)} := \sup_{x \in \Omega}\{|f(x)|\} \text{ for all } f \in C(\Omega).$$

The so-called natural ordering cone is given by

$$C_{C(\Omega)} := \{f \in C(\Omega) \mid f(x) \geq 0 \text{ for all } x \in \Omega\}.$$

In this case the unit ball coincides with the order interval $[-f, f]$ where $f \in C(\Omega)$ with

$$f(x) = 1 \text{ for all } x \in \Omega.$$

The ordering cone is closed and normal for the norm topology and it has a nonempty topological interior. Even $\text{int}(C_{C(\Omega)^*})$ is nonempty. The set of positive Radon measures of total mass 1 on Ω is a base for the cone $C_{C(\Omega)^*}$ of positive Radon measures on Ω (recall that a Radon measure on Ω is any continuous linear functional on $C(\Omega)$).

Example 1.50. Let Ω be any compact Hausdorff space. Let $M(\Omega)$ denote the linear space of all bounded Radon measures on Ω equipped with the norm $\|\cdot\|_{M(\Omega)}$ given by

$$\|\mu\|_{M(\Omega)} := \sup\left\{\int_\Omega f\, d\mu \,\bigg|\, f \in \mathcal{K}(\Omega) \text{ (linear space of real-valued continuous functions with compact support on } \Omega), \ |f(x)| \leq 1 \text{ for all } x \in \Omega\right\}$$

and partially ordered by the convex cone $C_{M(\Omega)}$ of positive Radon measures on Ω. Then $M(\Omega)$ is a Banach space and $C_{M(\Omega)}$ is closed and normal for the norm topology.

Example 1.51.

(a) For a nonempty subset Ω of $\mathbb{R}^n$ and any $p \in [1, \infty)$ $L_p(\Omega)$ denotes the real linear space of all (equivalence classes of) p-th power Lebesgue-integrable functions $f : \Omega \to \mathbb{R}$ with the norm $\|\cdot\|_{L_p(\Omega)}$ given by

$$\|f\|_{L_p(\Omega)} := \left(\int_\Omega |f(x)|^p\, dx\right)^{\frac{1}{p}} \text{ for all } f \in L_p(\Omega).$$

For every $p \in [1,\infty)$ the real linear spaces $L_p(\Omega)$ are separable Banach spaces. The so-called natural ordering cone is defined by
$$C_{L_p(\Omega)} := \{f \in L_p(\Omega) \mid f(x) \geq 0 \text{ almost everywhere on } \Omega\}.$$
For every $p \in [1,\infty)$ the topological interior of the ordering cone is empty. $C_{L_p(\Omega)}$ is normal for the norm topology for all $p \in [1,\infty)$ and it is weakly Daniell for all $p \in (1,\infty)$. $C_{L_1(\Omega)}$ has a bounded base. The linear space $L_2(\Omega)$ is a real Hilbert space and the quasi-interior $C_{L_2(\Omega)}^{\#}$ of its dual ordering cone is nonempty.

(b) The space $L_\infty(\Omega)$ is defined as the real linear space of all (equivalence classes of) essentially bounded functions $f : \Omega \to \mathbb{R}$ ($\emptyset \neq \Omega \subset \mathbb{R}^n$) with the norm $\|\cdot\|_{L_\infty(\Omega)}$ given by
$$\|f\|_{L_\infty(\Omega)} := \operatorname*{ess\,sup}_{x \in \Omega} \{|f(x)|\} \text{ for all } f \in L_\infty(\Omega).$$
The ordering cone $C_{L_\infty(\Omega)}$ is defined as
$$C_{L_\infty(\Omega)} := \{f \in L_\infty(\Omega) \mid f(x) \geq 0 \text{ almost everywhere on } \Omega\}.$$
It has a nonempty topological interior and it is weak* Daniell.

Example 1.52. Let $\mathcal{D}$ denote the real linear space of real-valued functions with compact support in $\mathbb{R}^n$ having derivatives of all orders. If $\mathcal{D}$ is equipped with the so-called Schwarz topology (e.g., see Peressini [249, p. 66]), then the topological dual space $\mathcal{D}^*$ is the space of distributions. Let $C_\mathcal{D}$ denote the ordering cone in $\mathcal{D}$ which consists of all non-negative functions in $\mathcal{D}$. Then the dual cone $C_{\mathcal{D}^*}$ is an ordering cone for $\mathcal{D}^*$ which is closed and normal.

Notes

Partially ordered linear spaces were investigated already about 60 years ago by Kantorovitch [166], Kakutani [164] and others. For a complete historical review of this mathematical area we refer to Nachbin [228]. Well-known books on partially ordered linear spaces were written by Nakano [229] (see also Fuchs [96]), Nachbin [228], Peressini [249], Vulikh [317] and Jameson [159]; convex cones are also examined by Fuchssteiner-Lusky [97]. But also books on topological linear spaces present several topics on partial orderings, e.g. Kelley-Namioka [170], Schaefer [273], Day [78], Holmes [125] and Cristescu [70]. In vector optimization partially ordered topological linear spaces are investigated by Hurwicz [127], Vogel [313], Kirsch-Warth-Werner [171], Penot [248] and in several papers of Borwein (e.g. [35]).

Vectorial norms were first introduced by Kantorovitch [167] who developed a theory of linear spaces equipped with a vectorial norm.

It should be noted that various notions presented in this book are used differently by some authors; for instance, "cone" and "quasi-interior" sometimes have another meaning.

Chapter 2

Maps on Linear Spaces

In this chapter various important classes of maps are considered for which one obtains interesting results in vector optimization. We especially consider convex maps and their generalizations and also several types of differentials. It is the aim of this chapter to present a brief survey on these maps.

2.1 Convex Maps

The importance of convex maps is based on the fact that the image set of such a map has useful properties. One of these properties is also valid for so-called convex-like maps which are investigated in this section as well.

First, recall the definition of a linear map.

Definition 2.1. Let X and Y be real linear spaces. A map $T : X \to Y$ is called *linear*, if for all $x, y \in X$ and all $\lambda, \mu \in \mathbb{R}$

$$T(\lambda x + \mu y) = \lambda T(x) + \mu T(y).$$

The set of continuous (bounded) linear maps between two real normed spaces $(X, \|\cdot\|_X)$ and $(Y, \|\cdot\|_Y)$ is a linear space as well and it is denoted $B(X,Y)$. With the norm $\|\cdot\| : B(X,Y) \to \mathbb{R}$ given by

$$\|T\| = \sup_{x \neq 0_X} \frac{\|T(x)\|_Y}{\|x\|_X} \text{ for all } T \in B(X,Y)$$

$(B(X,Y), \|\cdot\|)$ is even a normed space.

A linear map defines also a corresponding map as it may be seen in

Definition 2.2. Let X and Y be real separated locally convex linear spaces, and let $T : X \to Y$ be a linear map. A map $T^* : Y^* \to X^*$ given by

$$T^*(y^*)(x) = y^*(T(x)) \text{ for all } x \in X \text{ and all } y^* \in Y^*$$

is called the *adjoint* (or *conjugate* and *dual*, respectively) of T.

It is obvious that the adjoint T^* is also a linear map that it is uniquely determined. Adjoints are useful for the solution of linear functional equations.

Theorem 2.3. Let X and Y be real separated locally convex linear spaces, and let the elements $x \in X$, $x^* \in X^*$, $y \in Y$ and $y^* \in Y^*$ be given.

(a) If there is a linear map $T : X \to Y$ with $y = T(x)$ and $x^* = T^*(y^*)$, then $y^*(y) = x^*(x)$.

(b) If $x \neq 0_X$, $y^* \neq 0_{Y^*}$ and $y^*(y) = x^*(x)$, then there is a continuous linear map $T : X \to Y$ with $y = T(x)$ and $x^* = T^*(y^*)$.

Proof.

(a) Let a linear map $T : X \to Y$ with $y = T(x)$ and $x^* = T^*(y^*)$ be given. Then we get

$$y^*(y) = y^*(T(x)) = T^*(y^*)(x) = x^*(x)$$

which completes the proof.

2.1. Convex Maps

(b) Assume that for $x \neq 0_X$ and $y^* \neq 0_{Y^*}$ the functional equation

$$y^*(y) = x^*(x) \tag{2.1}$$

is satisfied. In the following we consider the two cases $x^*(x) \neq 0$ and $x^*(x) = 0$.

(i) First assume that $x^*(x) \neq 0$. Then we define a map $T : X \to Y$ by

$$T(z) = \frac{x^*(z)}{x^*(x)} y \text{ for all } z \in X. \tag{2.2}$$

Evidently, T is linear and continuous. From (2.1) and (2.2) we conclude $T(x) = y$ and

$$y^*(T(z)) = \frac{x^*(z)}{x^*(x)} y^*(y) = x^*(z) \text{ for all } z \in X$$

which means $x^* = T^*(y^*)$.

(ii) Now assume that $x^*(x) = 0$. Because of $y^* \neq 0_{Y^*}$ there is a $\tilde{y} \neq 0_Y$ with $y^*(\tilde{y}) = 1$. Since in a separated locally convex space X^* separates elements of X, $x \neq 0_X$ implies the existence of some $\tilde{x}^* \in X^*$ with $\tilde{x}^*(x) = 1$. Then we define the map $T : X \to Y$ as follows

$$T(z) = x^*(z)\tilde{y} + \tilde{x}^*(z)y \text{ for all } z \in X. \tag{2.3}$$

It is obvious that T is a continuous linear map. With (2.3) we conclude

$$T(x) = x^*(x)\tilde{y} + \tilde{x}^*(x)y = y.$$

Furthermore, we obtain with (2.3) and (2.1)

$$y^*(T(z)) = x^*(z)y^*(\tilde{y}) + \tilde{x}^*(z)y^*(y) = x^*(z) \text{ for all } z \in X$$

which implies $x^* = T^*(y^*)$. □

The class of linear maps is contained in the class of convex maps.

Definition 2.4. Let X and Y be real linear spaces, C_Y be a convex cone in Y, and let S be a nonempty convex subset of X. A map $f : S \to Y$ is called *convex* (or C_Y-*convex*), if for all $x, y \in S$ and all $\lambda \in [0, 1]$

$$\lambda f(x) + (1 - \lambda)f(y) - f(\lambda x + (1 - \lambda)y) \in C_Y \qquad (2.4)$$

(see Fig. 2.1 and 2.2). A map $f : S \to Y$ is called *concave* (or

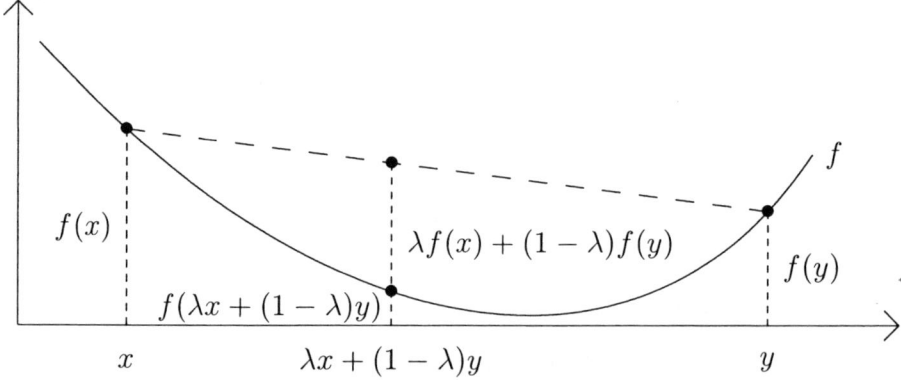

Figure 2.1: Convex functional.

C_Y-*concave*), if $-f$ is convex (see Fig. 2.3).

If $\leq_{C_Y}$ is the partial ordering in Y induced by C_Y, then the condition (2.4) can also be written as

$$f(\lambda x + (1 - \lambda)y) \leq_{C_Y} \lambda f(x) + (1 - \lambda)f(y).$$

If f is a linear map, then f and $-f$ are convex maps.

Definition 2.5. Let X and Y be real linear spaces, let C_Y be a convex cone in Y, let S be a nonempty subset of X, and let $f : S \to Y$ be a given map. The set

$$\operatorname{epi}(f) := \{(x, y) \mid x \in S, \ y \in \{(f(x)\} + C_Y\} \qquad (2.5)$$

2.1. Convex Maps 41

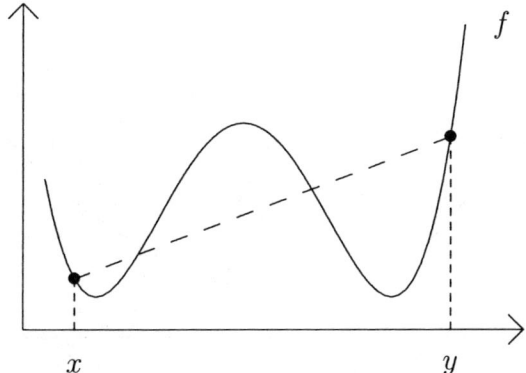

Figure 2.2: Non-convex functional.

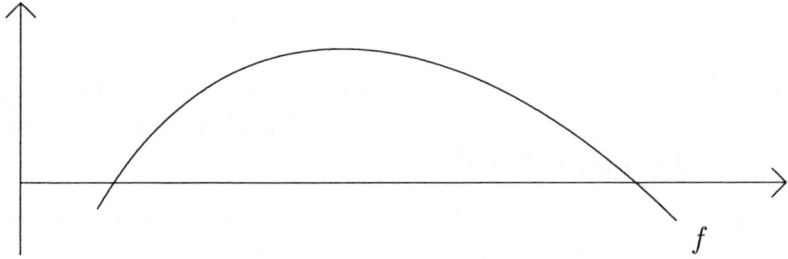

Figure 2.3: Concave functional.

is called the *epigraph* of f (see Fig. 2.4).

Notice that the epigraph in (2.5) can also be written as

$$\mathrm{epi}(f) = \{(x,y) \mid x \in S,\ f(x) \leq_{C_Y} y\}.$$

It turns out that a convex map can be characterized by its epigraph.

Theorem 2.6. *Let X and Y be real linear spaces, let C_Y be a convex cone in Y, let S be a nonempty subset of X and let $f : S \to Y$ be a given map. Then f is convex if and only if $\mathrm{epi}(f)$ is a convex set.*

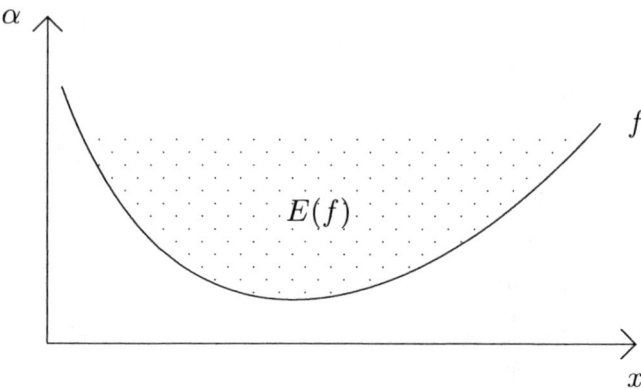

Figure 2.4: Epigraph of a functional.

Proof.

(a) Let f be a convex map (then S is a convex set). For arbitrary $z_1 = (x_1, y_1)$, $z_2 = (x_2, y_2) \in \text{epi}(f)$ and $\lambda \in [0,1]$ we obtain $\lambda x_1 + (1-\lambda)x_2 \in S$ and

$$\begin{aligned} \lambda y_1 + (1-\lambda)y_2 &\in \lambda(\{f(x_1)\} + C_Y) + (1-\lambda)(\{f(x_2)\} + C_Y) \\ &= \{\lambda f(x_1) + (1-\lambda)f(x_2)\} + C_Y \\ &\subset \{f(\lambda x_1 + (1-\lambda)x_2)\} + C_Y. \end{aligned}$$

Consequently, we have $\lambda z_1 + (1-\lambda)z_2 \in \text{epi}(f)$. Thus, $\text{epi}(f)$ is a convex set.

(b) If $\text{epi}(f)$ is a convex set, then S is convex as well. For arbitrary $x_1, x_2 \in S$ and $\lambda \in [0,1]$ we obtain $\lambda(x_1, f(x_1)) + (1-\lambda)(x_2, f(x_2)) \in \text{epi}(f)$ and

$$f(\lambda x_1 + (1-\lambda)x_2) \leq_{C_Y} \lambda f(x_1) + (1-\lambda)f(x_2).$$

Hence, f is a convex map. □

Next, we list some other properties of convex maps.

Lemma 2.7. *Let X, Y and Z be real linear spaces, let C_Y and C_Z be convex cones in Y and Z, respectively, and let S be a nonempty convex subset of X.*

2.1. Convex Maps

(a) *If $g : S \to Y$ is an affine linear map (i.e. there is a $b \in Y$ and a linear map $L : S \to Y$ with $g(x) = b + L(x)$ for all $x \in S$) and $f : Y \to Z$ is a convex map, then the composition $f \circ g$ is a convex map.*

(b) *If $g : S \to Y$ is a convex map and $f : Y \to Z$ is a convex and monotonically increasing map (that is: $y_1 \leq_{C_Y} y_2 \Rightarrow f(y_1) \leq_{C_Z} f(y_2)$), then the composition $f \circ g$ is a convex map.*

Proof. Take arbitrary $x_1, x_2 \in S$ and $\lambda \in [0, 1]$. Then we get for part (a)

$$\begin{aligned}
&\lambda(f \circ g)(x_1) + (1 - \lambda)(f \circ g)(x_2) - (f \circ g)(\lambda x_1 + (1 - \lambda)x_2) \\
&= \lambda f(g(x_1)) + (1 - \lambda)f(g(x_2)) - f(g(\lambda x_1 + (1 - \lambda)x_2)) \\
&= \lambda f(g(x_1)) + (1 - \lambda)f(g(x_2)) - f(\lambda g(x_1) + (1 - \lambda)g(x_2)) \\
&\in C_Z.
\end{aligned}$$

For the proof of part (b) we obtain with the convexity of g

$$\lambda g(x_1) + (1 - \lambda)g(x_2) - g(\lambda x_1 + (1 - \lambda)x_2) \in C_Y$$

and with the monotonicity of f

$$f(\lambda g(x_1) + (1 - \lambda)g(x_2)) - f(g(\lambda x_1 + (1 - \lambda)x_2)) \in C_Z.$$

Since f is also convex, we get

$$\lambda f(g(x_1)) + (1 - \lambda)f(g(x_2)) - f(\lambda g(x_1) + (1 - \lambda)g(x_2)) \in C_Z.$$

Consequently, we conclude

$$\lambda f(g(x_1)) + (1 - \lambda)f(g(x_2)) - f(g(\lambda x_1 + (1 - \lambda)x_2)) \in C_Z$$

and

$$\lambda(f \circ g)(x_1) + (1 - \lambda)(f \circ g)(x_2) - (f \circ g)(\lambda x_1 + (1 - \lambda)x_2) \in C_Z.$$

□

In vector optimization one is often merely concerned with the convexity of the set $f(S)+C_Y$ instead of epi(f). In this case the notion of convexity of f can be relaxed because the convexity of $f(S)+C_Y$ depends only on a property of the convex hull of $f(S)$.

Lemma 2.8. *Let X and Y be real linear spaces, let C_Y be a convex cone in Y, let S be a nonempty subset of X and let $f : S \to Y$ be a given map. Then the set $f(S) + C_Y$ is convex if and only if*

$$\mathrm{co}(f(S)) \subset f(S) + C_Y. \tag{2.6}$$

Proof.

(a) If the set $f(S) + C_Y$ is convex, then with Remark 7

$$\mathrm{co}(f(S)) \subset \mathrm{co}(f(S)) + C_Y = \mathrm{co}(f(S) + C_Y) = f(S) + C_Y.$$

(b) If the inclusion (2.6) is true, then

$$\mathrm{co}(f(S) + C_Y) = \mathrm{co}(f(S)) + C_Y \subset f(S) + C_Y$$

which implies that the set $f(S) + C_Y$ is convex. □

The inclusion (2.6) is used for the definition of convex-like maps.

Definition 2.9. *Let X and Y be real linear spaces, let C_Y be a convex cone, let S be a nonempty subset of X and let $f : S \to Y$ be a given map. Then f is called convex-like, if for every $x, y \in S$ and every $\lambda \in [0,1]$ there is an $s \in S$ with*

$$\lambda f(x) + (1-\lambda)f(y) - f(s) \in C_Y$$

(or: $f(s) \leq_{C_Y} \lambda f(x) + (1-\lambda)f(y)$).

Example 2.10.

(a) Obviously, every convex map is convex-like.

(b) Let the map $f : [\pi, \infty) \to \mathbb{R}^2$ be given by

$$f(x) = (x, \sin x) \text{ for all } x \in [\pi, \infty)$$

where $\mathbb{R}^2$ is partially ordered in the componentwise sense. The map f is convex-like but it is not convex.

Example 2.10, (b) shows that the class of convex-like maps is even much larger than the class of convex maps. With Lemma 2.8 we get immediately the following

Theorem 2.11. *Let X and Y be real linear spaces, let C_Y be a convex cone in Y, let S be a nonempty set and let $f : S \to Y$ be a given map. Then the map f is convex-like if and only if the set $f(S) + C_Y$ is convex (see Fig. 2.5).*

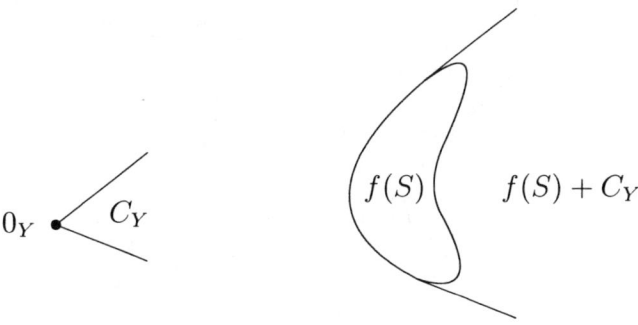

Figure 2.5: Convex-like map f.

2.2 Differentiable Maps

In the context with optimality conditions we have to work with generalized derivatives of maps. Therefore, we discuss various differentiability notions and we investigate the relationships among them.

Definition 2.12. Let X be a real linear space, let Y be a real topological linear space, let S be a nonempty subset of X, and let $f : S \to Y$ be a given map.

(a) If for two elements $\bar{x} \in S$ and $h \in X$ the limit
$$f'(\bar{x})(h) := \lim_{\lambda \to 0_+} \frac{1}{\lambda}(f(\bar{x} + \lambda h) - f(\bar{x}))$$
exists, then $f'(\bar{x})(h)$ is called the *directional derivative* of f at $\bar{x}$ in the direction h. If this limit exists for all $h \in X$, then f is called *directionally differentiable* at $\bar{x}$ (see Fig. 2.6).

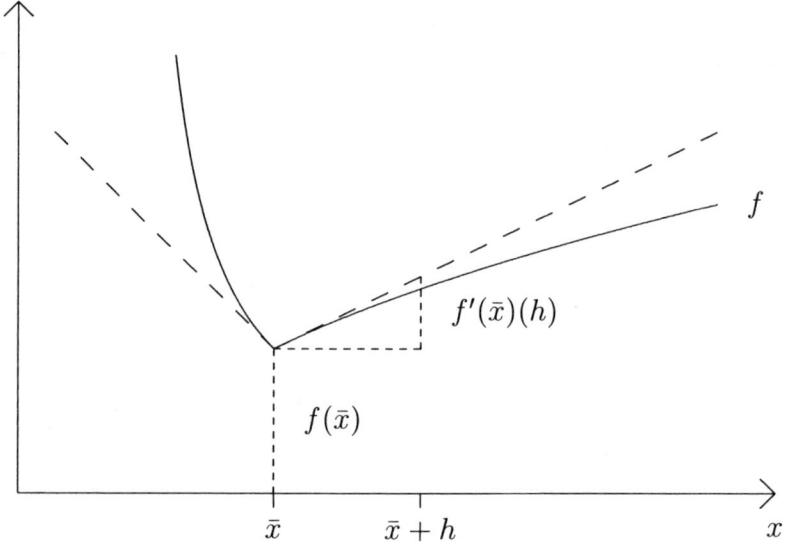

Figure 2.6: Directionally differentiable function.

(b) If for some $\bar{x} \in S$ and all $h \in X$ the limit
$$f'(\bar{x})(h) := \lim_{\lambda \to 0} \frac{1}{\lambda}(f(\bar{x} + \lambda h) - f(\bar{x}))$$
exists and if $f'(\bar{x})$ is a continuous linear map from X to Y, then $f'(\bar{x})$ is called the *Gâteaux derivative* of f at $\bar{x}$ and f is called *Gâteaux differentiable* at $\bar{x}$.

2.2. Differentiable Maps

Notice that for the limit defining the directional and Gâteaux derivative one considers arbitrary nets $(\lambda_i)_{i\in\mathbb{N}}$ converging to 0, $\lambda_i > 0$ for all $i \in \mathbb{N}$ in part (a), with the additional property that $\bar{x} + \lambda_i h$ belongs to the domain S for all $i \in I$. This restriction of the nets converging to 0 can be dropped, for instance, if S equals the whole space X.

Example 2.13. For the function $f : \mathbb{R}^2 \to \mathbb{R}$ with

$$f(x_1, x_2) = \left\{ \begin{array}{ll} x_1^2(1 + \frac{1}{x_2}) & \text{if } x_2 \neq 0 \\ 0 & \text{if } x_2 = 0 \end{array} \right\} \text{ for all } (x_1, x_2) \in \mathbb{R}^2$$

which is not continuous at $0_{\mathbb{R}^2}$, we obtain the directional derivative

$$f'(0_{\mathbb{R}^2})(h_1, h_2) = \lim_{\lambda \to 0_+} \frac{1}{\lambda} f(\lambda(h_1, h_2)) = \left\{ \begin{array}{ll} \frac{h_1^2}{h_2} & \text{if } h_2 \neq 0 \\ 0 & \text{if } h_2 = 0 \end{array} \right\}$$

in the direction $(h_1, h_2) \in \mathbb{R}^2$. Notice that $f'(0_{\mathbb{R}^2})$ is neither continuous nor linear.

Sometimes it is very useful to have a derivative notion which does not require any topology in Y. A possible generalization of a directional derivative which will be used in the second part of this book is given by

Definition 2.14. Let X and Y be real linear spaces, let S be a nonempty subset of X and let T be a nonempty subset of Y. Moreover, let a map $f : S \to Y$ and an element $\bar{x} \in S$ be given. A map $f'(\bar{x}) : S - \{\bar{x}\} \to Y$ is called a *directional variation* of f at $\bar{x}$ with respect to T, if the following holds: Whenever there is an element $x \in S$ with $x \neq \bar{x}$ and $f'(\bar{x})(x - \bar{x}) \in T$, then there is a $\lambda > 0$ with

$$\bar{x} + \lambda(x - \bar{x}) \in S \text{ for all } \lambda \in (0, \bar{\lambda}]$$

and

$$\frac{1}{\lambda}(f(\bar{x} + \lambda(x - \bar{x})) - f(\bar{x})) \in T \text{ for all } \lambda \in (0, \bar{\lambda}].$$

Example 2.15. Let X be a real linear space, let Y be a real topological linear space, and let S be a nonempty subset of X. Further, let $f : S \to Y$ be a given map, and let $x, \bar{x} \in S$ with $x \neq \bar{x}$ be fixed. Assume that there is a $\bar{\lambda} > O$ with

$$\bar{x} + \lambda(x - \bar{x}) \in S \text{ for all } \lambda \in (0, \bar{\lambda}].$$

(a) If $f'(\bar{x})$ is the directional derivative of f at $\bar{x}$ in the direction $x - \bar{x}$, then $f'(\bar{x})$ is a directional variation of f at $\bar{x}$ with respect to all nonempty open subsets of Y.

(b) Let f be an affine linear map, i.e. there is a $b \in Y$ and a linear map $L : S \to Y$ with

$$f(x) = b + L(x) \text{ for all } x \in S.$$

If for some nonempty set $T \subset Y$ $L(x - \bar{x}) \in T$, then

$$\frac{1}{\lambda}(f(\bar{x} + \lambda(x - \bar{x})) - f(\bar{x})) = L(x - \bar{x}) \in T \text{ for all } \lambda \in (0, \bar{\lambda}].$$

Consequently, L is the directional variation of f at $\bar{x}$ with respect to all nonempty sets $T \subset Y$.

A less general but more satisfying derivative notion may be obtained in normed spaces.

Definition 2.16. Let $(X, \|\cdot\|_X)$ and $(Y, \|\cdot\|_Y)$ be real normed spaces, let S be a nonempty open subset of X, and let $f : S \to Y$ be a given map. Furthermore let an element $\bar{x} \in S$ be given. If there is a continuous linear map $f'(\bar{x}) : X \to Y$ with the property

$$\lim_{\|h\|_X \to 0} \frac{\|f(\bar{x} + h) - f(\bar{x}) - f'(\bar{x})(h)\|_Y}{\|h\|_X} = 0,$$

then $f'(\bar{x})$ is called the *Fréchet derivative* of f at $\bar{x}$ and f is called *Fréchet differentiable* at $\bar{x}$.

According to this definition we obtain for Fréchet derivatives with the notations used above

$$f(\bar{x} + h) = f(\bar{x}) + f'(\bar{x})(h) + o(\|h\|_X)$$

2.2. Differentiable Maps

where the expression $o(\|h\|_X)$ of this Taylor series has the property
$$\lim_{\|h\|_X \to 0} \frac{o(\|h\|_X)}{\|h\|_X} = \lim_{\|h\|_X \to 0} \frac{f(\bar{x}+h) - f(\bar{x}) - f'(\bar{x})(h)}{\|h\|_X} = 0_Y.$$

With the next three assertions we present some known results on Fréchet differentiability.

Lemma 2.17. *Let $(X, \|\cdot\|_X)$ and $(Y, \|\cdot\|_Y)$ be real normed spaces, let S be a nonempty open subset of X, and let $f : S \to Y$ be a given map. If the Fréchet derivative of f at some $\bar{x} \in S$ exists, then the Gâteaux derivative of f at $\bar{x}$ exists as well and both are equal.*

Proof. Let $f'(\bar{x})$ denote the Fréchet derivative of f at $\bar{x}$. Then we have
$$\lim_{\lambda \to 0} \frac{\|f(\bar{x}+\lambda h) - f(\bar{x}) - f'(\bar{x})(\lambda h)\|_Y}{\|\lambda h\|_X} = 0 \text{ for all } h \in X \backslash \{0_X\}$$
implying
$$\lim_{\lambda \to 0} \frac{1}{|\lambda|} \|f(\bar{x}+\lambda h) - f(\bar{x}) - f'(\bar{x})(\lambda h)\|_Y = 0 \text{ for all } h \in X \backslash \{0_X\}.$$
Because of the linearity of $f'(\bar{x})$ we obtain
$$\lim_{\lambda \to 0} \frac{1}{\lambda}[f(\bar{x}+\lambda h) - f(\bar{x})] = f'(\bar{x})(h) \text{ for all } h \in X.$$
□

Corollary 2.18. *Let $(X, \|\cdot\|_X)$ and $(Y, \|\cdot\|_Y)$ be real normed spaces, let S be a nonempty open subset of X, and let $f : S \to Y$ be a given map. If f is Fréchet differentiable at some $\bar{x} \in S$, then the Fréchet derivative is uniquely determined.*

Proof. With Lemma 2.17 the Fréchet derivative coincides with the Gâteaux derivative. Since the Gâteaux derivative is as a limit uniquely determined, the Fréchet derivative is also uniquely determined.
□

The following lemma says that Fréchet differentiability implies continuity as well.

Lemma 2.19. *Let $(X, \|\cdot\|_X)$ and $(Y, \|\cdot\|_Y)$ be real normed spaces, let S be a nonempty open subset of X, and let $f : S \to Y$ be a given map. If f is Fréchet differentiable at some $\bar{x} \in S$, then f is continuous at $\bar{x}$.*

Proof. To a sufficiently small $\varepsilon > 0$ there is a ball around $\bar{x}$ so that for all $\bar{x} + h$ of this ball

$$\|f(\bar{x} + h) - f(\bar{x}) - f'(\bar{x})(h)\|_Y \leq \varepsilon \|h\|_X.$$

Then we conclude for some $\alpha > 0$

$$\begin{aligned}
&\|f(\bar{x} + h) - f(\bar{x})\|_Y \\
&= \|f(\bar{x} + h) - f(\bar{x}) - f'(\bar{x})(h) + f'(\bar{x})(h)\|_Y \\
&\leq \|f(\bar{x} + h) - f(\bar{x}) - f'(\bar{x})(h)\|_Y + \|f'(\bar{x})(h)\|_Y \\
&\leq \varepsilon \|h\|_X + \alpha \|h\|_X \\
&= (\varepsilon + \alpha) \|h\|_X.
\end{aligned}$$

Consequently f is continuous at $\bar{x}$. □

The following theorem gives a characterization of a convex Fréchet differentiable map.

Theorem 2.20. *Let $(X, \|\cdot\|_X)$ and $(Y, \|\cdot\|_Y)$ be real normed spaces, let S be a nonempty open convex subset of X, let C_Y be a closed convex cone in Y, and let a map $f : S \to Y$ be given which is Fréchet differentiable at every $x \in S$. Then the map f is convex if and only if*

$$f(y) + f'(y)(x - y) \leq_{C_Y} f(x) \text{ for all } x, y \in S$$

(see Fig. 2.7).

2.2. Differentiable Maps

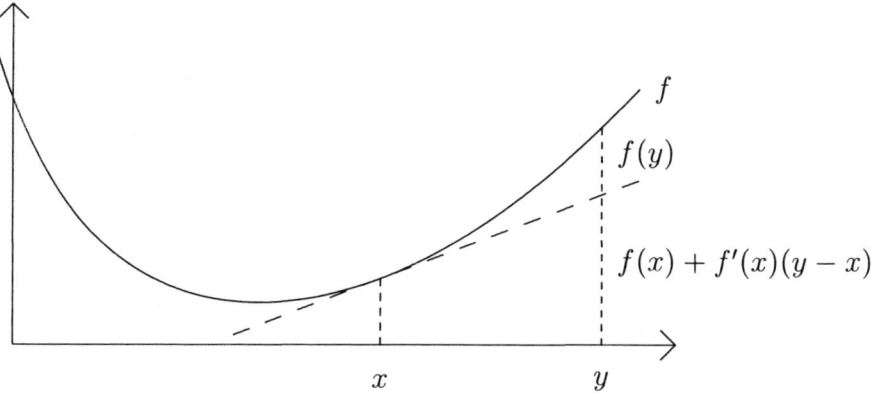

Figure 2.7: Illustration of the result of Thm. 2.20.

Proof.

(a) First, we assume that the map f is convex. Then it follows for all $x, y \in S$ and all $\lambda \in (0, 1]$

$$\lambda f(x) + (1 - \lambda) f(y) - f(\lambda x + (1 - \lambda) y) \in C_Y$$

and

$$f(x) - f(y) - \frac{1}{\lambda}(f(y + \lambda(x - y)) - f(y)) \in C_Y.$$

Since f is assumed to be Fréchet differentiable at y and C_Y is closed, we conclude

$$f(x) - f(y) - f'(y)(x - y) \in C_Y$$

or alternatively

$$f(y) + f'(y)(x - y) \leq_{C_Y} f(x).$$

(b) Next, we assume that

$$f(y) + f'(y)(x - y) \leq_{C_Y} f(x) \text{ for all } x, y \in S.$$

S is convex and, therefore, we obtain for all $x, y \in S$ and all $\lambda \in [0, 1]$

$$f(x) - f(\lambda x + (1 - \lambda) y) - f'(\lambda x + (1 - \lambda) y)((1 - \lambda)(x - y)) \in C_Y$$

and
$$f(y) - f(\lambda x + (1-\lambda)y) - f'(\lambda x + (1-\lambda)y)(-\lambda(x-y)) \in C_Y.$$

Since C_Y is a convex cone and Fréchet derivatives are linear maps, we get
$$\begin{aligned}&\lambda f(x) - \lambda f(\lambda x + (1-\lambda)y)\\&-\lambda(1-\lambda)f'(\lambda x + (1-\lambda)y)(x-y)\\&+(1-\lambda)f(y) - (1-\lambda)f(\lambda x + (1-\lambda)y)\\&-(1-\lambda)\lambda f'(\lambda x + (1-\lambda)y)(x-y)\\&\in C_Y\end{aligned}$$

which implies
$$\lambda f(x) + (1-\lambda)f(y) - f(\lambda x + (1-\lambda)y) \in C_Y.$$

Hence, f is a convex map.

□

The characterization of convex Fréchet differentiable maps presented in Theorem 2.20 is very helpful for the investigation of optimality conditions in vector optimization. This result leads to a generalization of the (Fréchet) derivative for convex maps which are not (Fréchet) differentiable.

Definition 2.21. Let X and Y be real topological linear spaces, let C_Y be a convex cone in Y, and let $f : X \to Y$ be a given map. For an arbitrary $\bar{x} \in X$ the set

$$\partial f(\bar{x}) := \{T \in B(X,Y) \mid f(\bar{x}+h) - f(\bar{x}) - T(h) \in C_Y \text{ for all } h \in X\}$$

(where $B(X,Y)$ denotes the linear space of the continuous linear maps from X to Y) is called the *subdifferential* of f at $\bar{x}$. Every $T \in \partial f(\bar{x})$ is called a *subgradient* of f at $\bar{x}$ (see Fig. 2.8).

Example 2.22. Let X and Y be real topological linear spaces, let C_Y be a pointed convex cone in Y, and let $\|\cdot\| : X \to Y$ be a vectorial norm. Then we have for every $\bar{x} \in X$

$$\partial \|\bar{x}\| = \{T \in B(X,Y) \mid T(\bar{x}) = \|\bar{x}\| \text{ and } T(x) \leq_{C_Y} \|x\| \text{ for all } x \in X\}.$$

2.2. Differentiable Maps

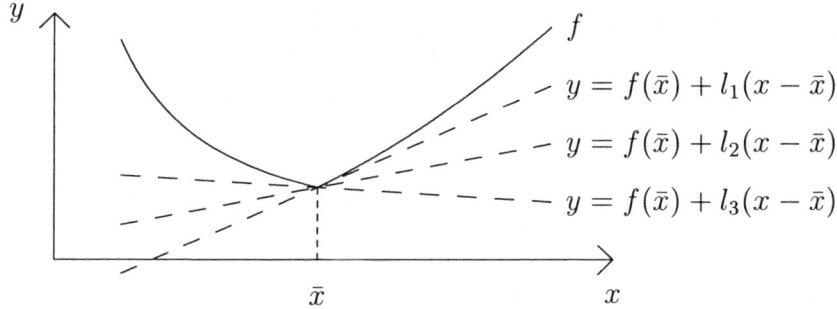

Figure 2.8: Subgradients of a convex functional.

Proof.

(a) First, choose an arbitrary $T \in B(X,Y)$ with $T(\bar{x}) = \|\bar{x}\|$ and
$$\|x\| - T(x) \in C_Y \text{ for all } x \in X.$$
Then we obtain for all $h \in X$
$$\begin{aligned}\|\bar{x} + h\| - \|\bar{x}\| - T(h) &= \|\bar{x} + h\| - T(\bar{x} + h) - \|\bar{x}\| + T(\bar{x}) \\ &\in C_Y\end{aligned}$$
which implies $T \in \partial \|\bar{x}\|$.

(b) Next, assume that any $T \in \partial \|\bar{x}\|$ is given. Then we get
$$\|\bar{x}\| - T(\bar{x}) = \|\bar{x} + \bar{x}\| - \|\bar{x}\| - T(\bar{x}) \in C_Y$$
and
$$-\|\bar{x}\| + T(\bar{x}) = \|\bar{x} - \bar{x}\| - \|\bar{x}\| - T(-\bar{x}) \in C_Y.$$
Since C_Y is pointed, we conclude
$$\|\bar{x}\| - T(\bar{x}) \in (-C_Y) \cap C_Y = \{0_Y\}$$
which means $T(\bar{x}) = \|\bar{x}\|$. Finally, we obtain
$$\begin{aligned}\|x\| - T(x) &\in \{\|x + \bar{x}\| - \|\bar{x}\| - T(x)\} + C_Y \\ &\subset C_Y + C_Y = C_Y \text{ for all } x \in X.\end{aligned}$$

This completes the proof. □

The next example is a special case of Example 2.22.

Example 2.23. Let $(X, \|\cdot\|_X)$ be a real normed space. Then we have for every $\bar{x} \in X$

$$\partial \|\bar{x}\|_X = \left\{ \begin{array}{l} \{x^* \in X^* \mid x^*(\bar{x}) = \|\bar{x}\|_X \text{ and } \|x^*\|_{X^*} = 1\} \text{ if } \bar{x} \neq 0_X \\ \{x^* \in X^* \mid \|x^*\|_{X^*} \leq 1\} \text{ if } \bar{x} = 0_X \end{array} \right\}.$$

Proof. The assertion follows directly from the preceding example for $Y = \mathbb{R}$ and $C_Y = \mathbb{R}_+$, if we notice that

$$\|x^*\|_{X^*} \leq 1 \iff x^*(x) \leq \|x\|_X \text{ for all } x \in X.$$

□

As a result of Example 2.23 the subdifferential of the norm at 0_X in a real normed space X coincides with the closed unit ball of the dual space.

With the following sequence of assertions it can be shown under appropriate assumptions that the subdifferential of a vectorial norm can be used in order to characterize the directional derivative of such a norm.

Lemma 2.24. Let X be a real linear space, let Y be a real topological linear space, let C_Y be a convex cone in Y which is Daniell, and let $\|\cdot\| : X \to Y$ be a vectorial norm. Then the directional derivative of the vectorial norm exists at every $\bar{x} \in X$ and in every direction $h \in X$.

Proof. Let $f : X \to Y$ be an arbitrary convex map with $f(0_X) = 0_Y$. Then we obtain for all $x \in X$ and all $\alpha, \beta \in \mathbb{R}$ with $0 < \alpha \leq \beta$

$$\frac{\alpha}{\beta} f(\beta x) - f(\alpha x) = \frac{\alpha}{\beta} f(\beta x) + \frac{\beta - \alpha}{\beta} f(0_X) - f\left(\frac{\alpha}{\beta}\beta x + \frac{\beta - \alpha}{\beta} 0_X\right) \in C_Y$$

resulting in

$$\frac{1}{\beta} f(\beta x) - \frac{1}{\alpha} f(\alpha x) \in C_Y.$$

2.2. Differentiable Maps

If we take especially

$$f(x) = \|\bar{x} + x\| - \|\bar{x}\| \text{ for all } x \in X,$$

then f is convex and $f(0_X) = 0_Y$. Hence, the above result applies to this special f, that is

$$\frac{1}{\beta}(\|\bar{x} + \beta x\| - \|\bar{x}\|) - \frac{1}{\alpha}(\|\bar{x} + \alpha x\| - \|\bar{x}\|) \in C_Y \quad (2.7)$$

for all $x \in X$ and all real numbers α, β with $0 < \alpha \leq \beta$.

Next, we show that the difference quotient which appears in the definition of the directional derivative is bounded. Since the vectorial norm is a convex map, we get for all $x \in X$ and all $\lambda > 0$

$$\frac{1}{1+\lambda}\|\bar{x} + \lambda x\| + \frac{\lambda}{1+\lambda}\|\bar{x} - x\| - \|\bar{x}\|$$

$$= \frac{1}{1+\lambda}\|\bar{x} + \lambda x\| + \frac{\lambda}{1+\lambda}\|\bar{x} - x\|$$

$$- \left\|\frac{1}{1+\lambda}(\bar{x} + \lambda x) + \frac{\lambda}{1+\lambda}(\bar{x} - x)\right\|$$

$$\in C_Y$$

implying

$$\frac{1}{\lambda}(\|\bar{x} + \lambda x\| - \|\bar{x}\|) \in \{\|\bar{x}\| - \|\bar{x} - x\|\} + C_Y.$$

This condition means that $\|\bar{x}\| - \|\bar{x} - x\|$ is, for every $\lambda > 0$, a lower bound of the difference quotient $\frac{1}{\lambda}(\|\bar{x} + \lambda x\| - \|\bar{x}\|)$. Since C_Y is assumed to be Daniell, we conclude with the condition (2.7) and the boundedness property that the directional derivative of the vectorial norm exists at every $\bar{x} \in X$ and in every direction $h \in X$. □

Lemma 2.25. *Let $(X, \|\cdot\|_X)$ and $(Y, \|\cdot\|_Y)$ be real reflexiv e Banach spaces, and let C_Y be a closed convex cone in Y which is Daniell and has a weakly compact base. If $\|\cdot\| : X \to Y$ is a vectorial norm* which is continuous at an $\bar{x} \in X$, then we have for the directional derivative at $\bar{x} \in X$ in every direction $h \in X$

$$T(h) \leq_{C_Y} \|\bar{x}\|'(h) \text{ for all } T \in \partial\|\bar{x}\|.$$

Proof. Notice that with Lemma 2.24 the directional derivative $\|\bar{x}\|'(h)$ exists for all $\bar{x}, h \in X$. By a result of Zowe [341] the subdifferential $\partial\|\bar{x}\|$ is nonempty. For every $\bar{x}, h \in X$ we get

$$\|\bar{x} + \lambda h\| - \|\bar{x}\| \in \{T(\bar{x} + \lambda h) - T(\bar{x})\} + C_Y$$
$$= \{\lambda T(h)\} + C_Y \text{ for all } \lambda > 0 \text{ and all } T \in \partial\|\bar{x}\|.$$

Consequently, we have

$$\frac{1}{\lambda}(\|\bar{x} + \lambda h\| - \|\bar{x}\|) \in \{T(h)\} + C_Y \text{ for all } \lambda > 0 \text{ and all } T \in \partial\|\bar{x}\|.$$

Since C_Y is closed, we conclude

$$\|\bar{x}\|'(h) \in \{T(h)\} + C_Y$$

which leads to the assertion. □

For the announced characterization result of the directional derivative of a vectorial norm we need a special lemma on subdifferentials.

Lemma 2.26. *Let $(X, \|\cdot\|_X)$ and $(Y, \|\cdot\|_Y)$ be real reflexive Banach spaces, and let C_Y be a convex cone in Y with a weakly compact base. If $f : X \to Y$ is a convex map which is continuous at some $\bar{x} \in X$, then*

$$t \circ \partial f(\bar{x}) = \partial(t \circ f)(\bar{x}) \text{ for all } t \in C_{Y^*}.$$

A proof of this lemma may be found in a paper of Zowe [341] even in a more general form (compare also Valadier [307] and Borwein [35, p. 437]).

Theorem 2.27. *Let $(X, \|\cdot\|_X)$ and $(Y, \|\cdot\|_Y)$ be real reflexive Banach spaces, and let C_Y be a closed convex cone in Y which is*

2.2. Differentiable Maps

Daniell and has a weakly compact base. If $\|\cdot\| : X \to Y$ is a vectorial norm which is continuous at an $\bar{x} \in X$, then the directional derivative of f at $\bar{x}$ in every direction h is given by

$$\|\bar{x}\|'(h) = \max \{T(h) \mid T \in B(X,Y), T(\bar{x}) = \|\bar{x}\|$$
$$\text{and } \|x\| - T(x) \in C_Y \text{ for all } x \in X\}$$

which means that there is a $\bar{T} \in B(X,Y)$ with $\bar{T}(\bar{x}) = \|\bar{x}\|$ and

$$\|x\| - \bar{T}(x) \in C_Y \text{ for all } x \in X$$

so that

$$\|\bar{x}\|'(h) = \bar{T}(h)$$

and

$$\|\bar{x}\|'(h) \in \{T(h)\} + C_Y \text{ for all } T \in B(X,Y) \text{ with } T(\bar{x}) = \|\bar{x}\|$$
$$\text{and } \|x\| - T(x) \in C_Y \text{ for all } x \in X.$$

Proof. Take any direction $h \in X$. From Example 2.22 and Lemma 2.25 we obtain immediately

$$\|\bar{x}\|'(h) \in \{T(h)\} + C_Y \text{ for all } T \in B(X,Y) \text{ with } T(\bar{x}) = \|\bar{x}\|$$
$$\text{and } \|x\| - T(x) \in C_Y \text{ for all } x \in X.$$

Therefore, we have only to show that there is a $\bar{T} \in \partial\|\bar{x}\|$ with $\|\bar{x}\|'(h) = \bar{T}(h)$.

With Corollary 3.19 (which will be stated later) there is a continuous linear functional $t \in C_{Y*}^{\#}$. Then we consider the functional $f := t \circ \|\cdot\| : X \to \mathbb{R}$. f is continuous at $\bar{x}$ and with Lemma 2.7, (b) it is even convex. With Lemma 2.25 we conclude

$$f'(\bar{x})(h) \geq \sup \{x^*(h) \mid x^* \in \partial f(\bar{x})\},$$

and since $\partial f(\bar{x}$ is weak*-compact in X^*, this supremum is actually attained, that is

$$f'(\bar{x})(h) \geq \max \{x^*(h) \mid x^* \in \partial f(\bar{x})\}.$$

In order to prove the equality we assume that there is an $\alpha \in \mathbb{R}$ with
$$f'(\bar{x})(h) > \alpha > \max\{x^*(h) \mid x^* \in \partial f(\bar{x})\}. \tag{2.8}$$

If S denotes the linear hull of $\{h\}$, we define a linear functional $l : S \to \mathbb{R}$ by
$$l(\lambda h) = \lambda \alpha \text{ for all } \lambda \in \mathbb{R}.$$

Then we get
$$l(\lambda h) \leq \lambda f'(\bar{x})(h) = f'(\bar{x})(\lambda h) \text{ for all } \lambda \in \mathbb{R}.$$

Since $f'(\bar{x})$ is sublinear, there is a continuous extension $\bar{l}$ of l on X with
$$\bar{l}(x) \leq f'(\bar{x})(x) \text{ for all } x \in X$$
which implies $\bar{l} \in \partial f(\bar{x})$. But with $\bar{l}(h) = \alpha$ we arrive at a contradiction to (2.8).

Summarizing these results we obtain
$$f'(\bar{x})(h) = \max\{x^*(h) \mid x^* \in \partial f(\bar{x})\}.$$

Consequently, there is an $x^* \in \partial f(\bar{x})$ with
$$f'(\bar{x})(h) = x^*(h).$$

With Lemma 2.26 there is a $\bar{T} \in \partial \|\bar{x}\|$ with $x^* = t \circ \bar{T}$ and we get
$$t \circ \|\bar{x}\|'(h) = (t \circ \|\bar{x}\|)'(h) = t \circ \bar{T}(h). \tag{2.9}$$

Assume that $\|\bar{x}\|'(h) \neq \bar{T}(h)$. Then we get from Lemma 2.25
$$\|\bar{x}\|'(h) - \bar{T}(h) \in C_Y \setminus \{0_Y\}$$

and, therefore,
$$t \circ \|\bar{x}\|'(h) - t \circ \bar{T}(h) > 0$$
which contradicts (2.9). Hence, $\|\bar{x}\|'(h) = \bar{T}(h)$ and this completes the proof. □

It should be noted that the assumptions of Theorem 27 are very restrictive (they are fulfilled, for instance, for $Y = \mathbb{R}^n$ and $C_Y = \mathbb{R}^n_+$). The assertion remains valid under even weaker conditions and for these investigations we refer to Borwein [35, p. 437].

Notes

A lot of material on convex functions may be found in the books of Rockafellar [260] and Roberts-Varberg [258]. For investigations on convex relations in analysis we refer to a paper of Borwein [32]. Convex-like maps were first introduced by Vogel [312, p. 165] who also formulated Theorem 2.11. In connection with a minisup theorem Aubin [8, § 13.3] presented a similar statement like Theorem 2.11 for so-called γ-convex functionals.

A survey on differentials in nonlinear functional analysis may be found in the extensive paper of Nashed [232]. The so-called directional variation was introduced by Kirsch-Warth-Werner [171, p. 33] in a more general form; they called it "B-Variation". The differentiability concept used in this book is based on a paper of Jahn-Sachs [156]. For a further generalized differentiability notion compare also the paper of Sachs [269]. The results on Fréchet differentiation can also be found in the books of Luenberger [217] and Jahn [149]. Subdifferentials were introduced by Moreaux and Rockafellar. We restrict ourselves to refer to the lecture notes of Rockafellar [262]. The books of Holmes [125], Ekeland-Temam [92] and Ioffe-Tihomirov [129] also present an interesting overview on subdifferentials and their use in optimization. Theorems on subdifferentials in partially ordered linear spaces may be found in the papers of Valadier [307], Zowe [341], Elster-Nehse [93], Penot [247] and Borwein [35].

Much of the work on vectorial norms described in the second section is based on various results of Holmes [125] and Borwein [35].

Chapter 3

Some Fundamental Theorems

For the investigation of vector optimization problems we need various fundamental theorems of convex analysis which are presented in this section. First, we formulate Zorn's lemma and the Hahn-Banach theorem and, as a consequence, we examine several types of separation theorems. Moreover, we discuss a James theorem on the characterization of weakly compact sets and we study two Krein-Rutman theorems on the extension of positive linear functionals and the existence of strictly positive linear functionals. Finally, we prove a Ljusternik theorem on certain tangent cones.

3.1 Zorn's Lemma and the Hahn-Banach Theorem

For the presentation of Zorn's lemma we need some useful definitions.

Definition 3.1. Let S be an arbitrary nonempty set which is partially ordered by a reflexive and transitive binary relation $\leq$ (since S is not assumed to have a linear structure, we do not require the conditions (iii) and (iv) in Definition 1.16, (b) to be satisfied).

(a) The set S is called *totally ordered*, if for all $x, y \in S$ either $x \leq y$ or $y \leq x$ is true.

(b) Let T be a nonempty subset of S. An element $\bar{x} \in S$ is called an *upper bound* of T, if
$$x \leq \bar{x} \text{ for all } x \in T.$$
$\bar{x} \in S$ is called a *lower bound* of T, if
$$\bar{x} \leq x \text{ for all } x \in T.$$

(c) An element $\bar{x} \in S$ is called a *maximal* element of S, if
$$x \in S, \bar{x} \leq x \implies x \leq \bar{x}.$$
$\bar{x} \in S$ is called a *minimal* element of S, if
$$x \in S, x \leq \bar{x} \implies \bar{x} \leq x.$$

(d) The set S is called *inductively ordered from above (from below)*, if every totally ordered subset of S has an upper (lower) bound.

With these notions we are able to formulate *Zorn's lemma*.

Lemma 3.2. *Let S be a nonempty set which is partially ordered by a reflexive and transitive binary relation. If S is inductively ordered from above (from below), then S has at least one maximal (minimal) element.*

Zorn's lemma may be derived from the axiom of choice. A first application of Zorn's lemma leads to a characterization of a base of a cone. This result refines Lemma 1.28. For the proof of the next lemma recall that a subset L of a real linear space is called a *linear manifold*, if
$$x, y \in L, \lambda \in \mathbb{R} \implies \lambda x + (1 - \lambda) y \in L.$$

Lemma 3.3. *Let C_X be a nontrivial convex cone in a real linear space X. A subset B of the ordering cone is a base for C_X if and only if there is a linear functional $x' \in C_{X'}^{\#}$ with*
$$B = \{ x \in C_X \mid x'(x) = 1 \}.$$

3.1. Zorn's Lemma and the Hahn-Banach Theorem

Proof. The "if" part of the assertion follows from Lemma 1.28, (a). Therefore, we assume that B is a base for C_X. Then we consider the set S of all linear manifolds in X containing B but not 0_X. The set S is partially ordered with respect to the set theoretical inclusion. By an application of Zorn's lemma there is a maximal linear manifold $\bar{L}$ in S. And this maximal linear manifold $\bar{L}$ is a hyperplane, that is, there is a linear functional $x' \in X'$ with

$$\bar{L} = \{x \in X \mid x'(x) = 1\}.$$

Then we conclude

$$x'(x) = 1 \text{ for all } x \in B$$

which implies $x' \in C_{X'}^{\#}$, and moreover, we obtain

$$B = \{x \in C_X \mid x'(x) = 1\}.$$

$\square$

The essential difference between Lemma 3.3 and Lemma 1.28 is that we do not need the assumption that the ordering cone C_X is reproducing.

Another important application of Zorn's lemma leads to the famous Hahn-Banach theorem.

Definition 3.4. Let X and Y be real linear spaces, and let C_Y be a convex cone in Y. A map $f : X \to Y$ is called *sublinear*, if (for all $x, y \in X$ and all $\lambda \geq 0$)

(a) $f(\lambda x) = \lambda f(x)$,

(b) $f(x + y) \leq_{C_Y} f(x) + f(y)$.

In the case of $Y = \mathbb{R}$ and $C_Y = \mathbb{R}_+$ we speak of a sublinear functional for which the condition (b) reads as

$$f(x + y) \leq f(x) + f(y).$$

For the proof of the Hahn-Banach theorem we need a couple of lemmas.

Lemma 3.5. *The set S of all sublinear functionals on a real linear space X is inductively ordered from below (with respect to a pointwise ordering).*

Proof. Let $\{f_i\}_{i \in I}$ be a totally ordered subset of S. If we restrict the functionals f_i to the one dimensional subspaces of X, we conclude
$$f(x) := \inf_{i \in I} f_i(x) > -\infty \text{ for all } x.$$
As an infimum of sublinear functionals the functional $f : X \to \mathbb{R}$ is sublinear and a lower bound of the f_i. □

Lemma 3.6. *Let S be a nonempty subset of a real linear space X. Let $g : X \to \mathbb{R}$ be a sublinear functional, and let $h : S \to \mathbb{R}$ be a given functional with*
$$h(x) \leq g(x) \text{ for all } x \in S.$$
Moreover, let $f : X \to \mathbb{R}$ denote a functional given by
$$f(x) = \inf_{\substack{y \in S \\ \lambda > 0}} (g(x + \lambda y) - \lambda h(y)) \text{ for all } x \in X.$$
Then the following holds:

(a) The functional f satisfies the inequality
$$f(x) \leq g(x) \text{ for all } x \in X. \tag{3.1}$$

(b) For a linear functional $l \in X'$ the conditions
$$l(x) \leq g(x) \text{ for all } x \in X. \tag{3.2}$$
and
$$h(x) \leq l(x) \text{ for all } x \in S \tag{3.3}$$
are equivalent to the inequality
$$l(x) \leq f(x) \text{ for all } x \in X. \tag{3.4}$$

3.1. Zorn's Lemma and the Hahn-Banach Theorem

(c) *If the set S is convex and h is concave, then f is a sublinear functional.*

Proof. First, we remark that the functional f is well-defined. For every $x \in X$, $y \in S$ and $\lambda > 0$ we get

$$\lambda h(y) \leq \lambda g(y) = g(\lambda y) \leq g(x + \lambda y) + g(-x)$$

implying

$$-g(-x) \leq g(x + \lambda y) - \lambda h(y).$$

Hence, the infimum exists and the functional f is well-defined.

(a) We show that f is bounded from above by g. With the inequality

$$g(x + \lambda y) - \lambda h(y) \leq g(x) + \lambda(g(y) - h(y))$$
$$\text{for all } x \in X, y \in S \text{ and } \lambda > 0$$

we obtain

$$f(x) \leq g(x) \text{ for all } x \in X.$$

(b) We assume that for some $l \in X'$ the inequalities (3.2) and (3.3) are satisfied. Then we have for all $x \in X$, $y \in S$ and $\lambda > 0$

$$l(x) = l(x + \lambda y) - \lambda l(y) \leq g(x + \lambda y) - \lambda h(y)$$

which implies the inequality (3.4). Conversely, let for some $l \in X'$ the inequality (3.4) be fulfilled. Then with (3.1) the inequality (3.2) holds trivially. For all $x \in S$ we get

$$-l(x) = l(-x) \leq f(-x) \leq g(-x + x) - h(x) = -h(x)$$

and

$$h(x) \leq \lambda(x).$$

Thus, the inequality (3.3) is true.

(c) Finally, we assume that S is convex and h is concave. For all $x \in X$ and all $\mu > 0$ we have

$$\begin{aligned}
\mu f(x) &= \inf_{\substack{y \in S \\ \lambda > 0}} \left(\mu g(x + \lambda y) - \mu \lambda h(y) \right) \\
&= \inf_{\substack{y \in S \\ \lambda > 0}} \left(g(\mu x + \mu \lambda y) - \mu \lambda h(y) \right) \\
&= f(\mu x)
\end{aligned}$$

which means that f is positively homogeneous. Since $f(0_X) = 0$, f is non-negatively homogeneous as well. In order to show that f is also subadditive we take arbitrary elements $u, v \in S$ and $\lambda, \mu > 0$. Then $w := \frac{\lambda}{\lambda+\mu} u + \frac{\mu}{\lambda+\mu} v \in S$ and

$$\begin{aligned}
f(x + y) &< g(x + y + (\lambda + \mu)w) - (\lambda + \mu)h(w) \\
&\leq g(x + \lambda u + y + \mu v) - \lambda h(u) - \mu h(v) \\
&\leq g(x + \mu u) - \lambda h(u) + g(y + \mu v) - \mu h(v).
\end{aligned}$$

Consequently, we conclude

$$f(x+y) \leq f(x) + f(y).$$

Hence, f is a sublinear functional. □

As a consequence of Lemma 3.6 we get the important result that the linear functionals are exactly the minimal elements of the set of all sublinear functionals.

Lemma 3.7. *Let S be the set of all sublinear functionals on a real linear space X which is partially ordered with respect to the pointwise ordering. Then $f \in S$ is a minimal element of S if and only if $f \in X'$.*

Proof.

(a) Let arbitrary $f \in X'$ and $g \in S$ with

$$g(x) \leq f(x) \text{ for all } x \in X$$

3.1. Zorn's Lemma and the Hahn-Banach Theorem

be given. Then we have

$$g(x) \leq f(x) = -f(-x) \leq -g(-x) \leq g(x) \text{ for all } x \in X$$

and, therefore, $f = g$. Consequently, f is a minimal element of S.

(b) Let an arbitrary minimal element g of the set S be given. For any fixed $y \in X$ we define the functional $f : X \to \mathbb{R}$ given by

$$f(x) = \inf_{\lambda > 0} \left(g(x + \lambda y) - \lambda g(y) \right) \text{ for all } x \in X.$$

By Lemma 3.6, (c) (where we set $S := \{y\}$) f is a sublinear functional. Since g is a minimal element of the set S and by Lemma 3.6, (a)

$$f(x) \leq g(x) \text{ for all } x \in X,$$

we conclude $f = g$. Then we get for all $x \in X$

$$g(x) = f(x) \leq g(x + y) - g(y)$$

and

$$g(x + y) \geq g(x) + g(y).$$

But g is also subadditive and, therefore, we conclude

$$g(x + y) = g(x) + g(y).$$

g is also homogeneous because for arbitrary $\mu > 0$ and $x \in X$ the equation

$$0 = g(\mu x - \mu x) = \mu g(x) + g(-\mu x)$$

implies

$$g(-\mu x) = -\mu g(x).$$

Thus, g is a linear functional. □

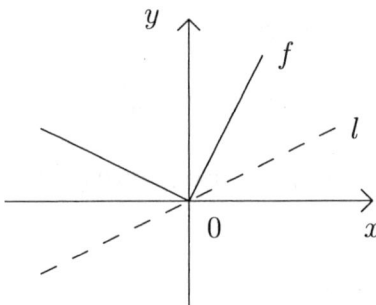

Figure 3.1: Illustration of the result of Thm. 3.8.

Now, we are able to formulate the *basic version of the Hahn-Banach theorem*.

Theorem 3.8. *For every sublinear functional g on a real linear space X there is a linear functional $f \in X'$ with*

$$f(x) \leq g(x) \text{ for all } x \in X$$

(see Fig. 3.1).

Proof. We consider the set S of all sublinear functionals $h : X \to \mathbb{R}$ with
$$h(x) \leq g(x) \text{ for all } x \in X.$$
With Lemma 3.5 the set S is inductively ordered from below (with respect to a pointwise ordering) and by Zorn's lemma S has at least one minimal element f which is, by Lemma 3.7, a linear functional. Finally, we conclude
$$f(x) \leq g(x) \text{ for all } x \in X.$$

□

Another consequence of Lemma 3.6 will be formulated as a *sandwich version of the Hahn-Banach theorem*.

Theorem 3.9. *Let S be a convex subset of a real linear space X. Let $g : X \to \mathbb{R}$ be a sublinear functional, and let $h : S \to \mathbb{R}$ be a*

3.1. Zorn's Lemma and the Hahn-Banach Theorem

concave functional with

$$h(x) \leq g(x) \text{ for all } x \in S.$$

Then there is a linear functional $l \in X'$ with

$$l(x) \leq g(x) \text{ for all } x \in X$$

and

$$h(x) \leq l(x) \text{ for all } x \in S.$$

Proof. We apply the basic version of the Hahn-Banach theorem to the functional f defined in Lemma 3.6. This is possible because, by Lemma 3.6, (c), f is a sublinear functional. Hence, there is a linear functional $l \in X'$ with

$$l(x) \leq f(x) \text{ for all } x \in S$$

and with Lemma 3.6, (b) we obtain directly the desired sandwich result. □

Next, we present the famous *extension theorem*. This theorem is weaker than the sandwich version of the Hahn-Banach theorem.

Theorem 3.10. *Let S be a subspace of a real linear space X. Let $g : X \to \mathbb{R}$ be a sublinear functional, and let $h : S \to \mathbb{R}$ be a linear functional with*

$$h(x) \leq g(x) \text{ for all } x \in S.$$

Then there is a linear functional $l \in X'$ with

$$l(x) \leq g(x) \text{ for all } x \in X$$

and

$$h(x) = l(x) \text{ for all } x \in S.$$

Proof. By Theorem 3.9 there is a linear functional $l \in X'$ with

$$l(x) \leq g(x) \text{ for all } x \in X$$

and
$$h(x) \leq l(x) \text{ for all } x \in S. \tag{3.5}$$
Since h is linear on S, we obtain with Lemma 3.7 that the inequality (3.5) implies
$$h(x) = l(x) \text{ for all } x \in S.$$
This completes the proof. □

Finally, we formulate another consequence of the sandwich version of the Hahn-Banach theorem. This result is a *convex version of the Hahn-Banach theorem*.

Theorem 3.11. *Let S be a nonempty convex subset of a real linear space X, and let $g : X \to \mathbb{R}$ be a sublinear functional. Then there is a linear functional $l \in X'$ with*
$$l(x) \leq g(x) \text{ for all } x \in X$$
and
$$\inf_{x \in S} l(x) = \inf_{x \in S} g(x).$$

Proof. We assume that $\alpha := \inf_{x \in S} g(x)$ is greater than $-\infty$, otherwise the assertion follows immediately from the basic version of the Hahn-Banach theorem. If we define the functional $h : S \to \mathbb{R}$ given by
$$h(x) = \alpha \text{ for all } x \in S,$$
then we obtain from the sandwich version of the Hahn-Banach theorem that there is a linear functional $l \in X'$ with
$$l(x) \leq g(x) \text{ for all } x \in X$$
and
$$\inf_{x \in S} g(x) = h(y) \leq l(y) \text{ for all } y \in S.$$
But that implies also
$$\inf_{x \in S} l(x) = \inf_{x \in S} g(x). \qquad \square$$

Until now, we studied only real-valued sublinear maps. But it is also possible to formulate various versions of the Hahn-Banach theorem for vector-valued sublinear maps. We restrict ourselves to the presentation of a *generalized basic version of the Hahn-Banach theorem*.

Definition 3.12. Let $(Y, \leq)$ be a partially ordered linear space. Then Y is said to have the *least upper bound property*, if every nonempty subset S of Y with an upper bound has a least upper bound, that is, if for every nonempty subset S of Y there is a $y \in Y$ with

$$s \leq y \text{ for all } s \in S,$$

then there is a $\bar{y} \in Y$ with

$$s \leq \bar{y} \text{ for all } s \in S$$

and

$$\bar{y} \leq \tilde{y} \text{ for every } \tilde{y} \in Y$$

with

$$s \leq \tilde{y} \text{ for all } s \in S.$$

Theorem 3.13. *Let X be a real linear space, and let $(Y, \leq)$ be a partially ordered linear space which has the least upper bound property. If $g : X \to Y$ is a sublinear map, then there is a linear map $l : X \to Y$ with*

$$l(x) \leq g(x) \text{ for all } x \in X.$$

For a proof of this generalized Hahn-Banach theorem we refer to Zowe [344, p. 18].

3.2 Separation Theorems

For various profound results in vector optimization separation theorems turn out to be most important. In this section we present a basic version of the separation theorem and we study several other versions which are of practical interest. The basic version of the separation theorem is a direct consequence of the convex version of the Hahn-Banach theorem.

First, we formulate the *basic version of the separation theorem*.

Theorem 3.14. *Let S and T be nonempty convex subsets of a real linear space X with $cor(S) \neq \emptyset$. Then $cor(S) \cap T = \emptyset$ if and only if there are a linear functional $l \in X' \setminus \{0_{X'}\}$ and a real number α with*

$$l(s) \leq \alpha \leq l(t) \text{ for all } s \in S \text{ and all } t \in T \qquad (3.6)$$

and

$$l(s) < \alpha \text{ for all } s \in cor(S) \qquad (3.7)$$

(see Fig. 3.2).

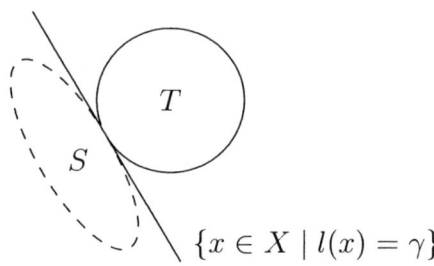

Figure 3.2: Illustration of the result of Thm. 3.14.

Proof.

(a) If there are an $l \in X' \setminus \{0_{X'}\}$ and an $\alpha \in \mathbb{R}$ with the properties (3.6) and (3.7), then it is evident that $cor(S) \cap T = \emptyset$.

(b) Now, we assume that $cor(S) \cap T = \emptyset$. For an arbitrary $\bar{x} \in cor(S)$ we define the translated sets $U := S - \{\bar{x}\}$ and $V := T - \{\bar{x}\}$. Since U is convex and $0_X \in cor(U)$, the Minkowski functional $p : X \to \mathbb{R}$ given by

$$p(x) = \inf_{\lambda > 0} \left\{ \lambda \mid \frac{1}{\lambda} x \in U \right\} \text{ for all } x \in X$$

is sublinear (e.g., see Dunford-Schwartz [84, p. 411]). Then by Theorem 3.11 there is a linear functional $l \in X'$ with

$$l(x) \leq p(x) \text{ for all } x \in X \qquad (3.8)$$

and

$$\inf_{x \in V} l(x) = \inf_{x \in V} p(x). \qquad (3.9)$$

Since

$$p(x) \leq 1 \text{ for all } x \in U,$$

we obtain with (3.8)

$$l(x) \leq 1 \text{ for all } x \in U.$$

Moreover, since

$$p(x) \geq 1 \text{ for } x \notin \text{cor}(U),$$

we conclude with (3.9) and the assumption $\text{cor}(U) \cap V = \emptyset$

$$l(y) \geq 1 \text{ for all } y \in V.$$

Consequently, we have

$$l(x) \leq 1 \leq l(y) \text{ for all } x \in U \text{ and all } y \in V$$

resulting in

$$l(s) \leq 1 + l(\bar{x}) \leq l(t) \text{ for all } s \in S \text{ and all } t \in T.$$

Obviously, l is not the zero functional. Hence, the first part of the assertion is shown. For the proof of the second part we observe only that

$$l(x) \leq p(x) < 1 \text{ for all } x \in \text{cor}(U)$$

which implies

$$l(s) < 1 + l(\bar{x}) \text{ for all } s \in \text{cor}(S).$$

$\square$

The basic version of the separation theorem is formulated in a non-topological setting. In order to get a topological version of the separation theorem we recall a result on the continuity of linear functionals.

Lemma 3.15. *Let X be a real topological linear space. A linear functional $l \in X'$ is discontinuous if and only if for every $\alpha \in \mathbb{R}$ the level set $\{x \in X \mid l(x) = \alpha\}$ is dense in X.*

A proof of this topological result can be found in the book of Holmes [125, p. 63].

With the last lemma we are now able to present the *topological version of the separation theorem* which is also known as *Eidelheit's separation theorem*.

Theorem 3.16. *Let S and T be nonempty convex subsets of a real topological linear space X with $\text{int}(S) \neq \emptyset$. Then $\text{int}(S) \cap T = \emptyset$ if and only if there are a continuous linear functional $l \in X^* \setminus \{0_X^*\}$ and a real number α with*

$$l(s) \leq \alpha \leq l(t) \text{ for all } s \in S \text{ and all } t \in T$$

and

$$l(s) < \alpha \text{ for all } s \in \text{int}(S). \tag{3.10}$$

Proof. With Lemma 1.32, (a) we have $\text{int}(S) = \text{cor}(S)$ and with the basic version of the separation theorem the assertion follows immediately, if we show the continuity of l. The inequality (3.10) implies that $\{x \in X \mid l(x) = \alpha\}$ is not dense in X. Consequently, with Lemma 3.15 the linear functional l is also continuous. This completes the proof. □

A well-known consequence of Eidelheit's separation theorem is that the dual space of a locally convex Hausdorff space X separates elements of X.

3.2. Separation Theorems

Corollary 3.17. *For every nonzero element x in a real separated locally convex space there is a continuous linear functional $l \in X^* \setminus \{0_{X^*}\}$ with $l(x) \neq 0$.*

Proof. Since x is nonzero, there is a convex 0_X-neighborhood that does not contain x. Then the assertion follows directly from Theorem 3.16. □

Next, we study two separation theorems which are helpful in locally convex spaces.

Theorem 3.18. *Let S be a nonempty closed convex subset of a real locally convex space X. Then $x \in X \setminus S$ if and only if there are a continuous linear functional $l \in X^* \setminus \{0_{X^*}\}$ and a real number α with*

$$l(x) < \alpha \leq l(s) \text{ for all } s \in S. \tag{3.11}$$

Proof.

(a) If for any $x \in X$ there are an $l \in X^* \setminus \{0_{X^*}\}$ and an $\alpha \in \mathbb{R}$ with the property (3.11), then we conclude immediately $x \notin S$.

(b) Take an arbitrary $x \in X \setminus S$. Since S is closed, there is a convex neighborhood N of x with $N \cap S = \emptyset$. By Theorem 3.16 there are a continuous linear functional $l \in X^* \setminus \{0_{X^*}\}$ and a real number α with
$$l(x) < \alpha \leq l(s) \text{ for all } s \in S.$$
□

With Lemma 3.3 we know that a base of a convex cone C_X in a real linear space X can be characterized by a linear functional $l \in C_{X'}^{\#}$. The question under which assumption the functional l is even continuous is answered in

Corollary 3.19. *Let C_X be a convex cone in a real locally convex space X. If C_X has a base, then the quasi-interior $C_{X^*}^{\#}$ of the topological dual cone for C_X is nonempty.*

Proof. Let B denote a base of the convex cone C_X. From the definition of B it follows that $0_X \notin \text{lin}(B)$ and with Lemma 1.32, (c) we conclude even $0_X \notin \bar{B}$. By Theorem 3.18 there are an $l \in X^* \setminus \{0_{X^*}\}$ and an $\alpha \in \mathbb{R}$ with

$$0 < \alpha \leq l(b) \text{ for all } b \in B.$$

Every $x \in C_X \setminus \{0_X\}$ can be uniquely represented as $x = \lambda b$ with a $\lambda > 0$ and a $b \in B$. Consequently, we get for every $x \in C_X \setminus \{0_X\}$

$$l(x) = \lambda l(b) > 0$$

which implies $l \in C_{X^*}^{\#}$. □

The next separation theorem is more general than Theorem 3.18.

Theorem 3.20. *Let S and T be nonempty convex subsets of a real locally convex space X where S is compact and T is closed. Then $S \cap T = \emptyset$ if and only if there is a continuous linear functional $l \in X^* \setminus \{0_{X^*}\}$ with*

$$\sup_{s \in S} l(s) < \inf_{t \in T} l(t). \tag{3.12}$$

Proof. Since S is compact and T is closed, by Lemma 1.34 the algebraic difference $T - S$ is closed. The set equation $S \cap T = \emptyset$ is equivalent to $0_X \notin T - S$. Since S and T are convex, the set $T - S$ is convex as well. Then, by Theorem 3.18, the set equation $S \cap T = \emptyset$ is equivalent to the existence of a continuous linear functional $l \in X^* \setminus \{0_{X^*}\}$ and a real number α with

$$0 < \alpha \leq l(t - s) \text{ for all } t \in T \text{ and all } s \in S.$$

This inequality is equivalent to

$$\begin{aligned} 0 &< \inf \{l(t) - l(s) \mid t \in T, s \in S\} \\ &= \inf \{l(t) \mid t \in T\} - \sup \{l(s) \mid s \in S\} \end{aligned}$$

implying

$$\sup_{s \in S} l(s) < \inf_{t \in T} l(t).$$

This completes the proof. □

It should be noticed that the last two separation theorems do not require that one of the considered sets has a nonempty interior. Instead we have a compactness assumption which is even stronger.

Before we present a special separation theorem for closed convex cones we list various useful results on convex cones.

Lemma 3.21. *Let C_X be a convex cone in a real linear space X.*

(a) If X is locally convex and C_X is closed, then

$$C_X = \{x \in X \mid x^*(x) \geq 0 \text{ for all } x^* \in C_{X^*}\}.$$

(b) If $cor(C_X) \neq \emptyset$, then

$$cor(C_X) = \{x \in X \mid x'(x) > 0 \text{ for all } x' \in C_{X'} \setminus \{0_{X'}\}\}.$$

(c) If X is a real topological linear space and $int(C_X) \neq \emptyset$, then

$$int(C_X) = \{x \in X \mid x^*(x) > 0 \text{ for all } x^* \in C_{X^*} \setminus \{0_{X^*}\}\}.$$

(d) Let X be locally convex and separated where the topology gives X as the topological dual space of X^*. Moreover, let C_X be closed and $int(C_{X^*}) \neq \emptyset$. Then we have

$$int(C_{X^*}) = C_{X^*}^{\#}.$$

Proof.

(a) We have only to show

$$C_X \supset \{x \in X \mid x^*(x) \geq 0 \text{ for all } x^* \in C_{X^*}\}$$

because the converse inclusion follows immediately from the definition of the dual cone C_{X^*}. Take any $x \in X$ with

$$x^*(x) \geq 0 \text{ for all } x^* \in C_{X^*} \qquad (3.13)$$

and assume that $x \notin C_X$. Since C_X is closed and convex, by Theorem 3.18 there are an $l \in X^* \setminus \{0_{X^*}\}$ and an $\alpha \in \mathbb{R}$ with

$$l(x) < \alpha \leq l(c) \text{ for all } c \in C_X. \tag{3.14}$$

Since C_X is a cone, we conclude

$$l(c) \geq 0 \text{ for all } c \in C_X \tag{3.15}$$

which implies $l \in C_{X^*}$. Consequently, with the inequality (3.13) we get $l(x) \geq 0$. But this contradicts the inequality $l(x) < 0$ which can be derived from (3.14) and (3.15).

(b) The inclusion

$$\text{cor}(C_X) \subset \{x \in X \mid x'(x) > 0 \text{ for all } x' \in C_{X'} \setminus \{0_{X'}\}\}$$

was already shown in Lemma 1.26. For the proof of the converse inclusion we take an arbitrary $x \in X$ with

$$x'(x) > 0 \text{ for all } x' \in C_{X'} \setminus \{0_{X'}\} \tag{3.16}$$

(we study only the non-trivial case $C_{X'} \neq \{0_{X'}\}$) and we assume that $x \notin \text{cor}(C_X)$. Then by the basic version of the separation theorem there are an $l \in X' \setminus \{0_{X'}\}$ and an $\alpha \in \mathbb{R}$ with

$$l(x) \leq \alpha \leq l(c) \text{ for all } c \in C_X.$$

Since C_X is a cone, we obtain $l \in C_{X'} \setminus \{0_{X'}\}$ and $l(x) \leq 0$ which contradicts the inequality (3.16).

(c) This assertion can be proved in analogy to the algebraic version under (b). We remark only that by Lemma 1.32, (a) $\text{int}(C_X) = \text{cor}(C_X)$.

(d) With Lemma 1.32, (a) and Lemma 1.25 we obtain $\text{int}(C_{X^*}) \subset C_{X^*}^{\#}$. For the proof of the converse inclusion we take an arbitrary $x^* \in C_{X^*}^{\#}$ and we assume that $x^* \notin \text{int}(C_{X^*})$. Then by Eidelheit's separation theorem and the fact that X is the dual space of X^* there are an $x \in X \setminus \{0_X\}$ and an $\alpha \in \mathbb{R}$ with

$$x^*(x) \leq \alpha \leq l(x) \text{ for all } l \in C_{X^*}. \tag{3.17}$$

3.2. Separation Theorems

This inequality implies

$$l(x) \geq 0 \text{ for all } l \in C_{X^*} \tag{3.18}$$

and with part (a) of this lemma we get $x \in C_X \setminus \{0_X\}$. Consequently, we have $x^*(x) > 0$ and from (3.17) and (3.18) we conclude $x^*(x) \leq 0$. But this is a contradiction. □

Now, we are able to present the promised *separation theorem for closed convex cones*.

Theorem 3.22. *Let X be a real separated locally convex space where the topology gives X as the topological dual space of X^*. Moreover, let S and T be closed convex cones in X with $\text{int}(S^*) \neq \emptyset$ (S^* denotes the dual cone for S). Then $(-S) \cap T = \{0_X\}$ if and only if there is a continuous linear functional $l \in X^* \setminus \{0_{X^*}\}$ with*

$$l(x) \leq 0 \leq l(y) \text{ for all } x \in -S \text{ and all } y \in T \tag{3.19}$$

and

$$l(x) < 0 \text{ for all } x \in -S \setminus \{0_X\} \tag{3.20}$$

(see Fig. 3.3).

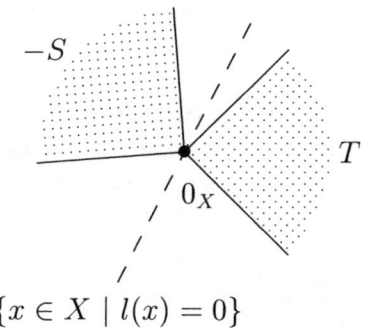

Figure 3.3: Illustration of the result of Thm. 3.22.

Proof.

(a) Let some $l \in X^* \setminus \{0_{X^*}\}$ be given with the properties (3.19) and

(3.20). If we assume that there is an $x \neq 0_X$ with $x \in (-S) \cap T$, then we get from (3.19) and (3.20)

$$l(x) < 0 \leq l(x)$$

which is a contradiction. Consequently, the set equation $(-S) \cap T = \{0_X\}$ is true.

(b) Now, assume that there is no $l \in X^* \setminus \{0_{X^*}\}$ with the properties (3.19) and (3.20). Then we obtain with Lemma 3.21, (d) that $\text{int}(S^*) \cap T^* = \emptyset$ where T^* denotes the dual cone for T. By Eidelheit's separation theorem and the fact that X is the topological dual space of X^* there are an $x \in X \setminus \{0_X\}$ and a real number α with

$$s^*(x) \leq \alpha \leq t^*(x) \text{ for all } s^* \in S^* \text{ and all } t^* \in T^*.$$

Since S^* and T^* are cones, we obtain even

$$s^*(x) \leq 0 \leq t^*(x) \text{ for all } s^* \in S^* \text{ and all } t^* \in T^*.$$

With Lemma 3.21, (a) this inequality implies $x \in (-S) \cap T$ which means that $(-S) \cap T \neq \{0_X\}$.

$\square$

We finish this section with a remark on weakly closed convex sets and with an additional strict separation theorem. This result is a simple application of Theorem 3.18. But first, we recall a characterization of weak convergence.

Lemma 3.23. *Let X be a real linear space and let Y be a subspace of X'. A net $(x_i)_{i \in I}$ in X converges to some $x \in X$ in the topology $\sigma(X, Y)$ if and only if $\lim_{i \in I} l(x_i) = l(x)$ for all $l \in Y$.*

Theorem 3.24. *Let S be a nonempty convex subset of a real locally convex space X. The set S is closed if and only if it is weakly closed.*

Proof. Lemma 3.23 implies that every convergent net in X is also weakly convergent and, therefore, every weakly closed set is also closed. Next, we show that every closed convex set is also weakly closed. Assume that S is closed and take an arbitrary $x \in X \backslash S$. By Theorem 3.18 there are a continuous linear functional $l \in X^* \backslash \{0_{X^*}\}$ and a real number α with

$$l(x) < \alpha \leq l(s) \text{ for all } s \in S.$$

Hence, by Lemma 3.23 no net in S can converge weakly to x. This implies that x does not belong to the weak closure of S. This completes the proof. □

With Theorem 3.24 and Theorem 3.18 it is also possible to formulate a strict separation theorem for reflexive Banach spaces where we do not need the assumption that at least one set has a nonempty interior.

Theorem 3.25. *Let S and T be nonempty closed convex subsets of a real reflexive Banach space $(X, \|\cdot\|)$ where S is bounded. Then $S \cap T = \emptyset$ if and only if there is a continuous linear functional $l \in X^* \backslash \{0_{X^*}\}$ with*

$$\sup_{s \in S} l(s) < \inf_{t \in T} l(t).$$

Proof. Since in a reflexive Banach space a bounded closed convex set is weakly compact, the set S is weakly compact. With Theorem 3.24 the set T is weakly closed, and with Lemma 1.34 we conclude that $T - S$ is weakly closed and, therefore, closed. Since $S \cap T = \emptyset$ is equivalent to $0_X \notin T - S$, we obtain the desired result with Theorem 3.18. □

3.3 A James Theorem

In this section we study reflexive Banach spaces and characterize weakly compact subsets. It is the aim to present a James theorem which is a famous and profound theorem of functional analysis. First, we begin with a general version of a well-known *Weierstraß theorem*.

Theorem 3.26. *Let S be a nonempty compact subset of a real topological linear space X, and let $f : S \to \mathbb{R}$ be a continuous functional. Then f attains its supremum on S.*

A consequence of this theorem is that every continuous linear functional on a real Banach space attains its supremum on a weakly compact set. The James theorem states that if every continuous linear functional attains its supremum on a bounded and weakly closed subset of a real Banach space, then this subset is weakly compact.

The *James theorem* reads as follows.

Theorem 3.27. *Let S be a nonempty bounded and weakly closed subset of a real quasi-complete locally convex space X. If every continuous linear functional $l \in X^*$ attains its supremum on S, then S is weakly compact.*

The proof of this theorem is rather complicated and technical. Therefore, we restrict ourselves only on a short discussion of this theorem under the additional assumption that X is a separable Banach space and S is the closed unit ball. For these investigations the following theorem on supporting hyperplanes (which is sometimes also called *bipolar theorem*) is essential.

Theorem 3.28. *Let X be a real linear space and let Y be a subspace of X'. For every nonempty subset S of X the $\sigma(X,Y)$-closed convex hull is*

$$cl(co(S))_{\sigma(X,Y)} = \{x \in X \mid l(x) \leq \sup_{s \in S} l(s) \text{ for all } l \in Y\}.$$

Proof. Let T denote the set

$$T := \{x \in X \mid l(x) \leq \sup_{s \in S} l(s) \text{ for all } l \in Y\}.$$

3.3. A James Theorem

(a) First, we show $\mathrm{cl}(\mathrm{co}(S))_{\sigma(X,Y)} \subset T$. It is evident that $S \subset T$, and since T is convex, we conclude $\mathrm{co}(S) \subset T$. But T is also $\sigma(X,Y)$-closed and, therefore, we get

$$\mathrm{cl}(\mathrm{co}(S))_{\sigma(X,Y)} \subset T.$$

(b) Now, we prove the inclusion $T \subset \mathrm{cl}(\mathrm{co}(S))_{\sigma(X,Y)}$. Take any $\bar{x} \in X$ with $\bar{x} \notin \mathrm{cl}(\mathrm{co}(S))_{\sigma(X,Y)}$. Since X equipped with the topology $\sigma(X,Y)$ is locally convex, by the separation theorem 3.18 there are a $\sigma(X,Y)$-continuous linear functional $l \in Y \setminus \{0_{X'}\}$ and a real number α with

$$l(\bar{x}) > \alpha \geq l(x) \text{ for all } x \in \mathrm{co}(S) \qquad (3.21)$$

(for this result observe that l is a $\sigma(X,Y)$-continuous linear functional on X if and only if $l \in Y$). With the inequality (3.21) we obtain

$$l(\bar{x}) > \sup_{s \in S} l(s)$$

which implies that $\bar{x} \notin T$. This completes the proof.

$\square$

The following lemma may be found in a paper of König [179, Korollar 4.4].

Lemma 3.29. *Let $(X, \|\cdot\|)$ be a real Banach space, and let S be a separable subset with the property that every continuous linear functional $l \in X^*$ attains its supremum on S. Then we have*

$$\mathrm{cl}(\mathrm{co}(S)) = \mathrm{cl}(\mathrm{co}(S))_{\sigma(X^{**}, X^*)}.$$

Now, we are able to prove a weaker version of the James theorem.

Theorem 3.30. *Let $(X, \|\cdot\|_X)$ be a real separable Banach space. If every continuous linear functional attains its supremum on the closed unit ball*

$$U(X) := \{x \in X \mid \|x\|_X \leq 1\},$$

then $U(X)$ is weakly compact (which is equivalent to the reflexivity of X).

Proof. With Lemma 3.29 we obtain
$$U(X) = \mathrm{cl}(U(X))_{\sigma(X^{**},X^*)}$$
and with Theorem 3.28 we conclude
$$\begin{aligned} U(x) &= \{x^{**} \in X^{**} \mid x^{**}(l) \leq \sup_{x \in U(X)} l(x) \text{ for all } l \in X^*\} \\ &= U(X^{**}) \end{aligned}$$
which implies that X is reflexive. Consequently, $U(x)$ is weakly compact. □

The usefulness of Theorem 3.27 is illustrated by

Example 3.31. Let Ω be a nonempty subset of $\mathbb{R}^n$. Then we consider the function space $L_1(\Omega)$ (compare Example 1.51) with the natural partial ordering $\leq$ and we assert that for arbitrary functions $f_1, f_2 \in L_1(\Omega)$ with $f_1 \leq f_2$ the order interval $[f_1, f_2]$ is weakly compact.

We prove this assertion with the James theorem (Theorem 3.27). Since $L_1(\Omega)^* = L_\infty(\Omega)$, we define for every $l \in L_\infty(\Omega)$ the function $g \in L_1(\Omega)$ with

$$g(x) = \begin{cases} f_1(x) & \text{almost everywhere on } \{x \in \Omega \mid l(x) < 0 \text{ on } \Omega\} \\ f_2(x) & \text{almost everywhere on } \{x \in \Omega \mid l(x) > 0 \text{ on } \Omega\} \\ 0 & \text{otherwise} \end{cases}$$

and we obtain

$$\sup_{f \in [f_1, f_2]} \int_\Omega l(x) f(x) \, dx = \int_\Omega l(x) g(x) \, dx.$$

Consequently, every continuous linear functional attains its supremum on $[f_1, f_2]$ and, therefore, $[f_1, f_2]$ is weakly compact.

3.3. A James Theorem

Next, we study some helpful consequences of the James theorem.

Definition 3.32. A nonempty subset S of a real normed space $(X, \|\cdot\|)$ is called *proximinal*, if every $x \in X$ has at least one best approximation from S, that is, for every $x \in X$ there is an $\bar{s} \in S$ with

$$\|x - \bar{s}\| \leq \|x - s\| \text{ for all } s \in S$$

(see Fig. 3.4).

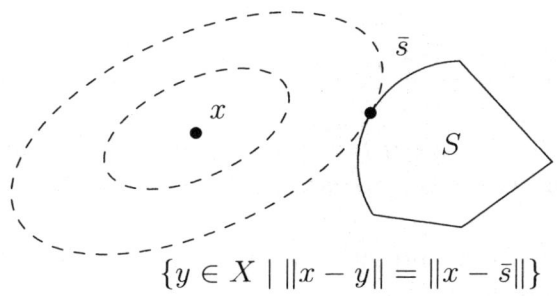

Figure 3.4: Best approximation.

It is evident that a proximinal set is necessarily closed, and every compact set is proximinal.

Definition 3.33. Let S be a nonempty subset of a real normed space $(X, \|\cdot\|)$. A functional $f : S \to \mathbb{R}$ is called *weakly lower semicontinuous*, if for every net $(x_i)_{i \in I}$ in S which converges weakly to some $x \in S$ the inequality

$$f(x) \leq \liminf_{i \in I} f(x_i)$$

is satisfied.

For instance, the norm $\|\cdot\|$ on X is weakly lower semicontinuous. Conditions ensuring that a set is proximinal are given by

Theorem 3.34. *Every nonempty weak*-closed subset S of the dual space X^* of a real normed space $(X, \|\cdot\|_X)$ is proximinal.*

Proof. Take any $x \in X^* \setminus S$ and any $y \in S$. Since every closed ball in X^* is weak*-compact, the set

$$S \cap \{x^* \in X \mid \|x^*\|_{X^*} \leq \|y\|_{X^*}\}$$

is weak*-compact as well. Notice that the functional $X^* \ni z \mapsto \|x - z\|_{X^*}$ is weakly* lower semicontinuous. Then the assertion follows immediately. □

The next corollary is a direct consequence of Theorem 3.34.

Corollary 3.35. *Every nonempty weakly closed subset of a real reflexive Banach space is proximinal.*

Now, we are able to present an interesting characterization of reflexive Banach spaces.

Theorem 3.36. *A real Banach space $(X, \|\cdot\|)$ is reflexive, if and only if*

(a) every nonempty weakly closed subset of X is proximinal

or

(b) every pair of disjoint nonempty closed convex subsets of X, one of which is bounded, can be strictly separated by a hyperplane.

Proof. If X is reflexive then the statements under (a) or (b) follow from Corollary 3.35 and Theorem 3.25. Therefore, we study only the case that X is not reflexive. In this case it follows that X contains a nonreflexive separable subspace M. Then the closed unit ball $U(M)$ in M is not weakly compact. Consequently, by Theorem 3.30 there is a continuous linear functional $l \in M^*$ with

$$l(x) < \sup_{u \in U(M)} l(u) \text{ for all } x \in U(M).$$

This implies that the closed convex sets

$$S := \{x \in M \mid l(x) \geq \sup_{u \in U(M)} l(u)\} \quad (3.22)$$

and $U(M)$ are disjoint.

(a) But then the weakly closed set S is not proximinal.

(b) Assume that the bounded closed convex set $U(M)$ and the closed convex set S can be strictly separated, i.e., there is a continuous linear functional $x^* \in M^*$ with

$$\sup_{u \in U(M)} x^*(u) < \inf_{s \in S} x^*(s). \quad (3.23)$$

The linear optimization problem

$$\inf \ x^*(x)$$
$$\text{subject to}$$
$$l(x) \geq \sup_{u \in U(M)} l(u)$$
$$x \in M$$

is solvable if and only if $x^* = \lambda l$ for some $\lambda > 0$. Hence, we get with (3.23) and (3.22)

$$\lambda \sup_{u \in U(M)} l(u) < \lambda \inf_{s \in S} l(s) = \lambda \sup_{u \in U(M)} l(u)$$

which is a contradiction. This completes the proof.

□

3.4 Two Krein-Rutman Theorems

In the literature one finds very often a popular Krein-Rutman theorem which states a result on the extension of positive linear functionals. Although this theorem will be presented in this section as well, our main aim is another Krein-Rutman theorem which is not so well-known. This theorem provides sufficient conditions under which strictly positive linear functionals exist or equivalently, it provides conditions which guarantee that the quasi-interior of the dual cone is nonempty. It turns out that this result has many applications in vector optimization.

First, we formulate the *extension theorem for positive linear functionals*.

Theorem 3.37. *Let X be a real linear space with a convex cone C_X which has a nonempty algebraic interior. Moreover, let M be a subspace of X which contains an element in the algebraic interior of C_X. Then for every linear functional $l \in C_{M'}$ (with $C_M := C_X \cap M$) there is a linear functional $f \in C_{X'}$ with*

$$f(x) = l(x) \text{ for all } x \in M.$$

Proof. Let S denote the span of M and C_X. For an arbitrary linear functional $l \in C_{M'}$ we define the sublinear functional $g : S \to \mathbb{R}$ given by

$$g(x) = \inf \{l(y) \mid y \in M \cap (\{x\} + C_X)\} \text{ for all } x \in S.$$

Then g is sublinear and

$$l(x) \leq g(x) \text{ for all } x \in M.$$

With the Hahn-Banach extension theorem there is a linear functional $f \in S'$ with

$$f(x) = l(x) \text{ for all } x \in M$$

and

$$f(x) \leq g(x) \text{ for all } x \in S.$$

In order to see that $f \in C_{S'}$ take any $\bar{x} \in C_S$. Then for an arbitrarily chosen $x \in M \cap C_X$ we get

$$\frac{1}{\lambda} x + \bar{x} \in C_X \text{ for all } \lambda > 0.$$

3.4. Two Krein-Rutman Theorems

Consequently, we obtain

$$f(-\bar{x}) \leq g(-\bar{x}) \leq l\left(\frac{1}{\lambda}x\right) = \frac{1}{\lambda}l(x)$$

and in the limit for $\lambda \to \infty$

$$f(\bar{x}) \geq 0.$$

Thus, $f \in C_{S'}$ and an additional extension argument completes the proof. □

Finally, we study the other announced Krein-Rutman theorem.

Theorem 3.38. *In a real separable normed space $(X, \|\cdot\|_X)$ with a closed pointed convex cone C_X the quasi-interior $C_{X^*}^{\#}$ of the topological dual cone is nonempty.*

Proof. The assertion is evident for a trivial cone C_X. Therefore, we assume that $C_X \neq \{0_X\}$. By Lemma 1.39 the unit ball $U(X^*)$ in X^* is weak*-metrizable. Since $U(X^*) \cap C_{X^*}$ is a weak*-compact subset of X^*, it is weak*-separable. Let $\{l_1, l_2, \ldots\}$ be a countable weak*-dense subset of $U(X^*) \cap C_{X^*}$ and consider the functional $l : X \to \mathbb{R}$ with

$$l(x) = \sum_{i=1}^{\infty} \frac{1}{2^i} l_i(x) \text{ for all } x \in X.$$

Since X^* is a Banach space and $\|l_i\|_{X^*} \leq 1$ for all $i \in \mathbb{N}$, the functional l exists and we get $l \in C_{X^*}$. Finally, we prove

$$l(x) > 0 \text{ for all } x \in C_X \backslash \{0_X\}. \tag{3.24}$$

Assume that there is some $x \in C_X \backslash \{0_X\}$ with $l(x) = 0$. Then we conclude

$$l_i(x) = 0 \text{ for all } i \in \mathbb{N}$$

and also

$$f(x) = 0 \text{ for all } f \in C_{X^*}. \tag{3.25}$$

Since C_X is pointed and $x \in C_X \backslash \{0_X\}$, we get $-x \notin C_X$. C_X is closed and, therefore, by Theorem 3.18 there is a continuous linear functional

$g \in C_{X^*} \setminus \{0_{X^*}\}$ with $g(x) > 0$. But this is a contradiction to (3.25). Hence, the inequality (3.24) is true which means that $l \in C_{X^*}^{\#}$. □

With Theorem 3.38 and Lemma 3.3 we obtain immediately

Corollary 3.39. *Every nontrivial closed pointed convex cone in a real separable normed space has a base.*

Example 3.40. Let Ω be a nonempty subset of $\mathbb{R}^n$, and let $C_{L_p(\Omega)}$ be the natural ordering cone of the function space $L_p(\Omega)$ with $p \in [1, \infty)$ (compare Example 1.51). It can be easily checked that the assumptions of Theorem 3.38 (and Corollary 3.39) are fulfilled in this setting. Consequently, $C_{L_p(\Omega)^*}^{\#}$ is nonempty and $C_{L_p(\Omega)}$ admits a base for all $p \in [1, \infty)$.

The separability assumption in Theorem 3.38 and Corollary 3.39 is essential and cannot be dropped. Krein-Rutman [186, p. 218] gave an interesting example which shows that the assertion fails in a non-separable space.

3.5 Contingent Cones and a Lyusternik Theorem

In this section we investigate contingent cones in normed spaces and present several important properties of these cones. A contingent cone to a set S at some $\bar{x} \in \mathrm{cl}(S)$ describes a local approximation of the set $S - \{\bar{x}\}$. This concept is very helpful for the investigation of optimality conditions. If the set S is given by equality constraints, then the contingent cone is related to a set which one obtains by "linearizing" the constraints. This is essentially the result of the Lyusternik theorem which will be formulated at the end of this section.

First, we introduce the helpful concept of contingent cones.

Definition 3.41. Let S be a nonempty subset of a real normed space $(X, \|\cdot\|)$.

(a) Let some $\bar{x} \in \mathrm{cl}(S)$ be given. An element $h \in X$ is called a

3.5. Contingent Cones and a Lyusternik Theorem

tangent to S at $\bar{x}$, if there are a sequence $(x_n)_{n \in \mathbb{N}}$ of elements $x_n \in S$ and a sequence $(\lambda_n)_{n \in \mathbb{N}}$ of positive real numbers λ_n so that

$$\bar{x} = \lim_{n \to \infty} x_n$$

and

$$h = \lim_{n \to \infty} \lambda_n (x_n - \bar{x}).$$

(b) The set $T(S, \bar{x})$ of all tangents to S at $\bar{x}$ is called the *contingent cone* (or the *Bouligand tangent cone*) to S at $\bar{x}$ (see Fig. 3.5).

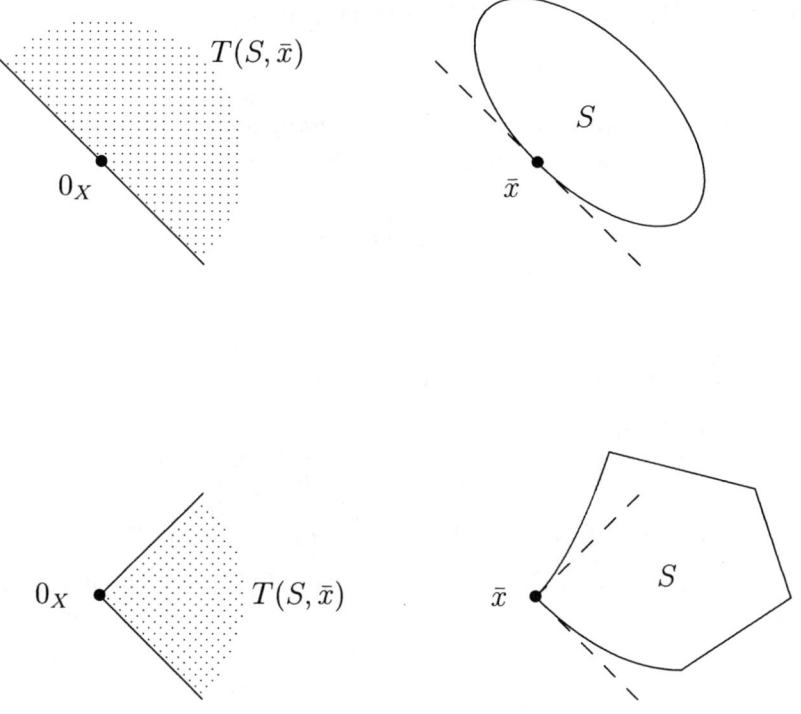

Figure 3.5: Two examples of contingent cones.

For the definition of $T(S, \bar{x})$ it is sufficient that $\bar{x}$ belongs to the closure of the set S. But later we will assume that $\bar{x}$ is an element of S. It is evident that the contingent cone is really a cone. If S is a subset of a real normed space with a nonempty interior, then for every $\bar{x} \in \text{int}(S)$ we have $T(S, \bar{x}) = X$.

The next lemma is easy to prove.

Lemma 3.42. *Let S_1 and S_2 be nonempty subsets of a real normed space. Then we have*

(a) $\bar{x} \in cl(S_1) \subset cl(S_2) \implies T(S_1, \bar{x}) \subset T(S_2, \bar{x}),$

(b) $\bar{x} \in cl(S_1 \cap S_2) \implies T(S_1 \cap S_2, \bar{x}) \subset T(S_1, \bar{x}) \cap T(S_2, \bar{x}).$

In the following we study some helpful properties of contingent cones.

Theorem 3.43. *Let S be a nonempty subset of a real normed space. If S is starshaped at some $\bar{x} \in S$, then*

$$\text{cone}(S - \{\bar{x}\}) \subset T(S, \bar{x}).$$

Proof. Take any $x \in S$. Then we have

$$x_n := \bar{x} + \frac{1}{n}(x - \bar{x}) = \frac{1}{n}x + \left(1 - \frac{1}{n}\right)\bar{x} \in S \text{ for all } n \in \mathbb{N}.$$

Hence, we get $\bar{x} = \lim_{n \to \infty} x_n$ and $x - \bar{x} = \lim_{n \to \infty} n(x_n - \bar{x})$. But this implies that $x - \bar{x}$ belongs to the contingent cone $T(S, \bar{x})$ and, therefore, we obtain

$$S - \{\bar{x}\} \subset T(S, \bar{x}).$$

Since $T(S, \bar{x})$ is a cone, it follows further

$$\text{cone}(S - \{\bar{x}\}) \subset T(S, \bar{x}).$$

□

3.5. Contingent Cones and a Lyusternik Theorem

Theorem 3.44. *Let S be a nonempty subset of a real normed space $(X, \|\cdot\|)$. For every $\bar{x} \in cl(S)$ we have*

$$T(S, \bar{x}) \subset cl(cone(S - \{\bar{x}\})).$$

Proof. Take an arbitrary tangent h to S at $\bar{x}$. Then there is a sequence $(x_n)_{n \in \mathbb{N}}$ of elements in X and a sequence $(\lambda_n)_{n \in \mathbb{N}}$ of positive real numbers with $\bar{x} = \lim_{n \to \infty} x_n$ and $h = \lim_{n \to \infty} \lambda_n (x_n - \bar{x})$. The last equation implies

$$h \in cl(cone(S - \{\bar{x}\})).$$

$\square$

With the next theorem we show that the contingent cone is always closed.

Theorem 3.45. *Let S be a nonempty subset of a real normed space $(X, \|\cdot\|)$. Then the contingent cone $T(S, \bar{x})$ is closed for every $\bar{x} \in cl(S)$.*

Proof. Let $(h_n)_{n \in \mathbb{N}}$ be an arbitrary sequence in $T(S, \bar{x})$ with $\lim_{n \to \infty} h_n = h \in X$. For every tangent h_n there are a sequence $(x_{n_i})_{i \in \mathbb{N}}$ of elements in S and a sequence $(\lambda_{n_i})_{i \in \mathbb{N}}$ of positive real numbers with $\bar{x} = \lim_{i \to \infty} x_{n_i}$ and $h_n = \lim_{i \to \infty} \lambda_{n_i}(x_{n_i} - \bar{x})$. Consequently, for every $n \in \mathbb{N}$ there is an $i(n) \in \mathbb{N}$ with

$$\|x_{n_i} - \bar{x}\| \leq \frac{1}{n} \text{ for all } i \geq i(n)$$

and

$$\|\lambda_{n_i}(x_{n_i} - \bar{x}) - h_n\| \leq \frac{1}{n} \text{ for all } i \geq i(n).$$

If we define

$$y_n := x_{n_{i(n)}} \in S \text{ for all } n \in \mathbb{N}$$

and

$$\mu_n := \lambda_{n_{i(n)}} > 0 \text{ for all } n \in \mathbb{N},$$

then we get $\bar{x} = \lim\limits_{n\to\infty} y_n$ and

$$\|\mu_n(y_n - \bar{x}) - h\| \leq \frac{1}{n} + \|h_n - h\| \text{ for all } n \in \mathbb{N}$$

which implies

$$h = \lim_{n\to\infty} \mu_n(y_n - \bar{x}).$$

Hence, h belongs to the contingent cone $T(S, \bar{x})$. □

With the last three theorems we get immediately the following

Corollary 3.46. *Let S be a nonempty subset of a real normed space. If S is starshaped at some $\bar{x} \in S$, then*

$$T(S, \bar{x}) = cl(cone(S - \{\bar{x}\})).$$

With the next theorem we answer the question under which conditions a contingent cone is even a convex cone.

Theorem 3.47. *Let S be a nonempty convex subset of a real normed space. Then the contingent cone $T(S, \bar{x})$ is convex for every $\bar{x} \in S$.*

Proof. Since S is convex, $S - \{\bar{x}\}$ and $cone(S - \{\bar{x}\})$ are convex as well. With Lemma 1.32 we conclude that the set $cl(cone(S - \{\bar{x}\}))$ is also convex. Finally, we get with Corollary 3.46 $T(S, \bar{x}) = cl(cone(S - \{\bar{x}\}))$. This completes the proof. □

The next theorem indicates already the importance of contingent cones in optimization theory.

Theorem 3.48. *Let S be a nonempty subset of a real normed space $(X, \|\cdot\|)$, and let $f : X \to \mathbb{R}$ be a given functional.*

(a) *If the functional f is continuous and convex, then for every $\bar{x} \in S$ with the property*

$$f(\bar{x}) \leq f(x) \text{ for all } x \in S$$

3.5. Contingent Cones and a Lyusternik Theorem

it follows

$$f(\bar{x}) \leq f(\bar{x} + h) \text{ for all } h \in T(S, \bar{x}).$$

(b) *If the set S is starshaped at some $\bar{x} \in S$ for which*

$$f(\bar{x}) \leq f(\bar{x} + h) \text{ for all } h \in T(S, \bar{x}),$$

then

$$f(\bar{x}) \leq f(x) \text{ for all } x \in S$$

(see Fig. 3.6).

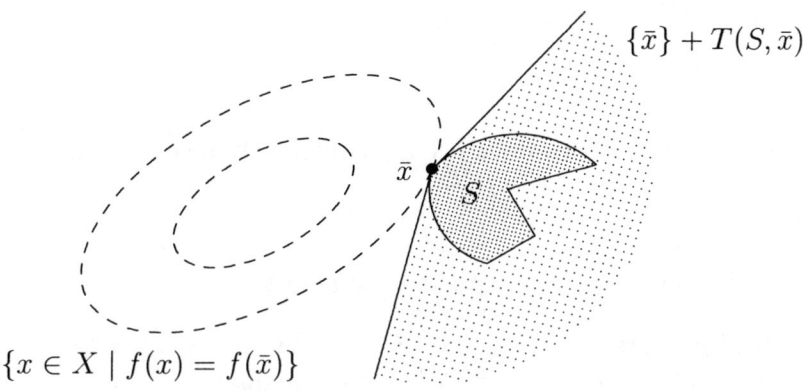

Figure 3.6: Illustration of the result of Thm. 3.48.

Proof.

(a) We choose an arbitrary $\bar{x} \in S$ and assume that the statement

$$f(\bar{x}) \leq f(\bar{x} + h) \text{ for all } h \in T(S, \bar{x})$$

does not hold. Then there are an $h \in T(S, \bar{x}) \setminus \{0_X\}$ and an $\alpha > 0$ with

$$f(\bar{x}) - f(\bar{x} + h) > \alpha > 0.$$

Since h is a tangent to S at $\bar{x}$, there is a sequence $(x_n)_{n\in\mathbb{N}}$ of elements in S and a sequence $(\lambda_n)_{n\in\mathbb{N}}$ of positive real numbers with $\bar{x} = \lim\limits_{n\to\infty} x_n$ and $h = \lim\limits_{n\to\infty} h_n$ where $h_n := \lambda_n(x_n - \bar{x})$ for all $n \in \mathbb{N}$. Since $h \neq 0_X$, we conclude $0 = \lim\limits_{n\to\infty} \dfrac{1}{\lambda_n}$. Then we get for sufficiently large $n \in \mathbb{N}$:

$$\begin{aligned} f(x_n) &= f\left(\frac{1}{\lambda_n}(\bar{x}+h_n) + \left(1 - \frac{1}{\lambda_n}\right)\bar{x}\right) \\ &\leq \frac{1}{\lambda_n} f(\bar{x}+h_n) + \left(1 - \frac{1}{\lambda_n}\right) f(\bar{x}) \\ &\leq \frac{1}{\lambda_n}(f(\bar{x}+h) + \alpha) + \left(1 - \frac{1}{\lambda_n}\right) f(\bar{x}) \\ &< \frac{1}{\lambda_n} f(\bar{x}) + \left(1 - \frac{1}{\lambda_n}\right) f(\bar{x}) \\ &= f(\bar{x}). \end{aligned}$$

Consequently, we obtain for a sufficiently large $n \in \mathbb{N}$

$$f(x_n) < f(\bar{x}).$$

This contraposition leads to the assertion.

(b) If S is starshaped at $\bar{x} \in S$, then by Theorem 3.43 it follows $S - \{\bar{x}\} \subset T(S, \bar{x})$. Therefore, the inequality

$$f(\bar{x}) \leq f(\bar{x} + h) \text{ for all } h \in T(S, \bar{x})$$

implies

$$f(\bar{x}) \leq f(x) \text{ for all } x \in S.$$

$\square$

From now on we study the contingent cone of a special subset of a Banach space which is the kernel of a given Fréchet differentiable map. Under suitable assumptions the kernel of the Fréchet derivative of this map is contained in the considered contingent cone. In essence, this is the result of the *Lyusternik theorem* which is formulated precisely in

3.5. Contingent Cones and a Lyusternik Theorem

Theorem 3.49. Let $(X, \|\cdot\|_X)$ and $(Z, \|\cdot\|_Z)$ be real Banach spaces, and let $h : X \to Z$ be a given map. Furthermore, let some $\bar{x} \in S$ with
$$S := \{x \in X \mid h(x) = 0_Z\}$$
be given. Let h be Fréchet differentiable on a neighborhood of $\bar{x}$, let $h'(\cdot)$ be continuous at $\bar{x}$, and let $h'(\bar{x})$ be surjective. Then it follows for the contingent cone
$$L(S, \bar{x}) := \{x \in X \mid h'(\bar{x})(x) = 0_Z\} \subset T(S, \bar{x}). \qquad (3.26)$$

The set $L(S, \bar{x})$ is also called the *linearizing cone* to S at $\bar{x}$. The proof of Theorem 3.49 is very technical and complicated. It may be found in the books of Ljusternik-Sobolew [208], Kirsch-Warth-Werner [171], Werner [323] and Jahn [149, p. 98–105].

With the following theorem we show that the inclusion (3.26) also holds in the opposite direction.

Theorem 3.50. Let $(X, \|\cdot\|_X)$ and $(Z, \|\cdot\|_Z)$ be real normed spaces, and let $h : X \to Z$ be a given map. Furthermore, let some $\bar{x} \in S$ with
$$S := \{x \in X \mid h(x) = 0_Z\}$$
be given. If h is Fréchet differentiable at $\bar{x}$, then it follows for the contingent cone
$$T(S, \bar{x}) \subset \{x \in X \mid h'(\bar{x})(x) = 0_Z\}.$$

Proof. Let $y \in T(S, \bar{x}) \setminus \{0_X\}$ be an arbitrary tangent vector (the assertion is evident for $y = 0_X$). Then there are a sequence $(x_n)_{n \in \mathbb{N}}$ of elements in S and a sequence $(\lambda_n)_{n \in \mathbb{N}}$ of positive real numbers with
$$\bar{x} = \lim_{n \to \infty} x_n$$
and
$$y = \lim_{n \to \infty} y_n$$

where
$$y_n := \lambda_n(x_n - \bar{x}) \text{ for all } n \in \mathbb{N}.$$

Consequently, by the definition of the Fréchet derivative we obtain:

$$\begin{aligned} h'(\bar{x})(y) &= h'(\bar{x})\bigl(\lim_{n\to\infty} \lambda_n(x_n - \bar{x})\bigr) \\ &= \lim_{n\to\infty} \lambda_n h'(\bar{x})(x_n - \bar{x}) \\ &= \lim_{n\to\infty} \lambda_n \Bigl[h(x_n) - h(\bar{x}) - (h(x_n) - h(\bar{x}) - h'(\bar{x})(x_n - \bar{x}))\Bigr] \\ &= -\lim_{n\to\infty} \|y_n\| \frac{h(x_n) - h(\bar{x}) - h'(\bar{x})(x_n - \bar{x})}{\|x_n - \bar{x}\|} \\ &= 0_Z. \end{aligned}$$

□

Since the assumptions of Theorem 3.50 are weaker than those of Theorem 3.49, we summarize the results of the two preceding theorems as follows: Under the assumptions of Theorem 3.49 we conclude for the contingent cone

$$T(S, \bar{x}) = \{x \in X \mid h'(\bar{x})(x) = 0_Z\}$$

(see Fig. 3.7).

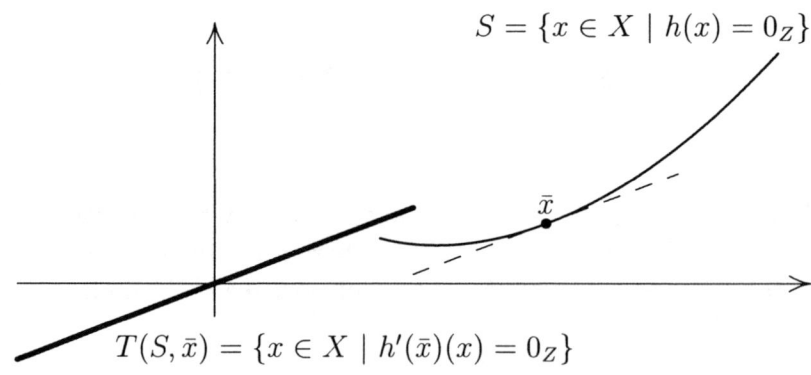

Figure 3.7: Illustration of the remark on page 98.

Notes

The characterization of a base of a convex cone in Lemma 3.3 can be found in the book of Peressini [249, p. 26]. The presentation of the different versions of the Hahn-Banach theorem and especially the key lemma 3.6 is due to König [181] ([178], [179]). For a very general version of the Hahn-Banach theorem we cite the paper of Rodé [263] (and König [180]). For a generalization of the Hahn-Banach theorem to vector-valued maps we refer to Zowe [342], [344], Elster-Nehse [94] and Borwein [32].

The basic version of the separation theorem is a nontopological version of Eidelheit's separation theorem. Eidelheit [90] presented a similar separation theorem in a real normed space. Theorem 3.23 was formulated by Mazur [222] in a normed setting. The results on convex cones (Lemma 3.20) may be found in the book of Vogel [313]. The separation theorem for closed convex cones (Theorem 3.21) was formulated by Borwein [29] and Vogel [313]. It can also be proved by using a theorem of the alternative which was formulated by Lehmann-Oettli [198] in a finite-dimensional setting (compare also Vogel [313, p. 80]).

The so-called James theorem is developed in a sequence of papers of James (e.g. [158]). The proof of Theorem 3.30 is due to König [179] and Example 3.31 is discussed in a paper of Rodé [264]. Theorem 3.34 and Theorem 3.36 are taken from the books of Holmes [124] and [125], respectively.

The two Krein-Rutman theorems were first published in 1948 in Russian. The extension theorem may also be found in the books of Day [78] and Holmes [125]. The proof of Theorem 3.38 is based on a proof given by Borwein [35, p. 425] who gave also a formulation of Corollary 3.39.

Contingent cones can also be formulated in separated topological linear spaces using nets instead of sequences. Notice that in a separated topological linear space the convergence of a net is unique in the sense that every net converges to at most one element. But, in general, in this setting the contingent cone is not always closed. If the space is metrizable, then the contingent cone is closed. In a normed space the presentation of this cone is simpler and we obtain the desired results.

The well-known properties of these contingent cones listed in the last section are discussed, for instance, in the books of Krabs [184] and Jahn [149]. The Lyusternik theorem is based on a result of Lyusternik [218]. A proof can be found in the books of Ljusternik-Sobolew [208], Kirsch-Warth-Werner [171], Ioffe-Tihomirov [129], Werner [323] and Jahn [149]. In the books of Girsanov [106] and Tichomirov [302] a formulation of the Lyusternik theorem is given without proof.

Part II

Theory of Vector Optimization

Vector optimization problems are those where we are looking for certain "optimal" elements of a nonempty subset of a partially ordered linear space. In this book we investigate optimal elements such as minimal, strongly minimal, properly minimal and weakly minimal elements. These notions are defined in chapter 4. Basic results concerning the connection of vector optimization problems with scalar optimization problems are studied in the fifth chapter. In chapter 6 we present existence theorems for these optima. The topic of chapter 7 is a generalized multiplier rule for abstract optimization problems. Finally, in the last chapter of this second part we discuss a duality theory for abstract optimization problems.

The first papers in this research area were published by Edgeworth [87] (1881) and Pareto [244] (1896) who were the initiators of vector optimization more than hundred years ago. The actual development of vector optimization begun with papers by Koopmans [182] (1951) and Kuhn-Tucker [187] (1951). An interesting and detailed article on the historical development of vector optimization was published by Stadler [286].

Chapter 4

Optimality Notions

For the investigation of "optimal" elements of a nonempty subset of a partially ordered linear space one is mainly interested in minimal or maximal elements of this set. But in certain situations it also makes sense to study several variants of these concepts; for example, strongly minimal, properly minimal and weakly minimal elements (or strongly maximal, properly maximal and weakly maximal elements). It is the aim of this first chapter of the second part to present the definition of these optimality notions together with some examples.

In Definition 3.1, (c) we defined already minimal and maximal elements of a partially ordered set S which is not assumed to have a linear structure. If S is a subset of a partially ordered linear space, Definition 3.1, (c) is equivalent to

Definition 4.1. Let S be a nonempty subset of a partially ordered linear space with an ordering cone C.

(a) An element $\bar{x} \in S$ is called a *minimal* element of the set S, if
$$(\{\bar{x}\} - C) \cap S \subset \{\bar{x}\} + C. \qquad (4.1)$$

(b) An element $\bar{x} \in S$ is called a *maximal* element of the set S, if
$$(\{\bar{x}\} + C) \cap S \subset \{\bar{x}\} - C. \qquad (4.2)$$

If the ordering cone C is pointed, then the inclusions (4.1) and (4.2) can be replaced by the set equations

$$(\{\bar{x}\} - C) \cap S = \{\bar{x}\} \quad (\text{ or: } x \leq_C \bar{x}, \, x \in S \Rightarrow x = \bar{x})$$

and

$$(\{\bar{x}\} + C) \cap S = \{\bar{x}\} \quad (\text{ or: } \bar{x} \leq_C x, \, x \in S \Rightarrow x = \bar{x}),$$

respectively (see Fig. 4.1). Since every maximal element of S is also minimal with respect to the partial ordering induced by the convex cone $-C$, without loss of generality it is sufficient to study the minimality notion.

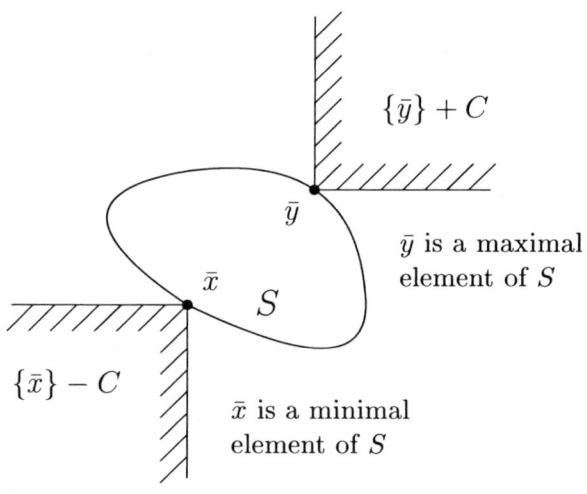

Figure 4.1: Minimal and maximal elements of a set S.

Example 4.2. Let X be the real linear space of functionals defined on a real linear space E and partially ordered by a pointwise ordering. Moreover, let S denote the subset of X which consists of all sublinear functionals on E. Then the algebraic dual space E' is the set of all minimal elements of S. This assertion is proved in Lemma 3.7 and is a key for the proof of the basic version of the Hahn-Banach theorem.

Example 4.3. Let X and Y be partially ordered linear spaces with the ordering cones C_X and C_Y, and let $T : X \to Y$ be a given linear map. We assume that there is a $q \in Y$ so that the set $S := \{x \in C_X \mid T(x) + q \in C_Y\}$ is nonempty. Then an *abstract complementary problem* leads to the problem of finding a minimal element of the set S (for further details see the paper of Cryer-Dempster [71] and Borwein [36]).

In the statistical decision theory and the theory of tests there are many prominent problems where one investigates minimal elements of a set (compare the book of Vogel [313]). The following example may be interpreted as a problem of finding minimal covariance matrices.

Example 4.4. Let X be the real linear space of real symmetric (n, n)-matrices, and let a partial ordering in X be given which is induced by the convex cone $C := \{A \in X \mid A \text{ is positive semidefinite}\}$. Then we are looking for minimal elements of a nonempty subset S of C. For example, if there is a matrix $A \in S$ which has a minimal trace among all matrices of S, then A is a minimal element of the set S.

Example 4.5. Let X and Y be real linear spaces, and let C_Y be a convex cone in Y. Furthermore, let S be a nonempty subset of X, and let $f : S \to Y$ be a given map. Then the *abstract optimization problem*

$$\min_{x \in S} f(x) \qquad (4.3)$$

is to be interpreted in the following way: Determine a minimal solution $x \in S$ which is defined as the inverse image of a minimal element $f(\bar{x})$ of the image set $f(S)$. If f is a vectorial norm (compare Definition 1.35), then the problem (4.3) is called a *vector approximation problem*. This kind of problems is studied in detail in chapter 9.

Now, we come to a vector optimization problem which arises in game theory.

Example 4.6. We consider a *cooperative n player game*. Let

X, Y_1,..., Y_n be real linear spaces, let S be a nonempty subset of X, and let C_{Y_1},..., C_{Y_n} be convex cones in Y_1,..., Y_n, respectively. Moreover, let for every player an objective map $f_i : S \to Y_i$ (for every $i \in \{1,\ldots,n\}$) be given. Every player tries to minimize its goal map f_i on S. But since they play exclusively cooperatively (and, therefore, this concept differs from that introduced by John von Neumann), they cannot hurt each other. In order to be able to introduce an optimality concept, it is convenient to define the product space $Y := \prod_{i=1}^{n} Y_i$, the product ordering cone $C := \prod_{i=1}^{n} C_{Y_i}$ and a map $f : X \to Y$ given by $f = (f_1, \ldots, f_n)$. Then an element $\bar{x} \in S$ is called a minimal solution (or an Edgeworth-Pareto optimal solution), if $\bar{x}$ is the inverse image of a minimal element of the image set $f(S)$. The product ordering allows an adequate description of the cooperation because an element $x \in S$ is preferred, if it is preferred by all players. Hence, cooperative n player games can be formulated as an abstract optimization problem. In chapter 10 special cooperative games, namely cooperative n player differential games, are discussed in detail.

The following lemma indicates that the minimal elements of a set S and the minimal elements of the set $S + C$ where C denotes the ordering cone are closely related.

Lemma 4.7. *Let S be a nonempty subset of a partially ordered linear space with an ordering cone C.*

(a) *If the ordering cone C is pointed, then every minimal element of the set $S + C$ is also a minimal element of the set S.*

(b) *Every minimal element of the set S is also a minimal element of the set $S + C$.*

Proof.

(a) Let $\bar{x} \in S + C$ be an arbitrary minimal element of the set $S + C$. If we assume that $\bar{x} \notin S$, then there is an element

$x \neq \bar{x}$ with $x \in S$ and $\bar{x} \in \{x\} + C$. Consequently, we get $x \in (\{\bar{x}\} - C) \cap (S + C)$ which contradicts the assumption that $\bar{x}$ is a minimal element of the set $S + C$. Hence, we obtain $\bar{x} \in S \subset S + C$ and, therefore, $\bar{x}$ is also a minimal element of the set S.

(b) Take an arbitrary minimal element $\bar{x} \in S$ of the set S, and choose any $x \in (\{\bar{x}\} - C) \cap (S + C) \neq \emptyset$. Then there are elements $s \in S$ and $c \in C$ so that $x = s + c$. Consequently, we obtain $s = x - c \in \{\bar{x}\} - C$, and since $\bar{x}$ is a minimal element of the set S, we conclude $s \in \{\bar{x}\} + C$. But then we get also $x \in \{\bar{x}\} + C$. This completes the proof.

$\square$

In some situations one is interested in an element of a set which is a lower bound of this set. Such an optimal element is called strongly minimal.

Definition 4.8. Let S be a nonempty subset of a partially ordered linear space with an ordering cone C.

(a) An element $\bar{x} \in S$ is called a *strongly minimal* element of the set S, if
$$S \subset \{\bar{x}\} + C \quad (\text{ or: } \bar{x} \leq_C x \text{ for all } x \in S)$$
(see Fig. 4.2).

(b) An element $\bar{x} \in S$ is called a *strongly maximal* element of the set S, if
$$S \subset \{\bar{x}\} - C \quad (\text{ or: } x \leq_C \bar{x} \text{ for all } x \in S).$$

In terms of lattice theory a strongly minimal element of a set S is also called *zero* element of S and a strongly maximal element of S is said to be *one* element of the set S.

The notion of strong minimality is very restrictive and is often not applicable in practice.

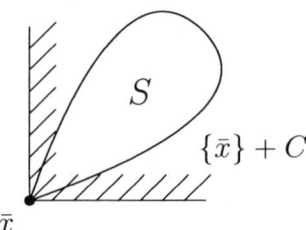

Figure 4.2: Strongly minimal element of a set S.

Example 4.9. Under the assumptions of Example 4.3 we consider again the set $S := \{x \in C_X \mid T(x) + q \in C_Y\}$. Obviously, if $q \in C_Y$, then 0_X is a strongly minimal element of the set S.

The next lemma which is easy to prove gives a relation between strongly minimal and minimal elements of a set.

Lemma 4.10. *Let S be a nonempty subset of a partially ordered linear space. Then every strongly minimal element of the set S is also a minimal element of S.*

Another refinement of the minimality notion is helpful from a theoretical point of view. These optima are called properly minimal. Until now there are various types of concepts of proper minimality. We present here a definition introduced by Borwein [29] and Vogel [313] in more general spaces.

Definition 4.11. Let S be a nonempty subset of a real normed space $(X, \|\cdot\|)$ whose partial ordering is induced by a convex cone C.

(a) An element $\bar{x} \in S$ is called a *properly minimal* element of the set S, if $\bar{x}$ is a minimal element of the set S and the zero element 0_X is a minimal element of the contingent cone $T(S+C, \bar{x})$ (see Fig. 4.3).

(b) An element $\bar{x} \in S$ is called a *properly maximal* element of the set

S, if $\bar{x}$ is a maximal element of the set S and the zero element 0_X is a maximal element of the contingent cone $T(S - C, \bar{x})$.

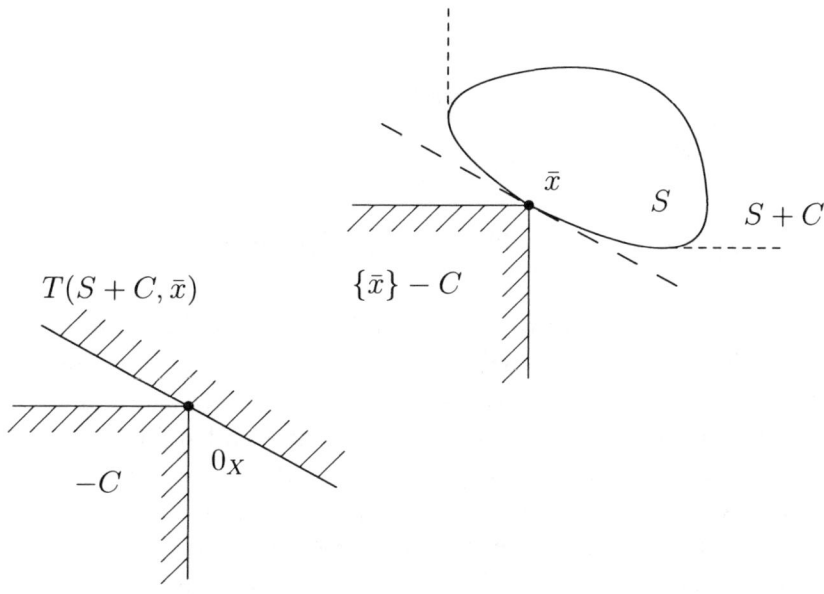

Figure 4.3: Properly minimal element of a set S.

It is evident that a properly minimal element of a set S is also a minimal element of S.

Finally, we come to an optimality notion which is weaker than all the considered notions.

Definition 4.12. Let S be a nonempty subset of a partially ordered linear space X with an ordering cone C which has a nonempty algebraic interior.

(a) An element $\bar{x} \in S$ is called a *weakly minimal* element of the set S, if $(\{\bar{x}\} - \text{cor}(C)) \cap S = \emptyset$ (see Fig. 4.4).

(b) An element $\bar{x} \in S$ is called a *weakly maximal* element of the set S, if $(\{\bar{x}\} + \text{cor}(C)) \cap S = \emptyset$.

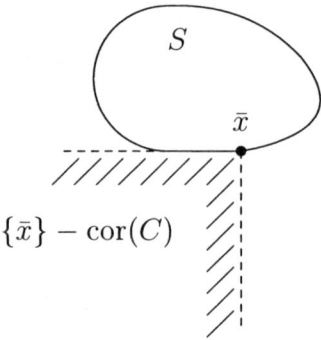

Figure 4.4: Weakly minimal element of a set S.

Notice that the notions "minimal" and "weakly minimal" are closely related. Take an arbitrary weakly minimal element $\bar{x} \in S$ of the set S, that is $(\{\bar{x}\} - \mathrm{cor}(C)) \cap S = \emptyset$. By Lemma 1.12, (a) the set $\hat{C} := \mathrm{cor}(C) \cup \{0_X\}$ is a convex cone and it induces another partial ordering in X. Consequently, $\bar{x}$ is also a minimal element of the set S with respect to the partial ordering induced by $\hat{C}$. But this observation is not very helpful from a practical point of view because a partial ordering induced by $\hat{C}$ leads to certain embarrassments (for instance, $\hat{C}$ is never algebraically closed). The concept of weak minimality is of theoretical interest, and it is not an appropriate notion for applied problems.

The next lemma is similar to Lemma 4.7.

Lemma 4.13. *Let S be a nonempty subset of a partially ordered linear space with an ordering cone C with a nonempty algebraic interior.*

(a) *Every weakly minimal element $\bar{x} \in S$ of the set $S + C$ is also a weakly minimal element of the set S.*

(b) *Every weakly minimal element $\bar{x} \in S$ of the set S is also a weakly minimal element of the set $S + C$.*

Proof.

(a) For an arbitrary weakly minimal element $\bar{x} \in S$ of the set $S + C$

we have
$$(\{\bar{x}\} - \operatorname{cor}(C)) \cap S \subset (\{\bar{x}\} - \operatorname{cor}(C)) \cap (S+C) = \emptyset$$
which implies that $\bar{x}$ is also a weakly minimal element of the set S.

(b) Take any element $\bar{x} \in S$ which is not a weakly minimal element of the set $S+C$. Then there is an element $x \in (\{\bar{x}\} - \operatorname{cor}(C)) \cap (S+C) \neq \emptyset$ and there is an $s \in S$ with $\bar{x} - x \in \operatorname{cor}(C)$ and $x - s \in C$. Consequently, we get with Lemma 1.12, (b)
$$\bar{x} - s \in \operatorname{cor}(C) + C = \operatorname{cor}(C)$$
or alternatively $s \in (\{\bar{x}\} - \operatorname{cor}(C)) \cap S$. Hence, $\bar{x}$ is not a weakly minimal element of the set S, and the assertion follows by contraposition.

□

With the next lemma we investigate again the connections between minimal and weakly minimal elements of a set.

Lemma 4.14. *Let S be a nonempty subset of a partially ordered linear space X with an ordering cone C for which $C \neq X$ and $\operatorname{cor}(C) \neq \emptyset$. Then every minimal element of the set S is also a weakly minimal element of the set S.*

Proof. The assumption $C \neq X$ implies $(-\operatorname{cor}(C)) \cap C = \emptyset$. Therefore, for an arbitrary minimal element $\bar{x}$ of S it follows
$$\begin{aligned}
\emptyset &= (\{\bar{x}\} - \operatorname{cor}(C)) \cap (\{\bar{x}\} + C) \\
&= (\{\bar{x}\} - \operatorname{cor}(C)) \cap (\{\bar{x}\} - C) \cap S \\
&= (\{\bar{x}\} - \operatorname{cor}(C)) \cap S
\end{aligned}$$
which means that $\bar{x}$ is also a weakly minimal element of S. □

In general, the converse statement of Lemma 4.14 is not true. This fact is illustrated by

Example 4.15. Consider the set
$$S := \{(x_1, x_2) \in [0,2] \times [0,2] \mid x_2 \geq 1 - \sqrt{1-(x_1-1)^2} \text{ for } x_1 \in [0,1]\}$$
in $X := \mathbb{R}^2$ (see Fig. 4.5) with the natural ordering cone $C := \mathbb{R}_+^2$. There are no strongly minimal elements of the set S. The set M of

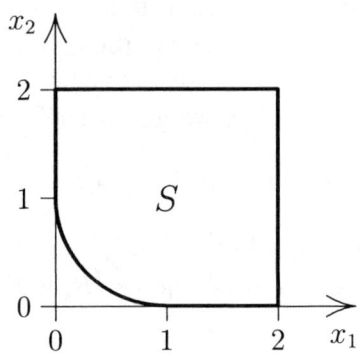

Figure 4.5: Illustration of the set S in Example 4.15.

all minimal elements of S is given as
$$M = \{(x_1, 1 - \sqrt{1-(x_1-1)^2}) \mid x_1 \in [0,1]\}.$$
The set M_p of all properly minimal elements of S reads as
$$M_p = M \backslash \{(0,1), (1,0)\},$$
and the set M_w of all weakly minimal elements of S is
$$M_w = M \cup \{(0, x_2) \in \mathbb{R}^2 \mid x_2 \in (1,2]\} \cup \{(x_1, 0) \in \mathbb{R}^2 \mid x_1 \in (1,2]\}.$$
Consequently, we have
$$M_p \underset{\neq}{\subset} M \underset{\neq}{\subset} M_w.$$

Notes

In engineering a vector optimization problem (like the one discussed in Example 4.5) is also called a multiobjective (or multi criteria or Edgeworth-Pareto) optimization problem, in economics one speaks also of a problem of multi criteria decision making, and sometimes the term polyoptimization has been used. In the applied sciences Edgeworth [87] and Pareto [244] were probably the first who introduced an optimality concept for such problems. In engineering and economics minimal or maximal elements of a set are often called efficient (Vogel [313]), Edgeworth-Pareto optimal (Stadler [283] and [286]) or nondominated (Yu [336]). Lemma 4.7 can also be found in the book of Vogel [313]. Strongly minimal elements are also investigated by Craven [69] and others. The notions of proper and weak minimality are especially qualified for the study of a generalized multiplier rule (see the paper of Borwein [29] and the book of Kirsch-Warth-Werner [171]).

The notion of proper minimality (or proper efficiency) was first introduced by Kuhn-Tucker [187] and modified by Geoffrion [103], and later it was formulated in a more general framework (Benson-Morin [24], Borwein [29], Vogel [313], Wendell-Lee [321], Wierzbicki [327], Hartley [116], Benson [23], Borwein [31], Nieuwenhuis [238], Henig [82a] and Zhuang [340]). The notion of proper minimality which is used in this book is due to Borwein. Proper efficiency plays an important role in the book of Kaliszewski [165] in a finite dimensional setting.

In the following we shortly present other definitions of proper minimality introduced for infinite dimensional spaces.

(a) Benson [23] gave the following definition:

Let S be a nonempty subset of a partially ordered linear space with an ordering cone C. An element $\bar{x} \in S$ is called a properly minimal element of the set S (in the sense of Benson), if $\bar{x}$ is a minimal element of the set S and the zero element 0_X is a minimal element of the set $\mathrm{cl}(\mathrm{cone}(S + C - \{\bar{x}\}))$.

If X is normed and S is starshaped at $\bar{x}$, then by Corollary 3.46 $T(S+C, \bar{x}) = \mathrm{cl}(\mathrm{cone}(S+C-\{\bar{x}\}))$ and the proper minimality notion of Borwein and Benson coincide.

(b) Wierzbicki [327], [329], [330] introduced a definition by using a second larger cone:

Let S be a nonempty subset of a partially ordered real normed space $(X, \|\cdot\|)$ with an ordering cone C, and assume that the cone
$$C_\varepsilon := \{x \in X \mid \inf_{\tilde{x} \in C} \|x - \tilde{x}\| \leq \varepsilon \|x\|\}.$$
is convex. An element $\bar{x} \in S$ is called a properly minimal element of the set S (in the sense of Wierzbicki), if $\bar{x}$ is a minimal element of S with respect to the partial ordering induced by C_ε.

Obviously, C_ε is always a cone and $C \subset C_\varepsilon$.

(c) Henig [118] used the same idea of enlarging the ordering cone and presented the following concept:

Let S be a nonempty subset of a topological real linear space X with an ordering cone C. An element $\bar{x} \in S$ is called a properly minimal element of the set S, if there is a convex cone $\tilde{C} \subset X$ with $C \setminus \{0_X\} \subset \text{int}(\tilde{C})$ so that $\bar{x}$ is a minimal element of S with respect to the partial ordering induced by $\tilde{C}$.

(d) Zhuang [340] created another type of proper minimality:

Let S be a nonempty subset of a real normed space $(X, \|\cdot\|)$ with an ordering cone C. An element $\bar{x} \in S$ is called super efficient (properly minimal in the sense of Zhuang), if there is a real number $\alpha > 0$ so that for the closed unit ball B
$$\text{cl}(\text{cone}(S - \{\bar{x}\})) \cap (B - C) \subset \alpha B.$$

Relationships between super efficiency and other optimality concepts are shown in [340] (for additional work on super efficiency see also [41]).

Chapter 5

Scalarization

In general, scalarization means the replacement of a vector optimization problem by a suitable scalar optimization problem which is an optimization problem with a real-valued objective functional. It is a fundamental principle in vector optimization that optimal elements of a subset of a partially ordered linear space can be characterized as optimal solutions of certain scalar optimization problems. Since the scalar optimization theory is widely developed scalarization turns out to be of great importance for the vector optimization theory.

We present families of scalar problems which fully describe the set of all optimal elements under suitable assumptions.

5.1 Necessary Conditions for Optimal Elements of a Set

In this section various necessary conditions for minimal, strongly minimal, properly minimal and weakly minimal elements are presented.

Before we discuss the minimality notion we introduce important monotonicity concepts.

Definition 5.1. Let S be a nonempty subset of a subset T of a partially ordered linear space with an ordering cone C.

(a) A functional $f : T \to \mathbb{R}$ is called *monotonically increasing* on

S, if for every $\bar{x} \in S$

$$x \in (\{\bar{x}\} - C) \cap S \implies f(x) \leq f(\bar{x})$$
$$(or: \quad x \leq_C \bar{x}, \ x \in S \implies f(x) \leq f(\bar{x})).$$

(b) A functional $f : T \to \mathbb{R}$ is called *strongly monotonically increasing* on S, if for every $\bar{x} \in S$

$$x \in (\{\bar{x}\} - C) \cap S, \ x \neq \bar{x} \implies f(x) < f(\bar{x})$$
$$(or: \quad x \leq_C \bar{x}, \ x \in S, \ x \neq \bar{x} \implies f(x) < f(\bar{x})).$$

(c) If $\text{cor}(C) \neq \emptyset$, then a functional $f : T \to \mathbb{R}$ is called *strictly monotonically increasing* on S, if for every $\bar{x} \in S$

$$x \in (\{\bar{x}\} - \text{cor}(C)) \cap S \implies f(x) < f(\bar{x}).$$

If $\text{cor}(C) \neq \emptyset$, then every functional which is strongly monotonically increasing on S is also strictly monotonically increasing on S.

Example 5.2.

(a) Let S be any subset of a partially ordered linear space X with the ordering cone C_X. Every linear functional $l \in C_{X'}$ is monotonically increasing on S. Furthermore, every linear functional $l \in C_{X'}^{\#}$ is strongly monotonically increasing on S. If $\text{cor}(C_X) \neq \emptyset$, then by Lemma 3.21, (b) every linear functional $l \in C_{X'} \setminus \{0_{X'}\}$ is strictly monotonically increasing on S.

(b) Consider the real linear space $L_p(\Omega)$ where $p \in [1, \infty)$ and Ω is a nonempty subset of $\mathbb{R}^n$. If we assume that this space is equipped with the natural ordering cone $C_{L_p(\Omega)}$ (compare Example 1.51, (a)), then the $L_p(\Omega)$-norm is strongly monotonically increasing on $C_{L_p(\Omega)}$. For the real linear space $L_\infty(\Omega)$ we obtain under the same assumptions that the $L_\infty(\Omega)$-norm is strictly monotonically increasing on $C_{L_\infty(\Omega)}$ (see Example 6.14, (a)).

(c) Let $(X, \langle ., .\rangle)$ be a partially ordered Hilbert space with an ordering cone C_X. Then the norm on X is strongly monotonically increasing on C_X if and only if $C_X \subset C_{X^*}$.

5.1. Necessary Conditions for Optimal Elements of a Set

Proof. First, we assume that the inclusion $C_X \subset C_{X^*}$ does not hold. Then there are elements $x, y \in C_X$ with $\langle x, y \rangle < O$. For an arbitrary $\alpha \in (0,1)$ define $z_\alpha := x + \alpha y$. Obviously, $z_\alpha \in C_X$ and $x \in (\{z_\alpha\} - C_X) \cap C_X$. But then we get for a sufficiently small $\alpha \in (0,1)$

$$\begin{aligned} \|z_\alpha\|^2 &= \langle x + \alpha y, x + \alpha y \rangle \\ &= \langle x, x \rangle + 2\alpha \langle x, y \rangle + \alpha^2 \langle y, y \rangle \\ &< \|x\|^2. \end{aligned}$$

Hence, the norm $\|\cdot\|$ is not strongly monotonically increasing on C_X.

Now, we assume that the inclusion $C_X \subset C_{X^*}$ holds. Choose arbitrary $y \in C_X$ and $x \in (\{y\} - C_X) \cap C_X$ with $x \neq y$. Since $y + x \in C_X$, $y - x \in C_X$ and $C_X \subset C_{X^*}$, we conclude

$$\|y\|^2 - \|x\|^2 = \langle y - x, y + x \rangle \geq O.$$

But this implies only the monotonicity of the norm on C_X. For the proof of the strong monotonicity assume that $\|x\| = \|y\|$. Because of the monotonicity of the norm on C_X we have

$$\|x\| \leq \|\lambda x + (1-\lambda)y\| \leq \|y\| \text{ for all } \lambda \in [0,1].$$

With the assumption $\|x\| = \|y\|$ we obtain

$$\|\lambda x + (1-\lambda)y\| = \lambda \|x\| + (1-\lambda)\|y\| \text{ for all } \lambda \in [0,1].$$

If we square this equation we get $\langle x, y \rangle = \|x\| \|y\|$. This Cauchy-Schwarz equality implies that there is a $\beta > 0$ with $x = \beta y$. Since we assumed $\|x\| = \|y\|$, in the case of $x \neq 0_X$ we get $\beta = 1$ and $x = y$, and in the case $x = 0_X$ we immediately obtain $x = y$. But this contradicts the assumption $x \neq y$. Consequently, the norm on X is strongly monotonically increasing on C_X. □

Now, we begin with the discussion of the minimality notion.

Theorem 5.3. *Let S be a nonempty subset of a partially ordered linear space X with a pointed, algebraically closed ordering cone C*

which has a nonempty algebraic interior. If $\bar{x} \in S$ is a minimal element of the set S, then for every $\hat{x} \in \{\bar{x}\} - \mathrm{cor}(C)$ there is a norm $\|\cdot\|$ on X which is monotonically increasing on C with the property

$$1 = \|\bar{x} - \hat{x}\| < \|x - \hat{x}\| \text{ for all } x \in S \setminus \{\bar{x}\}.$$

Proof. Take an arbitrary element $\hat{x} \in \{\bar{x}\} - \mathrm{cor}(C)$. As in the proof of Lemma 1.45, (b) we define a norm $\|\cdot\|$ on X by the Minkowski functional

$$\|x\| = \inf_{\lambda > 0} \left\{ \lambda \,\Big|\, \frac{1}{\lambda} x \in [\hat{x} - \bar{x}, \bar{x} - \hat{x}] \right\} \text{ for all } x \in X.$$

Then the order interval $[\hat{x} - \bar{x}, \bar{x} - \hat{x}]$ is the closed unit ball. Since $\bar{x}$ is assumed to be a minimal element of the set S, we conclude

$$[\hat{x} - \bar{x}, \bar{x} - \hat{x}] \cap (S - \{\hat{x}\}) = \{\bar{x} - \hat{x}\}$$

which implies

$$1 = \|\bar{x} - \hat{x}\| < \|x - \hat{x}\| \text{ for all } x \in S \setminus \{\bar{x}\}.$$

Finally, with the same arguments used in the proof of Lemma 1.45, (b) we obtain for all $c \in C$

$$x \in [0_X, c] \implies \|x\| \leq \|c\|.$$

This means that the norm $\|\cdot\|$ is monotonically increasing on C. $\square$

The preceding theorem states that under suitable assumptions every minimal element $\bar{x}$ of a set S is a unique best approximation from the set S to some element which is "strictly" less than $\bar{x}$. Hence, vector optimization problems lead to approximation problems.

A simpler necessary condition can be obtained, if the set $S + C$ is convex.

Theorem 5.4. *Let S be a nonempty subset of a partially ordered linear space X with a pointed nontrivial ordering cone C_X. If the set*

5.1. Necessary Conditions for Optimal Elements of a Set

$S + C_X$ is convex and has a nonempty algebraic interior, then for every minimal element $\bar{x} \in S$ of the set S there is a linear functional $l \in C_{X'} \setminus \{0_{X'}\}$ with the property

$$l(\bar{x}) \leq l(x) \text{ for all } x \in S.$$

Proof. If $\bar{x} \in S$ is a minimal element of the set S, then by Lemma 4.7, (b) $\bar{x}$ is also a minimal element of the set $S + C_X$, that is

$$(\{\bar{x}\} - C_X) \cap (S + C_X) = \{\bar{x}\}.$$

Since $\{\bar{x}\} - C_X$ and $S + C_X$ are convex, $\text{cor}(S + C_X) \neq \emptyset$ and $\bar{x} \notin \text{cor}(S + C_X)$, by the separation theorem 3.14 there are a linear functional $l \in X' \setminus \{0_{X'}\}$ and a real number α with

$$l(\bar{x} - c_1) \leq \alpha \leq l(x + c_2) \text{ for all } x \in S \text{ and all } c_1, c_2 \in C_X.$$

Since C_X is a cone, we immediately obtain $l \in C_{X'} \setminus \{0_{X'}\}$. Moreover, we get with $c_1 = c_2 = 0_X$

$$l(\bar{x}) \leq l(x) \text{ for all } x \in S.$$

□

If we consider an abstract optimization problem as defined in Example 4.5, the set S (in Theorem 5.4) equals the image set of the objective map. Then, by Theorem 2.11, the assumption that $S + C_X$ is convex is equivalent to the assumption that the objective map is convex-like. The result of Theorem 5.4 can also be interpreted in the following way: Under the stated assumptions for every minimal element $\bar{x}$ there is a linear functional $l \in C_{X'} \setminus \{0_{X'}\}$ so that $\bar{x}$ is a minimal solution of the scalar optimization problem

$$\min_{x \in S} l(x).$$

With the following theorem we formulate a necessary and sufficient condition with linear functionals but without the convexity assumption of Theorem 5.4.

Theorem 5.5. *Let S be a nonempty subset of a partially ordered locally convex linear space X with a pointed closed ordering cone C_X. An element $\bar{x} \in S$ is a minimal element of the set S if and only if for every $x \in S\setminus\{\bar{x}\}$ there is a continuous linear functional $l \in C_{X^*}\setminus\{0_{X^*}\}$ with $l(\bar{x}) < l(x)$.*

Proof. Let $\bar{x} \in S$ be a minimal element of the set S, i.e. $(\{\bar{x}\} - C_X) \cap S = \{\bar{x}\}$. This set equation can also be interpreted in the following way:

$$x \notin \{\bar{x}\} - C_X \text{ for all } x \in S\setminus\{\bar{x}\}. \tag{5.1}$$

Since C_X is closed and convex, the set $\{\bar{x}\} - C_X$ is closed and convex as well, and with Theorem 3.18 the statement (5.1) is equivalent to: For every $x \in S\setminus\{\bar{x}\}$ there is a continuous linear functional $l \in C_{X^*}\setminus\{0_{X^*}\}$ with $l(\bar{x}) < l(x)$. □

Roughly speaking, by Theorem 5.5 $\bar{x}$ is a minimal element of S if and only if $C_{X^*}\setminus\{0_{X^*}\}$ separates $\bar{x}$ from every other element in S. Theorem 5.5 is actually not a scalarization result. But with the same arguments we get a scalarization result for strongly minimal elements which is similar to that of Theorem 5.5.

Theorem 5.6. *Let S be a nonempty subset of a partially ordered locally convex linear space X with a closed ordering cone C_X. An element $\bar{x} \in S$ is a strongly minimal element of the set S if and only if for every $l \in C_{X^*}$*

$$l(\bar{x}) \leq l(x) \text{ for all } x \in S.$$

Proof. Let $\bar{x} \in S$ be a strongly minimal element of the set S, i.e.

$$S \subset \{\bar{x}\} + C_X. \tag{5.2}$$

Since C_X is a closed convex cone and X is locally convex, by Lemma 3.21, (a)

$$C_X = \{x \in X \mid l(x) \geq 0 \text{ for all } l \in C_{X^*}\}.$$

5.1. Necessary Conditions for Optimal Elements of a Set

Hence, the inclusion (5.2) is equivalent to

$$S - \{\bar{x}\} \subset \{x \in X \mid l(x) \geq 0 \text{ for all } l \in C_{X^*}\}$$

which can also be interpreted in the following way: For every $x \in S$ it follows

$$l(\bar{x}) \leq l(x) \text{ for all } l \in C_{X^*}.$$

□

Notice that we do not need any convexity assumption in Theorem 5.6. Thus, a strongly minimal element is a minimal solution for a whole class of scalar optimization problems. This shows that this optimality notion is indeed very strong.

Next, we turn our attention to the notion of proper minimality.

Theorem 5.7. *Let S be a nonempty subset of a partially ordered normed space $(X, \|\cdot\|_X)$ with an ordering cone C_X which has a weakly compact base. For some $\bar{x} \in S$ let $\text{cone}(T(S + C_X, \bar{x}) \cup (S - \{\bar{x}\}))$ be weakly closed. If $\bar{x}$ is a properly minimal element of the set S, then for every $\hat{x} \in \{\bar{x}\} - C_X$, $\hat{x} \neq \bar{x}$, there is an (additional) continuous norm $\|\cdot\|$ on X which is strongly monotonically increasing on C_X and which has the property*

$$1 = \|\bar{x} - \hat{x}\| < \|x - \hat{x}\| \text{ for all } x \in S\setminus\{\bar{x}\}.$$

Proof. The proof of this theorem is rather technical and, therefore, a short overview is given first in order to examine the geometry. In part (1) it is shown that the base of the convex cone $-C_X$ and the cone C generated by the set $T(S + C_X, \bar{x}) \cup (S - \{\bar{x}\})$ have a positive "distance" ε. This allows us to construct another cone $\hat{C}$ in the second part which is "larger" than the ordering cone C_X but for which $(-\hat{C}) \cap C = \{0_X\}$. It can be shown that $\hat{C}$ is convex, closed, pointed and that it has a nonempty interior. In part (3) we define the desired norm $\|\cdot\|$ as the Minkowski functional with respect to an appropriate order interval. Moreover, in part (4) several properties of the norm are proved. Notice for the following proof that the ordering cone C_X is pointed because it has a base (compare Lemma 3.3 and Lemma 1.27, (b)).

(1) In the following let B denote the weakly compact base of the ordering cone C_X and let C denote the cone generated by $T(S+C_X, \bar{x}) \cup (S - \{\bar{x}\})$, i.e.

$$C := \operatorname{cone}(T(S + C_X, \bar{x}) \cup (S - \{\bar{x}\})).$$

Since B is weakly compact and for every $x \in C$ the functional $\|x - \cdot\|_X : X \to \mathbb{R}$ is weakly lower semicontinuous, for every $x \in C$ the scalar optimization problem

$$\inf_{y \in -B} \|x - y\|_X$$

is solvable, i.e., there is a $y(x) \in -B$ with the property that

$$\|x - y(x)\|_X \leq \|x - y\|_X \text{ for all } y \in -B.$$

Next, we consider the scalar optimization problem

$$\varepsilon := \inf_{x \in C} \|x - y(x)\|_X.$$

If we assume $\varepsilon = 0$, then there is an infimal net

$$(\|x_i - y(x_i)\|_X)_{i \in I} \to 0 \text{ with } x_i \in C \text{ for all } i \in I. \qquad (5.3)$$

Since B is weakly compact and C is weakly closed, the set $C+B$ is weakly closed, and the condition (5.3) implies

$$0_X \in \operatorname{cl}(C + B) \subset \operatorname{cl}(C + B)_{\sigma(X, X^*)} = C + B. \qquad (5.4)$$

$\bar{x}$ is assumed to be a properly minimal element of the set S. Consequently, 0_X is a minimal element of the contingent cone $T(S + C_X, \bar{x})$ and a minimal element of the set $S - \{\bar{x}\}$, and we obtain

$$\begin{aligned} \{0_X\} &= (-C_X) \cap T(S + C_X, \bar{x}) \ \cup \ (-C_X) \cap (S - \{\bar{x}\}) \\ &= (-C_X) \cap (T(S + C_X, \bar{x}) \cup (S - \{\bar{x}\})) \end{aligned}$$

5.1. Necessary Conditions for Optimal Elements of a Set

and
$$\{0_X\} = (-C_X) \cap \text{cone}(T(S + C_X, \bar{x}) \cup (S - \{\bar{x}\}))$$
$$\supset (-B) \cap C.$$

Since $0_X \notin B$, we conclude $(-B) \cap C = \emptyset$ which contradicts the condition (5.4). Thus we get
$$0 < \varepsilon = \inf_{x \in C} \inf_{y \in -B} \|x - y\|_X,$$
i.e., the sets C and $-B$ have a positive "distance" ε.

(2) Now, we "separate" the sets $-B$ and C by a cone $-\hat{C}$. Since the base B is weakly compact and $0_X \notin B$ we obtain
$$0 < \delta := \inf_{y \in B} \|y\|_X.$$

For $\beta := \min\{\frac{\varepsilon}{2}, \frac{\delta}{2}\} > 0$ we define the set $U := B + N(0_X, \beta)$ ($N(0_X, \beta)$ denotes the closed ball around 0_X with radius β). It is evident that U is a convex set. Consequently, the cone generated by U and its closure $\hat{C} := \text{cl}(\text{cone}(U))$ is a convex cone. By definition, this cone has a nonempty topological interior. In order to see that $\hat{C}$ is pointed we investigate the cone $\tilde{C} := \text{cone}(B + N(0_X, \frac{3}{2}\beta))$ which is a superset of $\hat{C}$. If we assume that there is an $\tilde{x} \in (-\tilde{C}) \cap \tilde{C}$ with $\tilde{x} \neq 0_X$, then there are a $\lambda > 0$ and an $x \in B + N(0_X, \frac{3}{2}\beta)$ with $\tilde{x} = \lambda x$. Because of $-\tilde{x} = \lambda(-x) \in \tilde{C}$ we obtain for some $\mu > 0$
$$-\mu x \in B + N(0_X, \frac{3}{2}\beta).$$

Hence, x and $-\mu x$ are elements of the convex set $B + N(0_X, \frac{3}{2}\beta)$ which implies $0_X \in B + N(0_X, \frac{3}{2}\beta)$. But this is a contradiction to the choice of $\beta \leq \frac{\delta}{2}$. Consequently, $\tilde{C}$ is pointed and with $\hat{C} \subset \tilde{C}$ the cone $\hat{C}$ is pointed as well.

(3) Next, we choose an arbitrary $\hat{x} \in \{\bar{x}\} - C_X$ with $\hat{x} \neq \bar{x}$ and we define the order interval (with respect to the partial ordering induced by $\hat{C}$)
$$[\hat{x} - \bar{x}, \bar{x} - \hat{x}] := (\{\hat{x} - \bar{x}\} + \hat{C}) \cap (\{\bar{x} - \hat{x}\} - \hat{C}).$$

Because of the construction of $\hat{C}$ and the set U the element $\bar{x} - \hat{x}$ belongs to the interior of $\hat{C}$. Furthermore, $\hat{C}$ is closed and pointed. Consequently, the Minkowski functional $\|\cdot\| : X \to \mathbb{R}$ given by

$$\|x\| := \inf_{\lambda > 0} \left\{ \lambda \ \Big| \ \frac{1}{\lambda} x \in [\hat{x} - \bar{x}, \bar{x} - \hat{x}] \right\} \text{ for all } x \in X$$

is a norm on X and

$$[\hat{x} - \bar{x}, \bar{x} - \hat{x}] = \{x \in X \mid \|x\| \leq 1\}. \tag{5.5}$$

(4) We have to show several properties of the norm $\|\cdot\|$. Since 0_X belongs to the topological interior of the order interval $[\hat{x} - \bar{x}, \bar{x} - \hat{x}]$, there is an $\alpha > 0$ with

$$N(0_X, \alpha) \subset [\hat{x} - \bar{x}, \bar{x} - \hat{x}]$$

which implies with (5.5)

$$\|x\| \leq \alpha \|x\|_X \text{ for all } x \in X.$$

With this inequality it follows

$$|\|x\| - \|y\|| \leq \|x - y\| \leq \alpha \|x - y\|_X \text{ for all } x, y \in X.$$

Hence, the norm $\|\cdot\|$ is continuous.

In order to see that the norm $\|\cdot\|$ is strongly monotonically increasing on C_X, observe that the norm is monotonically increasing (with respect to $\hat{C}$) on $\hat{C}$, i.e.

$$\tilde{x} \in \hat{C}, \ x \in (\{\tilde{x}\} - \hat{C}) \cap \hat{C} \implies \|x\| \leq \|\tilde{x}\|.$$

For every $\tilde{x} \in C_X \subset \hat{C}$ and every $x \in (\{\tilde{x}\} - (C_X \setminus \{0_X\})) \cap C_X$ we have with $C_X \setminus \{0_X\} \subset \text{int}(\hat{C})$

$$\|x\| < \|\tilde{x}\|.$$

Hence, $\|\cdot\|$ is strongly monotonically increasing on C_X.

5.1. Necessary Conditions for Optimal Elements of a Set

Finally, we prove that $\bar{x}$ is a unique solution of a certain approximation problem. Since $\bar{x} - \hat{x}$ belongs to the closure of the unit ball, we obtain $\|\bar{x} - \hat{x}\| = 1$. Furthermore, we assert that

$$(-\hat{C}) \cap C = \{0_X\}. \tag{5.6}$$

Because of the construction of the set U and the choice of $\beta \leq \frac{\varepsilon}{2}$ for every $x \in C\setminus\{0_X\}$ there is an $\eta > 0$ with $N(x, \eta) \cap \text{cone}(U) = \emptyset$ which implies $x \notin \text{cl}(\text{cone}(U)) = \hat{C}$. Hence, $(-\hat{C}) \cap (C\setminus\{0_X\}) = \emptyset$ and the set equality (5.6) is evident. Moreover, with (5.6) and (5.5) we conclude

$$[\hat{x} - \bar{x}, \bar{x} - \hat{x}] \cap (\{\bar{x} - \hat{x}\} + C) = \{\bar{x} - \hat{x}\}$$

and

$$1 = \|\bar{x} - \hat{x}\| < \|\bar{x} - \hat{x} + x\| \text{ for all } x \in C\setminus\{0_X\}.$$

Since $S - \{\bar{x}\} \subset C$, we get

$$1 = \|\bar{x} - \hat{x}\| < \|x - \hat{x}\| \text{ for all } x \in S\setminus\{\bar{x}\}.$$

This completes the proof.

□

In the preceding theorem the ordering cone C_X is assumed to have a weakly compact base. Then C_X is necessarily nontrivial, pointed and closed. With the following lemmas we give sufficient conditions under which various assumptions of Theorem 5.7 are fulfilled.

Lemma 5.8. *Let $(X, \|\cdot\|_X)$ be a partially ordered reflexive Banach space with a nontrivial closed ordering cone C_X. The ordering cone C_X has a weakly compact base if and only if there is a continuous linear functional $l \in C_{X^*}^{\#}$ so that the set $\{x \in C_X \mid l(x) = 1\}$ is bounded.*

Proof. This lemma is a consequence of Lemma 3.3 (the continuity of l can be obtained with Lemma 3.13) and the fact that X is reflexive. □

Lemma 5.9. *Let S be a nonempty subset of a partially ordered normed space $(X, \|\cdot\|_X)$ with an ordering cone C_X. If the set $S + C_X$ is starshaped at some $\bar{x} \in S$ and the contingent cone $T(S + C_X, \bar{x})$ is weakly closed, then $\mathrm{cone}(T(S + C_X, \bar{x}) \cup (S - \{\bar{x}\}))$ is also weakly closed.*

Proof. Since the set $S + C_X$ is starshaped at $\bar{x}$, we conclude with Theorem 3.43

$$S - \{\bar{x}\} \subset S + C_X - \{\bar{x}\} \subset T(S + C_X, \bar{x}).$$

Hence, we obtain

$$\mathrm{cone}(T(S+C_X,\bar{x}) \cup (S - \{\bar{x}\})) = \mathrm{cone}(T(S+C_X,\bar{x})) = T(S+C_X,\bar{x})$$

which leads to the assertion. □

Lemma 5.10. *Let S be a nonempty subset of a partially ordered normed space $(X, \|\cdot\|_X)$ with an ordering cone C_X. If the set $S + C_X$ is convex, then for every $\bar{x} \in S$ the set $\mathrm{cone}(T(S+C_X,\bar{x}) \cup (S - \{\bar{x}\}))$ is weakly closed.*

Proof. The contingent cone $T(S+C_X, \bar{x})$ is closed (by Theorem 3.45) and also convex because of the convexity of the set $S + C_X$ (see Theorem 3.47). Consequently, by Theorem 3.24 the contingent cone $T(S + C_X, \bar{x})$ is weakly closed. Thus, the assertion follows from Lemma 5.9. □

The last lemma shows that the assumptions of Theorem 5.7 can be reduced, if the set $S + C_X$ is convex. But in this case a scalarization result which uses certain linear functionals is more interesting.

Theorem 5.11. *Let S be a nonempty subset of a partially ordered normed space X where the topology gives X as the topological dual space of X^*, and let C_X be a closed ordering cone in X with $\mathrm{int}(C_{X^*}) \neq \emptyset$. If the set $S + C_X$ is convex, then for every properly*

5.1. Necessary Conditions for Optimal Elements of a Set

minimal element $\bar{x} \in S$ of the set S there is a continuous linear functional $l \in C_{X^*}^{\#}$ with the property

$$l(\bar{x}) \leq l(x) \text{ for all } x \in S.$$

Proof. If $\bar{x} \in S$ is a properly minimal element of the set S, then the zero element 0_X is a minimal element of the contingent cone $T(S + C_X, \bar{x})$, i.e. (with Lemmas 3.21, (d) and 1.27, (b))

$$(-C_X) \cap T(S + C_X, \bar{x}) = \{0_X\}. \tag{5.7}$$

Since the set $S + C_X$ is convex, $\text{cone}(S + C_X - \{\bar{x}\})$ is also convex and by Lemma 1.32 the closure $\text{cl}(\text{cone}(S + C_X - \{\bar{x}\}))$ is convex as well. By Theorem 3.43 and Theorem 3.44 we get $T(S + C_X, \bar{x}) = \text{cl}(\text{cone}(S + C_X - \{\bar{x}\}))$ which is convex. Then, by Theorem 3.22, the set equation (5.7) is equivalent to the existence of a continuous linear functional $l \in X^* \backslash \{0_{X^*}\}$ with

$$l(-c) \leq 0 \leq l(t) \text{ for all } c \in C_X \text{ and all } t \in T(S + C_X, \bar{x}) \tag{5.8}$$

and

$$l(c) > 0 \text{ for all } c \in C_X \backslash \{0_X\}. \tag{5.9}$$

With the inequality (5.9) we conclude $l \in C_{X^*}^{\#}$. By Theorem 3.43 we obtain further

$$S - \{\bar{x}\} \subset S + C_X - \{\bar{x}\} \subset T(S + C_X, \bar{x}),$$

and, therefore, we get from the inequality (5.8)

$$l(\bar{x}) \leq l(x) \text{ for all } x \in S.$$

$\square$

The preceding theorem is comparable with Theorem 5.4 and Theorem 5.6. But now the linear functional l belongs to the quasi-interior of the dual ordering cone.

Finally, we present two necessary conditions for weakly minimal elements.

Theorem 5.12. *Let S be a nonempty subset of a partially ordered linear space X with an ordering cone C which has a nonempty algebraic interior. If $\bar{x} \in S$ is a weakly minimal element of the set S, then for every $\hat{x} \in \{\bar{x}\} - \text{cor}(C)$ there is a seminorm $\|\cdot\|$ on X which is strictly monotonically increasing on $\text{cor}(C)$ with the property*

$$1 = \|\bar{x} - \hat{x}\| \leq \|x - \hat{x}\| \text{ for all } x \in S.$$

Proof. Pick any element $\hat{x} \in \{\bar{x}\} - \text{cor}(C)$. As in the proof of Lemma 1.45, (a) we define a seminorm $\|\cdot\|$ on X by the Minkowski functional

$$\|x\| = \inf_{\lambda > 0} \left\{ \lambda \,\Big|\, \frac{1}{\lambda} x \in [\hat{x} - \bar{x}, \bar{x} - \hat{x}] \right\} \text{ for all } x \in X.$$

Then

$$\text{cor}([\hat{x} - \bar{x}, \bar{x} - \hat{x}]) = \{x \in X \mid \|x\| < 1\},$$

and since $\bar{x}$ is weakly minimal, we have

$$\{x \in X \mid \|x\| < 1\} \cap (S - \{\hat{x}\}) = \emptyset$$

which implies

$$1 \leq \|x - \hat{x}\| \text{ for all } x \in S.$$

$\bar{x} - \hat{x}$ belongs to the algebraic boundary of the order interval $[\hat{x} - \bar{x}, \bar{x} - \hat{x}]$ and, therefore, $\|\bar{x} - \hat{x}\| = 1$. With the same arguments as in Lemma 1.45, (a) we conclude for all $c \in \text{cor}(C)$

$$x \in \text{cor}([0_X, c]) \implies \|x\| < \|c\|$$

which means that the seminorm $\|\cdot\|$ is strictly monotonically increasing on $\text{cor}(C)$. □

It can be expected that for the weak minimality notion a special scalarization result can be formulated under a convexity assumption as well. This is done in

Theorem 5.13. *Let S be a nonempty subset of a partially ordered linear space X with an ordering cone C_X which has a nonempty algebraic interior. If the set $S + C_X$ is convex, then for every weakly minimal element $\bar{x} \in S$ of the set S there is a linear functional $l \in C_{X'} \setminus \{0_{X'}\}$ with the property*

$$l(\bar{x}) \leq l(x) \text{ for all } x \in S.$$

Proof. Let $\bar{x} \in S$ be a weakly minimal element of the set S. By Lemma 4.13, (b) $\bar{x}$ is also a weakly minimal element of the set $S + C_X$, i.e. $(\{\bar{x}\} - \text{cor}(C_X)) \cap (S + C_X) = \emptyset$. With Theorem 3.14 this set equation implies that there are a linear functional $l \in X' \setminus \{0_{X'}\}$ and a real number α with

$$l(\bar{x} - c_1) \leq \alpha \leq l(s + c_2) \text{ for all } c_1 \in C_X, \ s \in S \text{ and } c_2 \in C_X.$$

Since C_X is a cone, we get $l \in C_{X'} \setminus \{0_{X'}\}$ and the assertion is obvious. □

In this section we presented mainly two types of scalarization results: a nonconvex version via approximation problems (Theorem 5.3, 5.7, 5.12) and a (in general) convex version with the aid of linear functionals (Theorem 5.4, 5.11, 5.13). Only Theorem 5.5 and Theorem 5.6 are scalarization results with linear functionals without assuming that the set $S + C$ is convex.

5.2 Sufficient Conditions for Optimal Elements of a Set

It is the aim of this section to investigate under which assumptions the necessary conditions presented in section 5.1 are also sufficient for optimal elements of a set. We begin our discussion with the minimality notion.

Lemma 5.14. *Let S be a nonempty subset of a partially ordered linear space with a pointed ordering cone C. Moreover, let $f : S \to \mathbb{R}$*

be a given functional, and let an element $\bar{x} \in S$ be given with the property
$$f(\bar{x}) \leq f(x) \text{ for all } x \in S. \tag{5.10}$$

(a) If the functional f is monotonically increasing on S and if $\bar{x}$ is uniquely determined by (5.10), then $\bar{x}$ is a minimal element of the set S.

(b) If the functional f is strongly monotonically increasing on S, then $\bar{x}$ is a minimal element of the set S.

Proof. For the proof of both parts we assume that $\bar{x}$ is not a minimal element of the set S. Then there is an element $x \in (\{\bar{x}\}-C) \cap S$ with $x \neq \bar{x}$. This implies $f(x) \leq f(\bar{x})$ in part (a) which contradicts the unique solvability of the considered scalar optimization problem. In part (b) we obtain $f(x) < f(\bar{x})$ which is a contradiction to the minimality of f at $\bar{x}$. □

Next, we apply Lemma 5.14 to a special class of functionals f, namely certain seminorms and linear functionals.

Theorem 5.15. *Let S be a nonempty subset of a partially ordered linear space X with a pointed ordering cone C. Moreover, let $\|\cdot\|$ be a seminorm on X, and let elements $\hat{x} \in X$ and $\bar{x} \in S$ with*
$$S \subset \{\hat{x}\} + C \tag{5.11}$$
and
$$\|\bar{x} - \hat{x}\| \leq \|x - \hat{x}\| \text{ for all } x \in S$$
be given.

(a) *If the seminorm $\|\cdot\|$ is monotonically increasing on C and if $\bar{x}$ is the unique best approximation from the set S to $\hat{x}$, then $\bar{x}$ is a minimal element of the set S.*

(b) *If the seminorm $\|\cdot\|$ is strongly monotonically increasing on C, then $\bar{x}$ is a minimal element of the set S.*

5.2. Sufficient Conditions for Optimal Elements of a Set

Proof. We prove only part (a) of the assertion. The proof of the other part is similar. In order to be able to apply Lemma 5.14, (a) we show the monotonicity of the functional $\|\cdot - \hat{x}\|$ on S. For every $\bar{s} \in S$ we obtain with the inclusion (5.11)

$$(\{\bar{s}\} - C) \cap S \subset (\{\bar{s}\} - C) \cap (\{\hat{x}\} + C) = [\hat{x}, \bar{s}] = \{\hat{x}\} + [0_X, \bar{s} - \hat{x}].$$

Consequently, we have for every $x \in (\{\bar{s}\} - C) \cap S$

$$x - \hat{x} \in [0_X, \bar{s} - \hat{x}].$$

Hence, we conclude because of the monotonicity of the seminorm $\|\cdot\|$ on C

$$\|x - \hat{x}\| \leq \|\bar{s} - \hat{x}\|.$$

Consequently, the functional $\|\cdot - \hat{x}\|$ is monotonically increasing on S. This completes the proof. $\square$

Theorem 5.3 and Theorem 5.15, (a) lead to a characterization of minimal elements of a set.

Corollary 5.16. *Let S be a nonempty subset of a partially ordered linear space X with a pointed, algebraically closed ordering cone C which has a nonempty algebraic interior. Moreover, let an element $\hat{x} \in X$ with $S \subset \{\hat{x}\} + \text{cor}(C)$ be given. An element $\bar{x} \in S$ is a minimal element of the set S if and only if there is a norm $\|\cdot\|$ on X which is monotonically increasing on C with the property*

$$\|\bar{x} - \hat{x}\| < \|x - \hat{x}\| \text{ for all } x \in S\setminus\{\bar{x}\}.$$

If the set S has no lower bound $\hat{x}$, i.e., the inclusion (5.11) is not fulfilled, approximation problems are still qualified for the determination of minimal elements of the set S (this idea is due to Rolewicz [265]).

Theorem 5.17. *Let S be a nonempty subset of a partially ordered linear space X with a pointed ordering cone C. Moreover, let a seminorm $\|\cdot\|$ on X and an element $\tilde{x} \in S$ be given so that for some*

$\bar{x} \in S \cap (\{\tilde{x}\} - C)$

$$\|\bar{x} - \tilde{x}\| \geq \|x - \tilde{x}\| \text{ for all } x \in S \cap (\{\tilde{x}\} - C). \quad (5.12)$$

(a) *If the seminorm $\|\cdot\|$ is monotonically increasing on C and if $\bar{x}$ is uniquely determined by the inequality (5.12), then $\bar{x}$ is a minimal element of the set S.*

(b) *If the seminorm $\|\cdot\|$ is strongly monotonically increasing on C, then $\bar{x}$ is a minimal element of the set S.*

Proof. We proof only part (a) of this theorem. First, we show that the functional $-\|\cdot-\tilde{x}\|$ is monotonically increasing on $\{\tilde{x}\} - C$. For that purpose we take any arbitrary $\bar{y} \in \{\tilde{x}\} - C$ and choose any

$$x \in (\{\bar{y}\} - C) \cap (\{\tilde{x}\} - C) = \{\bar{y}\} - C.$$

Then we have $\tilde{x} - x \in \{\tilde{x} - \bar{y}\} + C$ and $\tilde{x} - \bar{y} \in \{\tilde{x} - x\} - C$. But we also have $\tilde{x} - \bar{y} \in C$. Because of the monotonicity of the seminorm on C we obtain

$$\|\tilde{x} - \bar{y}\| \leq \|\tilde{x} - x\|$$

implying

$$-\|\bar{y} - \tilde{x}\| \geq -\|x - \tilde{x}\|.$$

Then Lemma 5.14, (a) is applicable and $\bar{x}$ is a minimal element of the set $S \cap (\{\tilde{x}\} - C)$, i.e.

$$(\{\bar{x}\} - C) \cap S \cap (\{\tilde{x}\} - C) = \{\bar{x}\}.$$

Finally, the inclusion $(\{\bar{x}\} - C) \cap S \subset \{\tilde{x}\} - C$ leads to

$$(\{\bar{x}\} - C) \cap S = \{\bar{x}\},$$

i.e., $\bar{x}$ is a minimal element of the set S. □

Notice that in Theorem 5.15 we have to determine a minimal "distance" between $\hat{x}$ and the set S whereas in Theorem 5.17 a maximal "distance" between $\tilde{x}$ and elements in the set $S \cap (\{\tilde{x}\} - C)$ has to be determined.

5.2. Sufficient Conditions for Optimal Elements of a Set

Now, we study certain linear functionals.

Theorem 5.18. *Let S be a nonempty subset of a partially ordered linear space X with a pointed ordering cone C_X.*

(a) If there are a linear functional $l \in C_{X'}$ and an element $\bar{x} \in S$ with
$$l(\bar{x}) < l(x) \text{ for all } x \in S\setminus\{\bar{x}\},$$
then $\bar{x}$ is a minimal element of the set S.

(b) If there are a linear functional $l \in C_{X'}^{\#}$ and an element $\bar{x} \in S$ with
$$l(\bar{x}) \leq l(x) \text{ for all } x \in S,$$
then $\bar{x}$ is a minimal element of the set S.

Proof. The proof follows directly from Lemma 5.14 and the remark in Example 5.2, (a). □

Notice that the Krein-Rutman theorem 3.38 gives conditions under which the set $C_{X^*}^{\#}$ is nonempty. If we compare Theorem 5.4 and Theorem 5.18 we see that one cannot prove the sufficiency of the necessary condition formulated in Theorem 5.4. Hence, one cannot present a complete characterization like Corollary 5.16 for linear functionals instead of norms.

Since we already characterized strongly minimal elements of a set in Theorem 5.6, we study now the proper minimality notion.

Theorem 5.19. *Let S be a nonempty subset of a partially ordered normed space $(X, \|\cdot\|_X)$ with a pointed ordering cone C which has a nonempty algebraic interior. Let $\|\cdot\|$ be any (additional) continuous norm on X which is strongly monotonically increasing on C. Moreover, let an element $\hat{x} \in X$ with $S \subset \{\hat{x}\} + cor(C)$ be given. If there is an element $\bar{x} \in S$ with the property*

$$\|\bar{x} - \hat{x}\| \leq \|x - \hat{x}\| \text{ for all } x \in S, \qquad (5.13)$$

then $\bar{x}$ is a properly minimal element of the set S.

Proof. Since the norm $\|\cdot\|$ is strongly monotonically increasing on C and $S - \{\hat{x}\} \subset \text{cor}(C)$, by Lemma 5.14, (b) $\bar{x}$ is a minimal element of the set S. Next, we prove that 0_X is a minimal element of the contingent cone $T(S+C,\bar{x})$.

Since the norm $\|\cdot\|$ is assumed to be strongly monotonically increasing on C, we obtain from (5.13)

$$\|\bar{x} - \hat{x}\| \leq \|x - \hat{x}\| \leq \|x + c - \hat{x}\| \text{ for all } x \in S \text{ and all } c \in C$$

resulting in

$$\|\bar{x} - \hat{x}\| \leq \|x - \hat{x}\| \text{ for all } x \in S + C. \tag{5.14}$$

It is evident that the functional $\|\cdot - \hat{x}\|$ is both convex and continuous in the topology generated by the norm $\|\cdot\|_X$. Then by Theorem 3.48, (a) the inequality (5.14) implies

$$\|\bar{x} - \hat{x}\| \leq \|\bar{x} - \hat{x} + h\| \text{ for all } h \in T(S+C,\bar{x}). \tag{5.15}$$

With $T := T(S+C,\bar{x}) \cap (\{\hat{x} - \bar{x}\} + C)$ the inequality (5.15) is also true for all $h \in T$, and by Lemma 5.14, (b) 0_X is a minimal element of the set T.

Now, we assume that 0_X is not a minimal element of the contingent cone $T(S+C,\bar{x})$. Then there is an $x \in (-C_X) \cap T(S+C,\bar{x})$ with $x \neq 0_X$. Because of the inclusion $S \subset \{\hat{x}\} + \text{cor}(C)$ there is a $\lambda > 0$ with $\lambda x \in \{\hat{x} - \bar{x}\} + C$. Consequently, we get

$$\lambda x \in (-C) \cap T(S+C,\bar{x}) \cap (\{\hat{x} - \bar{x}\} + C)$$

and, therefore, we have $\lambda x \in (-C) \cap T$ which contradicts the fact that 0_X is a minimal element of the set T. Hence, 0_X is a minimal element of the contingent cone $T(S+C,\bar{x})$, and the assertion is obvious. □

In Theorem 5.7 we do not need the assumptions $\text{cor}(C) \neq \emptyset$ and $\hat{x} \in \{\bar{x}\} - \text{cor}(C)$ which play an important role in Theorem 5.19. On the other hand in Theorem 5.19 it is not required that $\bar{x}$ is uniquely determined by the inequality (5.13). With Theorem 5.7 and Theorem 5.19 we get immediately a characterization of properly minimal elements.

5.2. Sufficient Conditions for Optimal Elements of a Set

Corollary 5.20. *Let S be a nonempty subset of a partially ordered normed space $(X, \|\cdot\|_X)$ with an ordering cone C which has a nonempty algebraic interior and a weakly compact base. Moreover, let an element $\hat{x} \in X$ with $S \subset \{\hat{x}\} + cor(C)$ be given, and for some $\bar{x} \in S$ let the set $cone(T(S+C, \bar{x}) \cup (S - \{\bar{x}\}))$ be weakly closed. Then $\bar{x}$ is a properly minimal element of the set S if and only if there is an (additional) continuous norm $\|\cdot\|$ on X which is strongly monotonically increasing on C and which has the property*

$$1 = \|\bar{x} - \hat{x}\| < \|x - \hat{x}\| \text{ for all } x \in S \setminus \{\bar{x}\}.$$

In the preceding corollary we assume that the ordering cone C has a nonempty algebraic interior and a weakly compact base. Then by Lemma 1.45, (c) there is an (additional) norm $\|\cdot\|$ on X so that the real normed space $(X, \|\cdot\|)$ is also reflexive. This shows that the assumptions of Corollary 5.20 are very restrictive.

Another sufficient condition for properly minimal elements is given by

Theorem 5.21. *Let S be a nonempty subset of a partially ordered normed space X with a pointed ordering cone C_X which has a nonempty quasi-interior $C_{X^*}^{\#}$ of the topological dual ordering cone. If there are a continuous linear functional $l \in C_{X^*}^{\#}$ and an element $\bar{x} \in S$ with the property*

$$l(\bar{x}) \leq l(x) \text{ for all } x \in S, \tag{5.16}$$

then $\bar{x}$ is a properly minimal element of the set S.

Proof. With Theorem 5.18, (b) we conclude that $\bar{x}$ is a minimal element of the set S. Take any tangent $h \in T(S + C_X, \bar{x})$. Then there are a sequence $(x_n)_{n \in \mathbb{N}}$ of elements in $S + C_X$ and a sequence $(\lambda_n)_{n \in \mathbb{N}}$ of positive real numbers with $\bar{x} = \lim_{n \in \mathbb{N}} x_n$ and $h = \lim_{n \in \mathbb{N}} \lambda_n(x_n - \bar{x})$. The linear functional l is continuous and, therefore, we get $l(\bar{x}) = \lim_{n \in \mathbb{N}} l(x_n)$. Since the functional l is also strongly monotonically increasing on X, the inequality (5.16) implies

$$l(\bar{x}) \leq l(x) \text{ for all } x \in S + C_X.$$

Then it follows

$$\begin{aligned} l(h) &= \lim_{n \in \mathbb{N}} l(\lambda_n(x_n - \bar{x})) \\ &= \lim_{n \in \mathbb{N}} \lambda_n(l(x_n) - l(\bar{x})) \\ &\geq 0. \end{aligned}$$

Hence, we obtain

$$0 \leq l(h) \text{ for all } h \in T(S + C_X, \bar{x}).$$

Consequently, by Theorem 5.18, (b) 0_X is a minimal element of the contingent cone $T(S + C_X, \bar{x})$. This completes the proof. □

With Theorem 5.11 and Theorem 5.21 we are able to formulate the following characterization of properly minimal elements under a convexity assumption.

Corollary 5.22. *Let S be a nonempty subset of a partially ordered normed space X where the topology gives X as the topological dual space of X^*, and let C_X be a closed ordering cone in X with $\text{int}(C_{X^*}) \neq \emptyset$. Moreover, let the set $S + C_X$ be convex. An element $\bar{x} \in S$ is a properly minimal element of the set S if and only if there is a continuous linear functional $l \in C_{X^*}^{\#}$ with the property*

$$l(\bar{x}) \leq l(x) \text{ for all } x \in S.$$

Notice that by Lemma 3.21, (d) we have $\text{int}(C_{X^*}) = C_{X^*}^{\#}$ under the assumptions of Corollary 5.22. Even though the characterization result in Corollary 5.22 is very important for the theory in the following sections, the assumptions are very restrictive and, therefore, we modify the notion of proper minimality.

Definition 5.23. *Let S be a nonempty subset of a partially ordered topological linear space X with an ordering cone C_X which has a nonempty quasi-interior $C_{X^*}^{\#}$ of the topological dual ordering cone.*

5.2. Sufficient Conditions for Optimal Elements of a Set

An element $\bar{x} \in S$ is called an *almost properly minimal* element of the set S, if there is a linear functional $l \in C_{X*}^{\#}$ with the property

$$l(\bar{x}) \leq l(x) \text{ for all } x \in S.$$

Recall that, by the Krein-Rutman theorem 3.38, in a partially ordered separable normed space $(X, \|\cdot\|)$ with a closed and pointed ordering cone C_X the set $C_{X*}^{\#}$ is nonempty. Obviously, under the assumptions of Corollary 5.22 the notions "properly minimal" and "almost properly minimal" coincide. Moreover, by Theorem 5.18, (b) every almost properly minimal element is a minimal element as well.

Finally, we turn our attention to the weak minimality notion. For the following results we need a basic lemma.

Lemma 5.24. *Let S be a nonempty subset of a partially ordered linear space with an ordering cone C which has a nonempty algebraic interior. Moreover, let $f : S \to \mathbb{R}$ be a given functional which is strictly monotonically increasing on S. If there is an element $\bar{x} \in S$ with the property*

$$f(\bar{x}) \leq f(x) \text{ for all } x \in S,$$

then $\bar{x}$ is a weakly minimal element of the set S.

Proof. If $\bar{x} \in S$ is not a weakly minimal element of the set S, then we have $f(x) < f(\bar{x})$ for some $x \in (\{\bar{x}\} - \text{cor}(C)) \cap S$ which is a contradiction to the minimality of f at $\bar{x}$. □

Theorem 5.25. *Let S be a nonempty subset of a partially ordered linear space X with an ordering cone C which has a nonempty algebraic interior. Moreover, let $\|\cdot\|$ be a seminorm on X which is strictly monotonically increasing on C, and let an element $\hat{x} \in X$ with $S \subset \{\hat{x}\} + C$ be given. If there is an element $\bar{x} \in S$ with*

$$\|\bar{x} - \hat{x}\| \leq \|x - \hat{x}\| \text{ for all } x \in S,$$

then $\bar{x}$ is a weakly minimal element of the set S.

Proof. The proof of this theorem is analogous to the proof of Theorem 5.15. □

With Theorem 5.12 and Theorem 5.25 we get the following characterization of weakly minimal elements of a set.

Corollary 5.26. *Let S be a nonempty subset of a partially ordered linear space X with an ordering cone C which has a nonempty algebraic interior. Moreover, let an element $\hat{x} \in X$ with $S \subset \{\hat{x}\} + cor(C)$ be given. An element $\bar{x} \in S$ is a weakly minimal element of the set S if and only if there is a seminorm $\|\cdot\|$ on X which is strictly monotonically increasing on $cor(C)$ with the property*

$$\|\bar{x} - \hat{x}\| \leq \|x - \hat{x}\| \text{ for all } x \in S.$$

In contrast to Corollary 5.16 concerning minimal elements we do not require in Corollary 5.26 that $\bar{x}$ is a unique best approximation from S at $\hat{x}$. For the following result we do not need the assumption that a "strict" lower bound $\hat{x}$ exists.

Theorem 5.27. *Let S be a nonempty subset of a partially ordered linear space X with an ordering cone C which has a nonempty algebraic interior. Moreover, let an element $\tilde{x} \in S$ and a seminorm $\|\cdot\|$ on X be given which is strictly monotonically increasing on C. If there is an element $\bar{x} \in S$ with the property*

$$\|\bar{x} - \tilde{x}\| \geq \|x - \tilde{x}\| \text{ for all } x \in S \cap (\{\tilde{x}\} - C),$$

then $\bar{x}$ is a weakly minimal element of the set S.

The proof of Theorem 5.27 is similar to that of Theorem 5.17. The next theorem is evident using Lemma 5.24.

Theorem 5.28. *Let S be a nonempty subset of a partially ordered linear space X with an ordering cone C_X which has a nonempty algebraic interior. If for some $\bar{x} \in S$ there is a linear functional*

$l \in C_{X'} \backslash \{0_{X'}\}$ with the property

$$l(\bar{x}) \leq l(x) \text{ for all } x \in S,$$

then $\bar{x}$ is a weakly minimal element of the set S.

Although we cannot formulate a complete characterization of minimal elements with the aid of linear functionals (compare Theorem 5.4 and Theorem 5.18), this can be done for weakly minimal elements.

Corollary 5.29. *Let S be a nonempty subset of a partially ordered linear space X with an ordering cone C_X which has a nonempty algebraic interior. Moreover, let the set $S + C_X$ be convex. An element $\bar{x} \in S$ is a weakly minimal element of the set S if and only if there is a linear functional $l \in C_{X'} \backslash \{0_{X'}\}$ with the property*

$$l(\bar{x}) \leq l(x) \text{ for all } x \in S.$$

The preceding corollary follows from Theorem 5.13 and Theorem 5.28.

5.3 Parametric Approximation Problems

The results on norm scalarization presented in the two preceding sections are now extended. We introduce special parametric norms for scalarization which can be used for a complete characterization of minimal and weakly minimal elements in the general nonconvex case. The only assumption formulated is that the considered set has a strict lower bound. It turns out that these parametric norms are well-known norms in special cases arising in applications.

The parametric norms are introduced as follows:

Definition 5.30. *Let Y be a real topological linear space partially ordered by a closed pointed convex cone C with a nonempty interior* int(C). For every $a \in \text{int}(C)$ let $\|\cdot\|_a$ denote the Minkowski functional of the order interval $[-a, a]$, i.e.

$$\|y\|_a := \inf\left\{\lambda > 0 \,\Big|\, \frac{1}{\lambda} \in [-a, a]\right\} \text{ for all } y \in Y.$$

Since a belongs to the interior of the closed pointed convex cone C, the order interval $[-a, a]$ is an absolutely convex and absorbing (i.e., $0_Y \in \text{cor}([-a, a])$) set which is algebraically bounded. Therefore, the Minkowski functional of the order interval $[-a, a]$ is indeed a norm for every $a \in \text{int}(C)$ (see [125]). Consequently, the parametric norm is well defined and we have

$$[-a, a] = \{y \in Y \mid \|y\|_a \leq 1\} \text{ for all } a \in \text{int}(C) \qquad (5.17)$$

([125, p. 40]). In other words: The parametric norm $\|\cdot\|_a$ is chosen in such a way that its unit ball equals the order interval $[-a, a]$.

The following result gives a complete characterization of minimal and weakly minimal elements with the aid of the parametric norm $\|\cdot\|_a$. This theorem clarifies the relationship between vector optimization and approximation theory.

Theorem 5.31. *Let S be a nonempty subset of a real topological linear space Y partially ordered by a closed pointed convex cone C with a nonempty interior $\text{int}(C)$. Moreover, let an element $\hat{y} \in Y$ be given with the property*

$$S \subset \{\hat{y}\} + \text{int}(C). \qquad (5.18)$$

(a) An element $\bar{y} \in S$ is a minimal element of the set S if and only if there is an element $a \in \text{int}(C)$ so that

$$\|\bar{y} - \hat{y}\|_a < \|y - \hat{y}\|_a \text{ for all } y \in S\setminus\{\bar{y}\} \qquad (5.19)$$

(see Fig. 5.1).

(b) An element $\bar{y} \in S$ is a weakly minimal element of the set S if and only if there is an element $a \in \text{int}(C)$ so that

$$\|\bar{y} - \hat{y}\|_a \leq \|y - \hat{y}\|_a \text{ for all } y \in S. \qquad (5.20)$$

5.3. Parametric Approximation Problems

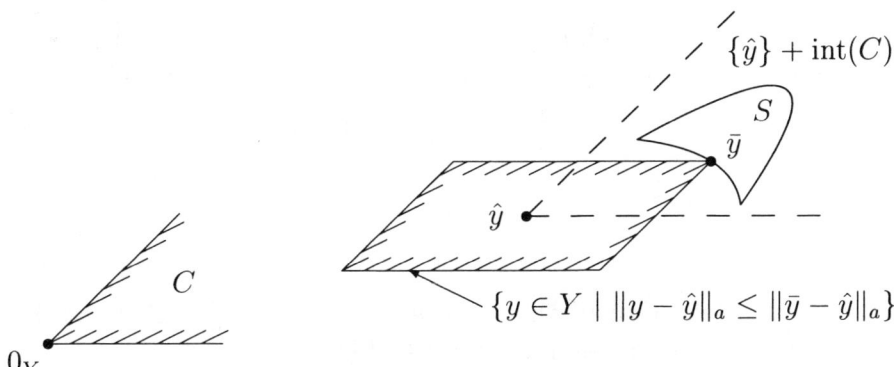

Figure 5.1: Illustration of the result in Thm. 5.31, (a).

Proof. Let an arbitrary $\hat{y} \in Y$ with the property (5.18) be chosen.

(a) If $\bar{y}$ is a minimal element of the set S, then we have
$$(\{\bar{y}\} - C) \cap S = \{\bar{y}\}$$
implying
$$(\{\bar{y} - \hat{y}\} - C) \cap (S - \{\hat{y}\}) = \{\bar{y} - \hat{y}\}. \qquad (5.21)$$
With the inclusion (5.18) we obtain $\hat{y} - \bar{y} \in -\text{int}(C)$ and we conclude that
$$S - \{\hat{y}\} \subset \text{int}(C) \subset \{\hat{y} - \bar{y}\} + \text{int}(C) \subset \{\hat{y} - \bar{y}\} + C. \qquad (5.22)$$
Consequently, the set equation (5.21) implies
$$(\{\hat{y} - \bar{y}\} + C) \cap (\{\bar{y} - \hat{y}\} - C) \cap (S - \{\hat{y}\}) = \{\bar{y} - \hat{y}\}$$
and
$$[-(\bar{y} - \hat{y}), \bar{y} - \hat{y}] \cap (S - \{\hat{y}\}) = \{\bar{y} - \hat{y}\}. \qquad (5.23)$$
If we notice the set equation (5.17), then (5.23) is equivalent to the inequality (5.19) for $a := \bar{y} - \hat{y}$.

For the converse implication let for an arbitrary $a \in \text{int}(C)$ a solution $\bar{y} \in S$ of the inequality (5.19) be given, and assume that $\bar{y}$ is not a minimal element of the set S. Then there is a $y \neq \bar{y}$ with $y \in (\{\bar{y}\} - C) \cap S$. Consequently, we have

$$y - \hat{y} \in \{\bar{y} - \hat{y}\} - C$$

which implies that

$$\|y - \hat{y}\|_a \leq \|\bar{y} - \hat{y}\|_a$$

by the definition of the parametric norm $\|\cdot\|_a$. But this is a contradiction to the inequality (5.19).

(b) Assume that $\bar{y}$ is a weakly minimal element of the set S. Then the set equation

$$(\{\bar{y}\} - \text{int}(C)) \cap S = \emptyset$$

is satisfied, and with (5.22) we get

$$(\{\hat{y} - \bar{y}\} + \text{int}(C)) \cap (\{\bar{y} - \hat{y}\} - \text{int}(C)) \cap (S - \{\hat{y}\}) = \emptyset$$

and

$$\text{int}([-(\bar{y} - \hat{y}), \bar{y} - \hat{y}]) \cap (S - \{\hat{y}\}) = \emptyset.$$

But this set equation implies

$$\{y \in Y \mid \|y - \hat{y}\|_a < 1\} \cap (S - \{\hat{y}\}) = \emptyset$$

for $a := \bar{y} - \hat{y}$. Hence, the inequality (5.20) is satisfied.

Finally, we prove the converse statement. Let an arbitrary $a \in \text{int}(C)$ be given, and assume that a $\bar{y} \in S$ solves the inequality (5.20) which is not a weakly minimal element of the set S. Then there is a

$$y \in (\{\bar{y}\} - \text{int}(C)) \cap S,$$

and we get

$$y - \hat{y} \in \{\bar{y} - \hat{y}\} - \text{int}(C).$$

By the definition of the parametric norm $\|\cdot\|_a$ this implies that

$$\|y - \hat{y}\|_a < \|\bar{y} - \hat{y}\|_a$$

which contradicts the inequality (5.20). □

5.3. Parametric Approximation Problems

Notice that by Theorem 5.31 every minimal and weakly minimal element of a set can be characterized as a solution of a certain approximation problem with a parametric norm. This result is even true for a nonconvex set. The only requirement formulated by the inclusion (5.18) says that the set S must have a strictly lower bound $\hat{y}$.

In the following lemmas we point out that the parametric norm $\|\cdot\|_a$ is well-known in special spaces.

Lemma 5.32. *Let the linear space $\mathbb{R}^n$ be partially ordered in a componentwise sense. Then for every vector $a \in \mathbb{R}^n$ with positive components the parametric norm $\|\cdot\|_a$ is given as*

$$\|y\|_a = \max_{1 \leq i \leq n} \left\{ \frac{|y_i|}{a_i} \right\} \text{ for all } y \in \mathbb{R}^n.$$

Proof. The proof of this lemma is obvious with the equation (5.17), if we notice that for every $a \in \mathbb{R}^n$ with positive components we have

$$[-a, a] = \{y \in \mathbb{R}^n \mid |y_i| \leq a_i \text{ for all } i \in \{1, \ldots, n\}\}.$$

$\square$

The parametric norm in Lemma 5.32 is the weighted maximum norm (or the weighted Chebyshev norm). A similar result is obtained in the space of continuous functions.

Lemma 5.33. *Let the linear space $C^n([t_0, t_1])$ (linear space of continuous functions defined on $[t_0, t_1]$ with $0 < t_0 < t_1 < \infty$ and having values in $\mathbb{R}^n$) be equipped with the usual maximum norm and partially ordered by the natural ordering cone*

$$C := \{y \in C^n([t_0, t_1]) \mid y_i(t) \geq 0 \text{ for all } t \in [t_0, t_1] \text{ and all } i \in \{1, \ldots, n\}\}.$$

Then for every function $a \in C^n([t_0, t_1])$ with

$$a_i(t) > 0 \text{ for all } t \in [t_0, t_1] \text{ and all } i \in \{1, \ldots, n\}$$

the parametric norm $\|\cdot\|_a$ is given as

$$\|y\|_a = \max_{\substack{t\in[t_0,t_1]\\ 1\leq i\leq n}} \left\{\frac{|y_i(t)|}{a_i(t)}\right\} \text{ for all } y \in C^n([t_0,t_1]).$$

Proof. Let $a \in C^n([t_0,t_1])$ be an arbitrary function which is componentwise and pointwise positive. Then we get

$$[-a,a] = \{y \in C^n([t_0,t_1]) \mid |y_i(t)| \leq a_i(t)$$
$$\text{for all } t \in [t_0,t_1] \text{ and all } i \in \{1,\ldots,n\}\},$$

and the assertion is obvious with the equation (5.17). □

The next lemma shows that in the space of continuous linear operators the parametric norm equals the weighted operator norm.

Lemma 5.34. *Let $(X, \langle \cdot, \cdot \rangle)$ be a real Hilbert space, let $B(X,X)$ denote the linear space of continuous linear operators $T : X \to X$, and let the linear space*

$$Y := \{T \in B(X,X) \mid T \text{ is self-adjoint}\}$$

be given which is equipped with the operator norm $\|\cdot\|$ given as

$$\|T\| = \sup_{x \neq 0_X} \left\{\frac{|\langle Tx, x\rangle|}{\langle x, x\rangle}\right\} \text{ for all } T \in Y$$

and partially ordered by the natural ordering cone

$$C := \{T \in Y \mid \langle Tx, x\rangle \geq 0 \text{ for all } x \in X\}.$$

Then for every positive definite operator $A \in Y$ the parametric norm $\|\cdot\|_A$ is given as

$$\|T\|_A = \sup_{x \neq 0_X} \left\{\frac{|\langle Tx, x\rangle|}{\langle Ax, x\rangle}\right\} \text{ for all } T \in Y.$$

5.3. Parametric Approximation Problems

Proof. For this proof we remark only that for every positive definite operator $A \in Y$ we obtain

$$[-A, A] = \{T \in Y \mid |\langle Tx, x \rangle| \leq \langle Ax, x \rangle \text{ for all } x \in X\}.$$

Then the assertion follows from the equation (5.17). □

Using the preceding lemmas we can specialize the formulation of Theorem 5.31 for concrete applications. First we present a result for a multiobjective optimization problem.

Corollary 5.35. *Let M be a nonempty set, and let $f : M \to \mathbb{R}^n$ be a given vector function. The linear space $\mathbb{R}^n$ is assumed to be partially ordered in a componentwise sense. Assume that there is a $\hat{y} \in \mathbb{R}^n$ with the property that*

$$\hat{y}_i < f_i(x) \text{ for all } x \in M \text{ and all } i \in \{1, \ldots, n\}.$$

(a) A vector $\bar{x} \in M$ is a minimal solution of the multiobjective optimization problem $\min_{x \in M} f(x)$ (i.e., $f(\bar{x})$ is a minimal element of $f(M)$) if and only if there are positive real numbers $a_1, \ldots, a_n$ so that

$$\max_{1 \leq i \leq n} \left\{ \frac{f_i(\bar{x}) - \hat{y}_i}{a_i} \right\} < \max_{1 \leq i \leq n} \left\{ \frac{f_i(x) - \hat{y}_i}{a_i} \right\}$$
for all $x \in M$ with $f(x) \neq f(\bar{x})$.

(b) A vector $\bar{x} \in M$ is a weakly minimal solution of the multiobjective optimization problem $\min_{x \in M} f(x)$ (i.e., $f(\bar{x})$ is a weakly minimal element of $f(M)$) if and only if there are positive real numbers $a_1, \ldots, a_n$ so that

$$\max_{1 \leq i \leq n} \left\{ \frac{f_i(\bar{x}) - \hat{y}_i}{a_i} \right\} \leq \max_{1 \leq i \leq n} \left\{ \frac{f_i(x) - \hat{y}_i}{a_i} \right\} \text{ for all } x \in M.$$

Proof. This corollary is a direct consequence of Theorem 5.31 and Lemma 5.32. □

Hence, optimal solutions of a general multiobjective optimization problem can be characterized as solutions of certain Chebyshev approximation problems. This result is even true for nonconvex problems, if the objective functions $f_1, \ldots, f_n$ have a lower bound.

A well-known problem in statistics is the problem of the determination of minimal covariance matrices. In this context we consider covariance operators defined on a real Hilbert space.

Corollary 5.36. *Let the assumptions in Lemma 5.34 be satisfied, and let S (set of covariance operators) be an arbitrary subset of Y for which $S \subset C$.*

(a) *A covariance operator $\bar{T} \in S$ is a minimal element of the set S if and only if there is a positive definite operator $A \in Y$ (i.e., there is an $\alpha > 0$ with $\langle Ax, x \rangle \geq \alpha \langle x, x \rangle$ for all $x \in X$) so that*

$$\sup_{x \neq 0_X} \left\{ \frac{\langle (\bar{T} + I)x, x \rangle}{\langle Ax, x \rangle} \right\} < \sup_{x \neq 0_X} \left\{ \frac{\langle (T + I)x, x \rangle}{\langle Ax, x \rangle} \right\}$$

for all $T \in S$ with $T \neq \bar{T}$

(I denotes the identity operator).

(b) *A covariance operator $\bar{T} \in S$ is a weakly minimal element of the set S if and only if there is a positive definite operator $A \in Y$ (i.e., there is an $\alpha > 0$ with $\langle Ax, x \rangle \geq \alpha \langle x, x \rangle$ for all $x \in X$) so that*

$$\sup_{x \neq 0_X} \left\{ \frac{\langle (\bar{T} + I)x, x \rangle}{\langle Ax, x \rangle} \right\} \leq \sup_{x \neq 0_X} \left\{ \frac{\langle (T + I)x, x \rangle}{\langle Ax, x \rangle} \right\} \text{ for all } T \in S$$

(I denotes the identity operator).

Proof. The cone C is pointed, convex, closed, and it has a nonempty interior. This interior consists exactly of all positive definite operators of Y. Since $S \subset C$ and $I \in \text{int}(C)$, we conclude that

$$S - \{-I\} = S + \{I\} \subset C + \text{int}(C) = \text{int}(C).$$

Hence, the inclusion (5.18) is fulfilled for $\hat{y} := -I$. With Theorem 5.31 and Lemma 5.34 we then obtain the desired result. $\square$

Since, in general, covariance matrices are positive semidefinite, the inclusion $S \subset C$ is always satisfied in this case. Therefore, it makes sense to assume that $S \subset C$.

It is important to note that Corollary 5.36 is valid without any assumption on the set S of covariance operators. Therefore, this result is of practical interest. In the case of covariance matrices, it is known from statistics that every covariance matrix which has the smallest trace or for which its maximal eigenvalue is uniquely the smallest, is a minimal covariance matrix. This is one possibility in order to determine at least one minimal covariance matrix. With Corollary 5.36 we know that every minimal covariance matrix can be obtained by determining the matrix for which the sum with the identity matrix has a uniquely smallest weighted spectral norm.

Notes

Example 5.2, (c) is taken from a paper of Rolewicz [265] where the mentioned inclusion $C_X \subset C_{X^*}$ is given by Wierzbicki [326]. Theorem 5.4 is perhaps the oldest necessary condition for minimal elements (e.g., compare Arrow-Barankin-Blackwell [7], Karlin [168], Dinkelbach [80], Fandel [95], Vogel [313]). Theorem 5.3 and Theorem 5.12 extend a result given by Wierzbicki [327], [328], [329] for so-called order preserving and order presenting functionals (compare also a paper of Vogel [314] for convex problems). The necessary conditions which are given with the aid of approximation problems can also be found in the papers of Jahn [141], [142]. Theorem 5.11 is due to Borwein [29].

Theorem 5.18 presents probably the best-known sufficient condition for minimal elements (e.g., see Arrow-Barankin-Blackwell [7], Hurwicz [127], Dinkelbach [80], Fandel [95] and Vogel [313]). The results of the second section are based on the papers of Jahn [141], [142]. Theorem 5.15, (b) also generalizes corresponding results of Rolewicz [265] and Vogel [314]. For $Y = \mathbb{R}^n$ and the natural ordering cone approximation problems (like these in the second section) are also investigated by Dinkelbach [81], Salukvadze [270], Dinkelbach-Dürr [82], Huang [126], Yu [335], Yu-Leitmann [337], Gearhart [101] and others. In the case of $X = \mathbb{R}^n$ and $C = \mathbb{R}^n_+$ one can show that in Corollary 5.16 the norm is actually a weighted Chebyshev norm (compare the papers of Steuer-Choo [294] and Jahn [140]). The sufficiency condition for properly minimal elements formulated in Theorem 5.19 generalizes a corresponding result of Dinkelbach-Dürr [82] in the case of $X = \mathbb{R}^n$ and $C = \mathbb{R}^n_+$. A similar abstract result is shown by Vogel [314] under assumptions on C which are hard to check. Theorem 5.21 and Corollary 5.22 are taken from a paper of Borwein [29]. An overview on the results of this chapter applied to vector optimization problems in Operations Research is given in a paper of Jahn [143] (compare also Section 11.2). Scalarization results for the notion of proper minimality are also studied by Henig [118], [119] and Zhuang [340].

The results of section 5.3 are based on investigations of Jahn in [146] (see also [148]). A result similar to Corollary 5.35 can also be found in papers of Bowman [43] and Steuer-Choo [294]. Related results on Chebyshev approximation were obtained by Dinkelbach [81], Dinkelbach-Dürr [82], Yu [335], Yu-Leitmann [337] and Gearhart [101]. The problem of the determination of minimal covariance matrices was already investigated by Vogel [313].

Chapter 6

Existence Theorems

In this chapter we study assumptions which guarantee that at least one optimal element of a subset of a partially ordered linear space exists. These investigations will be done for the minimality, proper minimality and weak minimality notions. Strongly minimal elements are not considered because this optimality notion is too restrictive.

Zorn's lemma is the most important result which provides a sufficient condition for the existence of a minimal element of a set. Recall that we already used this lemma in order to prove some special existence results. From Lemma 3.5 and Zorn's lemma it follows that the set of sublinear functionals partially ordered in the natural way has minimal elements. This fact is used in the proof of the basic version of the Hahn-Banach theorem. Moreover, recall the proof of Lemma 3.3 where we show that a base of a cone is contained in a maximal linear manifold which does not contain the zero element. This is also a consequence of Zorn's lemma.

In order to get existence results under weak assumptions on a set we introduce the following

Definition 6.1. Let S be a nonempty subset of a partially ordered linear space X with an ordering cone C. If for some $x \in X$ the set $S_x = (\{x\} - C) \cap S$ is nonempty, S_x is called a *section* of the set S (see Fig. 6.1).

The assertion of the following lemma is evident.

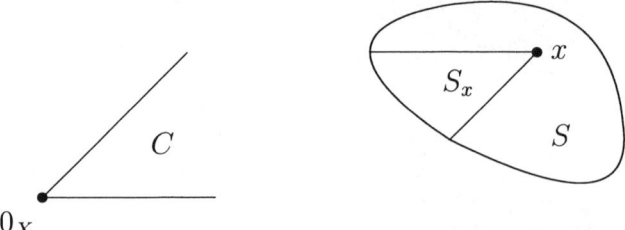

Figure 6.1: Section S_x of a set S.

Lemma 6.2. *Let S be a nonempty subset of a partially ordered linear space X with an ordering cone C.*

(a) Every minimal element of a section of the set S is also a minimal element of the set S.

(b) If $\operatorname{cor}(C) \neq \emptyset$, then every weakly minimal element of a section of the set S is also a weakly minimal element of the set S.

It is important to remark that for the notion of proper minimality a similar statement is not true in general.

We begin now with a discussion of the minimality notion. The following existence result is a consequence of Zorn's lemma.

Theorem 6.3. *Let S be a nonempty subset of a partially ordered topological linear space X with a closed ordering cone C. Then we have:*

(a) If the set S has a closed section which has a lower bound and the ordering cone C is Daniell, then there is at least one minimal element of the set S.

(b) If the set S has a closed and bounded section and the ordering cone C is Daniell and boundedly order complete, then there is at least one minimal element of the set S.

(c) If the set S has a compact section, then there is at least one minimal element of the set S.

Proof. Let S_x (for some $x \in X$) be an appropriate section of the set S. If we show that the section S_x is inductively ordered from below, then by Zorn's lemma (Lemma 3.2) S_x has at least one minimal element which is, by Lemma 6.2, (a), also a minimal element of the set S.

Let $\{s_i\}_{i \in I}$ be any totally ordered subset of the section S_x. Let $\mathcal{F}$ denote the set of all finite subsets of I which are partially ordered with respect to the set theoretical inclusion. Then for every $F \in \mathcal{F}$ the minimum

$$x_F := \min \{s_i \mid i \in F\}$$

exists and belongs to S_x. Consequently, $(x_F)_{F \in \mathcal{F}}$ is a decreasing net in S_x. Next, we consider several cases.

(a) S_x is assumed to have a lower bound so that $(x_F)_{F \in \mathcal{F}}$ has an infimum. Since S_x is closed and C is Daniell, $(x_F)_{F \in \mathcal{F}}$ converges to its infimum which belongs to S_x. This implies that S_x is inductively ordered from below.

(b) Since S_x is bounded and C is boundedly order complete, the net $(x_F)_{F \in \mathcal{F}}$ has an infimum. The ordering cone C is Daniell and, therefore, $(x_F)_{F \in \mathcal{F}}$ converges to its infimum. And since S_x is closed, this infimum belongs to S_x. Hence, S_x is inductively ordered from below.

(c) Now, S_x is assumed to be compact. The family of compact subsets S_{s_i} $(i \in I)$ has the finite intersection property, i.e., every finite subfamily has a nonempty intersection. Since S_x is compact, the family of subsets S_{s_i} $(i \in I)$ has a nonempty intersection (see Dunford-Schwartz [84, p. 17]), that is, there is an element

$$\hat{x} \in \bigcap_{i \in I} S_{s_i} = \bigcap_{i \in I} (\{s_i\} - C) \cap S_x.$$

Hence, $\hat{x}$ is a lower bound of the subset $\{s_i\}_{i \in I}$ and belongs to S_x. Consequently, the section S_x is inductively ordered from below.

□

Notice that the preceding theorem remains valid, if "section" is replaced by the set itself.

Example 6.4. We consider again the problem formulated in Example 4.3. Let X and Y be partially ordered topological linear spaces with the closed ordering cones C_X and C_Y where C_Y is also assumed to be Daniell. Moreover, let $T : X \to Y$ be a continuous linear map and let $q \in Y$ be given so that the set $S := \{x \in C_X \mid T(x) + q \in C_Y\}$ is nonempty. Clearly the set S is closed and has a lower bound (namely 0_X). Then by Theorem 6.3, (a) the set S has at least one minimal element.

The next result follows from Theorem 6.3, (c) and the James theorem.

Theorem 6.5. *Let S be a nonempty subset of a real locally convex space X.*

(a) *If S is weakly compact, then for every closed convex cone C in X the set S has at least one minimal element with respect to the partial ordering induced by C.*

(b) *In addition, let X be quasi-complete. If S is bounded and weakly closed and for every closed convex cone C in X the set S has at least one minimal element with respect to the partial ordering induced by C, then S is weakly compact.*

Proof.

(a) By Lemma 3.24 every closed convex cone C is also weakly closed. Since S is weakly compact, then by Theorem 6.3, (c) S has at least one minimal element with respect to the partial ordering induced by C.

(b) It is evident that the functional 0_{X^*} attains its supremum on the set S. Therefore, take an arbitrary continuous linear functional $l \in X^* \setminus \{0_{X^*}\}$ (if it exists) and define the set $C := \{x \in X \mid l(x) \leq 0\}$ which is a closed convex cone. Let $\bar{x} \in S$ be a minimal element of the set S with respect to the partial ordering induced by C, i.e.

$$(\{\bar{x}\} - C) \cap S \subset \{\bar{x}\} + C. \tag{6.1}$$

Since
$$\{\bar{x}\} - C = \{x \in X \mid l(x) \geq l(\bar{x})\}$$
and
$$\{\bar{x}\} + C = \{x \in X \mid l(x) \leq l(\bar{x})\},$$
the inclusion (6.1) is equivalent to the implication

$$x \in S, \; l(x) \geq l(\bar{x}) \implies l(x) = l(\bar{x}).$$

This implication can also be written as

$$l(\bar{x}) \geq l(x) \text{ for all } x \in S.$$

This means that the functional l attains its supremum on S at $\bar{x}$. Then by the James theorem (Theorem 3.27) the set S is weakly compact.

$\square$

The preceding theorem shows that the weak compactness assumption on a set plays an important role for the existence of minimal elements. This theorem is immediately applicable to a closed unit ball of a Banach space.

Corollary 6.6. *A real Banach space is reflexive if and only if the closed unit ball has at least one minimal element with respect to every partial ordering induced by a closed convex cone.*

Proof. The assertion is a direct consequence of Theorem 6.5, if we observe that a real Banach space is reflexive if and only if the closed unit ball is weakly compact (see Lemma 1.41). $\square$

Corollary 6.6 presents an interesting characterization of the reflexivity of Banach spaces where the reflexivity is related to the existence of certain minimal elements. Recall that in Theorem 3.36, (a) the reflexivity of a Banach space is already characterized by the existence of a solution of certain approximation problems. Hence, there is a close connection between these two types of characterization.

Next, we study existence theorems which follow from scalarization results presented in section 5.2.

Theorem 6.7. *Assume that either assumption (a) or assumption (b) below holds:*

(a) *Let S be a nonempty subset of a partially ordered normed space $(X, \|\cdot\|_X)$ with a pointed ordering cone C, and let X be the topological dual space of a real normed space $(Y, \|\cdot\|_Y)$. Moreover, for some $x \in X$ let a weak*-closed section S_x be given.*

(b) *Let S be a nonempty subset of a partially ordered reflexive Banach space $(X, \|\cdot\|_X)$ with a pointed ordering cone C. Furthermore, for some $x \in X$ let a weakly closed section S_x be given.*

If, in addition, the section S_x has a lower bound $\hat{x} \in X$, i.e. $S_x \subset \{\hat{x}\} + C$, and the norm $\|\cdot\|_X$ is strongly monotonically increasing on C, then the set S has at least one minimal element.

Proof. If the assumption (a) is satisfied, then by Theorem 3.34 the section S_x is proximinal. On the other hand, if the assumption (b) is satisfied, then by Corollary 3.35 the section S_x is proximinal as well. Consequently, there is an $\bar{x} \in S_x$ with

$$\|\bar{x} - \hat{x}\|_X \leq \|s - \hat{x}\|_X \text{ for all } s \in S_x.$$

Since the norm $\|\cdot\|_X$ is strongly monotonically increasing on C, by Theorem 5.15, (b) $\bar{x}$ is a minimal element of S_x. Finally, an application of Lemma 6.2, (a) completes the proof. □

Example 6.8.

(a) As in Example 5.2, (b) we consider again the real linear space $L_p(\Omega)$ where $p \in (1, \infty)$ and Ω is a nonempty subset of $\mathbb{R}^n$.

Assume that this space is partially ordered in a natural way (compare Example 1.51, (a)). We know from Example 5.2, (b) that the $L_p(\Omega)$-norm is strongly monotonically increasing on the ordering cone. Consequently, by Theorem 6.7 every subset of $L_p(\Omega)$ which has a weakly closed section bounded from below has at least one minimal element.

(b) Let S be a nonempty subset of a partially ordered Hilbert space $(X, \langle .,. \rangle)$ with an ordering cone C_X which has the property $C_X \subset C_{X^*}$ (see Example 5.2, (c)). If S has a weakly closed section bounded from below, then S has at least one minimal element.

For the minimality notion a scalarization result concerning positive linear functionals leads to an existence theorem which is contained in Theorem 6.5, (a). But for the proper minimality notion such a scalarization result is helpful.

Theorem 6.9. *Every weakly compact subset of a partially ordered separable normed space with a closed pointed ordering cone has at least one properly minimal element.*

Proof. By a Krein-Rutman theorem (Theorem 3.38) the quasi-interior of the topological dual cone is nonempty. Then every continuous linear functional which belongs to that quasi-interior attains its infimum on a weakly compact set, and Theorem 5.21 leads to the assertion. □

A further existence theorem for properly minimal elements is given by

Theorem 6.10. *Assume that either assumption (a) or assumption (b) below holds:*

(a) *Let S be a nonempty subset of a partially ordered normed space $(X, \|\cdot\|_X)$ with a pointed ordering cone C which has a nonempty algebraic interior, and let X be the topological dual space of a* *real normed space* $(Y, \|\cdot\|_Y)$. *Moreover, let the set S be weak*-closed.*

(b) *Let S be a nonempty subset of a partially ordered reflexive Banach space $(X, \|\cdot\|_X)$ with a pointed ordering cone C which has a nonempty algebraic interior. Furthermore, let the set S be weakly closed.*

If, in addition, there is an $\hat{x} \in X$ with $S \subset \{\hat{x}\} + \operatorname{cor}(C)$ and the norm $\|\cdot\|_X$ is strongly monotonically increasing on C, then the set S has at least one properly minimal element.

Proof. The proof is similar to that of Theorem 6.7 where we use now the scalarization result in Theorem 5.19. □

Example 6.11. Let S be a nonempty subset of a partially ordered Hilbert space $(X, \langle \cdot, \cdot \rangle)$ with an ordering cone C_X which has a nonempty algebraic interior and for which $C_X \subset C_{X^*}$ (compare Example 5.2, (c)). If S is weakly closed and there is an $\hat{x} \in X$ with $S \subset \{\hat{x}\} + \operatorname{cor}(C_X)$, then the set S has at least one properly minimal element.

Finally, we turn our attention to the weak minimality notion. Using Lemma 4.14 we can easily extend the existence theorems for minimal elements to weakly minimal elements, if we assume additionally that the ordering cone $C \subset X$ does not equal X and that it has a nonempty algebraic interior. This is one possibility in order to get existence results for the weak minimality notion. In the following theorems we use directly appropriate scalarization results for this optimality notion.

Theorem 6.12. *Let S be a nonempty subset of a partially ordered locally convex space X with a closed ordering cone $C_X \neq X$ which has a nonempty algebraic interior. If S has a weakly compact section, then the set S has at least one weakly minimal element.*

Proof. Since the ordering cone C_X is closed and does not equal

X, there is at least one continuous linear functional $l \in C_{X^*} \setminus \{0_{X^*}\}$ (compare Theorem 3.18). This functional attains its infimum on a weakly compact section of S which is, by Theorem 5.28, a weakly minimal element of this section. An application of Lemma 6.2, (b) completes the proof. □

Notice that Theorem 6.12 could also be proved using Theorem 6.5, (a) and Lemma 4.14.

Theorem 6.13. *Assume that either assumption (a) or assumption (b) below holds:*

(a) *Let S be a nonempty subset of a partially ordered normed space $(X, \|\cdot\|_X)$ with an ordering cone C which has a nonempty algebraic interior, and let X be the topological dual space of a real normed space $(Y, \|\cdot\|_Y)$. Moreover, for some $x \in X$ let a weak*-closed section S_x be given.*

(b) *Let S be a nonempty subset of a partially ordered reflexive Banach space $(X, \|\cdot\|_X)$ with an ordering cone C which has a nonempty algebraic interior. Furthermore, for some $x \in X$ let a weakly closed section S_x be given.*

If, in addition, the section S_x has a lower bound $\hat{x} \in X$, i.e. $S_x \subset \{\hat{x}\} + C$, and the norm $\|\cdot\|_X$ is strictly monotonically increasing on C, then the set S has at least one weakly minimal element.

Proof. The proof is similar to that of Theorem 6.7 where we now use the scalarization result in Theorem 5.25. □

Example 6.14.

(a) Let S be a nonempty subset of $L_\infty(\Omega)$ (compare Example 1.51, (b)) which is assumed to be partially ordered in the natural way. If the set S has a weak*-closed section bounded from below, then S has at least one weakly minimal element.

Proof. If we consider the linear space $L_\infty(\Omega)$ as the topological dual space of $L_1(\Omega)$, then the assertion follows from Theorem

6.13, if we show that the norm $\|\cdot\|_{L_\infty(\Omega)}$ is strictly monotonically increasing on the ordering cone C. It is evident that

$$\operatorname{int}(C) = \{f \in L_\infty(\Omega) \mid \text{there is an } \alpha > 0 \text{ with}$$
$$f(x) \geq \alpha \text{ almost everywhere on } \Omega\} \neq \emptyset.$$

By Lemma 1.32, (a) $\operatorname{int}(C)$ equals the algebraic interior of C. Take any functions $f, g \in C$ with $f \in \{g\} - \operatorname{int}(C)$. Then we have $g - f \in \operatorname{int}(C)$ which implies that there is an $\alpha > 0$ with

$$g(x) - f(x) \geq \alpha \text{ almost everywhere on } \Omega$$

and

$$g(x) \geq \alpha + f(x) \text{ almost everywhere on } \Omega.$$

Consequently, we get

$$\operatorname*{ess\,sup}_{x \in \Omega} \{g(x)\} \geq \alpha + \operatorname*{ess\,sup}_{x \in \Omega} \{f(x)\}$$

and

$$\|g\|_{L_\infty(\Omega)} > \|f\|_{L_\infty(\Omega)}.$$

Hence, the norm $\|\cdot\|_{L_\infty(\Omega)}$ is strictly monotonically increasing on C. □

(b) Let $C(\Omega)$ be the partially ordered linear space of real-valued continuous functions on a compact Hausdorff space Ω with the natural ordering cone and the maximum norm (compare Example 1.49). If S is a nonempty subset of $C(\Omega)$ which has a weakly closed section in a reflexive subspace of $C(\Omega)$ and a lower bound in this subspace, then the set S has at least one weakly minimal element.

Proof. As in the proof of part (a) one can show that the maximum norm is strictly monotonically increasing on the ordering cone. Then the assertion follows from Theorem 6.13. □

Notes

Theorem 6.3, Theorem 6.5 and Corollary 6.6 are due to Borwein [36]. But it should be mentioned that Theorem 6.3, (c) was first proved by Vogel [313] and Theorem 6.3, (a) can essentially be found, without proof, in a survey article of Penot [248]. For further existence results we refer to the papers of Bishop-Phelps [25] (for the Bishop and Phelps lemma see also Holmes [125, p. 164]), Cesari-Suryanarayana [51], [52], [53], Corley [63], Isac [130] and Chew [59]. Example 6.4 is also discussed by Borwein [36]. The application of certain scalarization results in order to get existence theorems is also described in a paper of Jahn [144].

In functional analysis existence theorems play an important role for the proof of known results like Ekeland's variational principle (see Ekeland [91], Ekeland-Temam [92, p. 29–30] and Borwein [36, p. 72]). For further information we cite the papers of Phelps [250] and the thesis of Landes [197].

Next, we give a short presentation of some of the results of Bishop-Phelps [25]:

(a) Let $(X, \|\cdot\|)$ be a real normed space, let $l \in X^*$ be an arbitrary continuous linear functional, and let an arbitrary $\gamma \in (0,1)$ be given. Then the cone

$$C(l,\gamma) := \{x \in X \mid \gamma\|x\| \leq l(x)\}$$

is called the *Bishop-Phelps cone*.

Notice that this cone is convex and pointed and, therefore, it can be used as an ordering cone in the space X.

(b) The following *Bishop-Phelps lemma* is a special type of an existence result for maximal elements:

Let S be a nonempty closed subset of a real Banach space $(X, \|\cdot\|_X)$, and let a continuous linear functional $l \in X^*$ be given with $\|l\|_{X^*} = 1$ and $\sup_{x \in S} l(x) < \infty$. Then for every $x \in S$ and every $\gamma \in (0,1)$ there is a maximal element $\bar{x} \in \{x\} + C(l,\gamma)$ of the set S with respect to the Bishop-Phelps ordering cone $C(l,\gamma)$.

(c) The so-called *Bishop-Phelps theorem* is an important consequence of this lemma:

Let S be a nonempty closed bounded and convex subset of a real Banach space $(X, \|\cdot\|)$. Then the set of support functionals of S is dense in X^*.

Finally, we present *Ekeland's variational principle* which is also a consequence of a special existence argument for minimal elements:

Let (X, d) be a complete metric space, and let $\varphi : X \to \mathbb{R} \cup \{+\infty\}$ be a lower semicontinuous function bounded from below. Moreover, let some $\varepsilon > 0$ and some $\bar{x} \in X$ be arbitrarily given where

$$\varphi(\bar{x}) \leq \inf_{x \in X} \varphi(x) + \varepsilon.$$

Then there is an $\hat{x} \in X$ with

$$\varphi(\hat{x}) \leq \varphi(\bar{x}),$$

$$d(\bar{x}, \hat{x}) \leq 1$$

and

$$\varphi(x) - \varphi(\hat{x}) > -\varepsilon d(x, \hat{x}) \text{ for all } x \in X \setminus \{\hat{x}\}.$$

It was shown by Landes [197] that Ekeland's variational principle is closely related with the Bishop-Phelps lemma: Both results can be deduced from a Brézis-Browder theorem [45].

Chapter 7

Generalized Lagrange Multiplier Rule

In this chapter we present a generalization of the famous and well-known Lagrange multiplier rule published in 1797. Originally, Lagrange formulated his rule for the optimization of a real-valued function under side-conditions in the form of equalities. In this context we investigate an abstract optimization problem (introduced in Example 4.5) with equality and inequality constraints. For this problem we derive a generalized multiplier rule as a necessary optimality condition and we show under which assumptions this multiplier rule is also sufficient for optimality. The results are also applied to multiobjective optimization problems.

7.1 Necessary Conditions for Minimal and Weakly Minimal Elements

The derivation of necessary optimality conditions for minimal and weakly minimal elements can be restricted to the weak minimality notion. If the ordering cone does not equal the whole space and if it has a nonempty algebraic interior, then, by Lemma 4.14, every minimal element of a set is also a weakly minimal element of this set. Hence, under this assumption a necessary condition for weakly minimal elements is a necessary condition for minimal elements as well.

In this section we derive the generalized multiplier rule for Fréchet differentiable maps, although this can be done for more general differentiability notions. For an extensive presentation of these generalizations the reader is referred to the book of Kirsch-Warth-Werner [171].

The standard assumption for this section reads as follows:

$$\left.\begin{array}{l}\text{Let } (X, \|\cdot\|_X) \text{ and } (Z_2, \|\cdot\|_{Z_2}) \text{ be real Banach spaces;} \\ \text{let } (Y, \|\cdot\|_Y) \text{ and } (Z_1, \|\cdot\|_{Z_1}) \text{ be partially ordered} \\ \text{normed spaces;} \\ \text{let } C_Y \text{ and } C_{Z_1} \text{ denote the ordering cones in } Y \text{ and } Z_1, \\ \text{respectively, which are assumed to have a nonempty} \\ \text{interior;} \\ \text{let } \hat{S} \text{ be a nonempty convex subset of } X \text{ which has a} \\ \text{nonempty interior;} \\ \text{let } f : X \to Y, \ g :\to Z_1 \text{ and } h : X \to Z_2 \text{ be given} \\ \text{maps.}\end{array}\right\} \quad (7.1)$$

Under this assumption we define the *constraint set*

$$S := \{x \in \hat{S} \mid g(x) \in -C_{Z_1} \text{ and } h(x) = 0_{Z_2}\}$$

(which is assumed to be nonempty) and we consider the *abstract optimization problem*

$$\min_{x \in S} f(x). \quad (7.2)$$

The map f is also called the *objective map*. As indicated in Example 4.5 we define a solution of the problem (7.2) in the following way:

Definition 7.1. Let the abstract optimization problem (7.2) be given under the assumption (7.1).

(a) An element $\bar{x} \in S$ is called a *minimal solution* of the problem (7.2), if $f(\bar{x})$ is a minimal element of the image set $f(S)$.

(b) An element $\bar{x} \in S$ is called a *weakly minimal solution* of the problem (7.2), if $f(\bar{x})$ is a weakly minimal element of the image set $f(S)$.

7.1. Necessary Conditions for Minimal and Weakly Minimal Elements

In order to obtain a necessary condition for a weakly minimal solution of the abstract optimization problem (7.2), we need a basic lemma on contingent cones.

Lemma 7.2. *Let $(X, \|\cdot\|_X)$ be a real normed space, and let $(Y, \|\cdot\|_Y)$ be a partially ordered normed space with an ordering cone C_Y which has a nonempty interior. Moreover, let S be a nonempty subset of X and let a map $r : X \to Y$ be given. If the map r is Fréchet differentiable at some $\bar{x} \in S$ with $r(\bar{x}) \in -C_Y$, then*

$$\{h \in T(S, \bar{x}) \mid r(\bar{x}) + r'(\bar{x})(h) \in -int(C_Y)\}$$
$$\subset T(\{x \in S \mid r(x) \in -int(C_Y)\}, \bar{x})$$

(where $T(.,.)$ denotes the contingent cone introduced in Definition 3.41).

Proof. We choose an arbitrary $h \in T(S, \bar{x})$ with the property $r(\bar{x}) + r'(\bar{x})(h) \in -int(C_Y)$. For $h = 0_X$ the assertion is trivial. Therefore, we assume that $h \neq 0_X$. Then there is a sequence $(x_n)_{n \in \mathbb{N}}$ of elements $x_n \in S$ and a sequence $(\lambda_n)_{n \in \mathbb{N}}$ of positive real numbers λ_n so that

$$\bar{x} = \lim_{n \to \infty} x_n$$

and

$$h = \lim_{n \to \infty} \lambda_n (x_n - \bar{x}).$$

If we set

$$h_n := \lambda_n(x_n - \bar{x}) \text{ for all } n \in \mathbb{N},$$

we get

$$r(x_n) = \frac{1}{\lambda_n}\Big[\lambda_n(r(x_n) - r(\bar{x}) - r'(\bar{x})(x_n - \bar{x})) + r'(\bar{x})(h_n - h)$$
$$+ r(\bar{x}) + r'(\bar{x})(h)\Big] + \left(1 - \frac{1}{\lambda_n}\right) r(\bar{x}) \text{ for all } n \in \mathbb{N} \quad (7.3)$$

and

$$\lim_{n \to \infty} \lambda_n(r(x_n) - r(\bar{x}) - r'(\bar{x})(x_n - \bar{x})) + r'(\bar{x})(h_n - h) = 0_Y. \quad (7.4)$$

By assumption we have
$$r(\bar{x}) + r'(\bar{x})(h) \in -\text{int}(C_Y)$$
and, therefore, it follows with (7.4)
$$\begin{aligned} y_n &:= \lambda_n(r(x_n) - r(\bar{x}) - r'(\bar{x})(x_n - \bar{x})) + r'(\bar{x})(h_n - h) + r(\bar{x}) \\ &\quad + r'(\bar{x})(h) \\ &\in -\text{int}(C_Y) \text{ for sufficiently large } n \in \mathbb{N} \end{aligned}$$
and
$$\frac{1}{\lambda_n} y_n \in -\text{int}(C_Y) \text{ for sufficiently large } n \in \mathbb{N}.$$
Since
$$\left(1 - \frac{1}{\lambda_n}\right) r(\bar{x}) \in -C_Y \text{ for sufficiently large } n \in \mathbb{N},$$
we conclude with (7.3), Lemma 1.12, (b) and Lemma 1.32, (a)
$$\begin{aligned} r(x_n) &= \frac{1}{\lambda_n} y_n + \left(1 - \frac{1}{\lambda_n}\right) r(\bar{x}) \\ &\in -\text{int}(C_Y) - C_Y \\ &= -\text{int}(C_Y) \text{ for sufficiently large } n \in \mathbb{N}. \end{aligned}$$
But this leads to
$$h \in T(\{x \in S \mid r(x) \in -\text{int}(C_Y)\}, \bar{x}).$$

$\square$

With the preceding lemma and the Lyusternik theorem we obtain a first necessary condition for a weakly minimal solution of the problem (7.2).

Theorem 7.3. *Let the abstract optimization problem (7.2) be given under the assumption (7.1), and let $\bar{x} \in S$ be a weakly minimal solution of the problem (7.2). Moreover, let f and g be Fréchet differentiable at $\bar{x}$ and let h be continuously Fréchet differentiable at*

7.1. Necessary Conditions for Minimal and Weakly Minimal Elements 165

$\bar{x}$ where $h'(\bar{x})$ is assumed to be surjective. Then there is no $x \in int(\hat{S})$ with
$$f'(\bar{x})(x - \bar{x}) \in -int(C_Y),$$
$$g(\bar{x}) + g'(\bar{x})(x - \bar{x}) \in -int(C_{Z_1})$$
and
$$h'(\bar{x})(x - \bar{x}) = 0_{Z_2}.$$

Proof. Assume that there is an $x \in int(\hat{S})$ with $f'(\bar{x})(x - \bar{x}) \in -int(C_Y)$, $g(\bar{x}) + g'(\bar{x})(x - \bar{x}) \in -int(C_{Z_1})$ and $h'(\bar{x})(x - \bar{x}) = 0_{Z_2}$. Then we get with the Lyusternik theorem (Theorem 3.49)
$$x - \bar{x} \in T(\{s \in X \mid h(s) = 0_{Z_2}\}, \bar{x}).$$
Since $\hat{S}$ is convex and $x \in int(\hat{S})$, we obtain $x - \bar{x} \in T(\tilde{S}, \bar{x})$ where
$$\tilde{S} := \{s \in \hat{S} \mid h(s) = 0_{Z_2}\}.$$
Next, we define the map $r : X \to Y \times Z_1$ by
$$r(x) = \begin{pmatrix} f(x) - f(\bar{x}) \\ g(x) \end{pmatrix} \text{ for all } x \in X.$$
Obviously we have
$$r(\bar{x}) = \begin{pmatrix} 0_Y \\ g(\bar{x}) \end{pmatrix} \in (-C_Y) \times (-C_{Z_1})$$
and, therefore, we conclude with Lemma 7.2
$$\{h \in T(\tilde{S}, \bar{x}) \mid f'(\bar{x})(h) \in -int(C_Y), \ g(\bar{x}) + g'(\bar{x})(h) \in -int(C_{Z_1})\}$$
$$\subset T(\{s \in \tilde{S} \mid f(s) - f(\bar{x}) \in -int(C_Y), \ g(s) \in -int(C_{Z_1})\}, \bar{x}).$$
Because of $x - \bar{x} \in T(\tilde{S}, \bar{x})$, $f'(\bar{x})(x - \bar{x}) \in -int(C_Y)$ and $g(\bar{x}) + g'(\bar{x})(x - \bar{x}) \in -int(C_{Z_1})$ we conclude
$$x - \bar{x} \in T(\{s \in \hat{S} \mid f(s) - f(\bar{x}) \in -int(C_Y),$$
$$g(s) \in -int(C_{Z_1}), \ h(s) = 0_{Z_1}\}, \bar{x}).$$

But this implies that $\bar{x}$ is no weakly minimal solution of the problem (7.2). □

Now, we are ready to present the promised multiplier rule which generalizes a corresponding result of Lagrange. This necessary optimality condition is based on the previous theorem and a separation theorem.

Theorem 7.4. *Let the abstract optimization problem (7.2) be given under the assumption (7.1), and let $\bar{x} \in S$ be a weakly minimal solution of the problem (7.2). Moreover, let f and g be Fréchet differentiable at $\bar{x}$, let h be continuously Fréchet differentiable at $\bar{x}$, and let the image set $h'(\bar{x})(X)$ be closed. Then there are continuous linear functionals $t \in C_{Y^*}$, $u \in C_{Z_1^*}$ and $v \in Z_2^*$ with $(t, u, v) \neq 0_{Y^* \times Z_1^* \times Z_2^*}$ so that*

$$(t \circ f'(\bar{x}) + u \circ g'(\bar{x}) + v \circ h'(\bar{x}))(x - \bar{x}) \geq 0 \text{ for all } x \in \hat{S} \quad (7.5)$$

and

$$(u \circ g)(\bar{x}) = 0. \quad (7.6)$$

If, in addition, there is an $\hat{x} \in \text{int}(\hat{S})$ with $g(\bar{x}) + g'(\bar{x})(\hat{x} - \bar{x}) \in -\text{int}(C_{Z_1})$ and $h'(\bar{x})(\hat{x} - \bar{x}) = 0_{Z_2}$ and if the map $h'(\bar{x})$ is surjective, then $t \neq 0_{Y^*}$.

Proof. First, we assume that $h'(\bar{x})$ is not surjective. Then, by an application of a separation theorem (Theorem 3.18), there is a continuous linear functional $v \in Z_2^* \setminus \{0_{Z_2^*}\}$ with $v \circ h'(\bar{x}) = 0_{X^*}$. If we set $t = 0_{Y^*}$ and $u = 0_{Z_1^*}$, we get immediately the conditions (7.5) and (7.6). In this case the first part of the assertion is shown.

In the following assume that the map $h'(\bar{x})$ is surjective. In this case we define the set

$$M := \{(f'(\bar{x})(x - \bar{x}) + y, \ g(\bar{x}) + g'(\bar{x})(x - \bar{x}) + z_1, \ h'(\bar{x})(x - \bar{x})) $$
$$\in Y \times Z_1 \times Z_2 \mid x \in \text{int}(\hat{S}), \ y \in \text{int}(C_Y), \ z_1 \in \text{int}(C_{Z_1})\}$$

which can also be written as

$$M = (f'(\bar{x}), g'(\bar{x}), h'(\bar{x}))(\text{int}(\hat{S}) - \{\bar{x}\})$$
$$+ \text{int}(C_Y) \times (\{g(\bar{x})\} + \text{int}(C_{Z_1})) \times \{0_{Z_2}\}.$$

7.1. Necessary Conditions for Minimal and Weakly Minimal Elements

The map $h'(\bar{x})$ is continuous, linear and surjective. Then, by the open map theorem, $h'(\bar{x})$ maps every open subset of X onto an open subset of Z_2, and it is evident that the set M equals its interior. The set M is a convex set because $(f'(\bar{x}), g'(\bar{x}), h'(\bar{x}))$ is a linear map and $\text{int}(\hat{S}) - \{\bar{x}\}$ is a convex set. Since $\bar{x} \in S$ is a weakly minimal solution of the problem (7.2), by the necessary condition given in Theorem 7.3 the zero element $0_{Y \times Z_1 \times Z_2}$ does not belong to the set M, i.e. we get

$$M \cap \{0_{Y \times Z_1 \times Z_2}\} = \emptyset.$$

The set M is convex and open and, therefore, by Eidelheit's separation theorem (Theorem 3.16), the preceding set equation implies the existence of continuous linear functionals $t \in Y^*$, $u \in Z_1^*$ and $v \in Z_2^*$ with $(t, u, v) \neq 0_{Y^* \times Z_1^* \times Z_2^*}$ and

$$t(f'(\bar{x})(x - \bar{x}) + y) + u(g(\bar{x}) + g'(\bar{x})(x - \bar{x}) + z_1) + v(h'(\bar{x})(x - \bar{x}))$$
$$> 0 \text{ for all } x \in \text{int}(\hat{S}),\ y \in \text{int}(C_Y) \text{ and } z_1 \in \text{int}(C_{Z_1}). \tag{7.7}$$

With Lemma 1.32, (b) and the continuity of the arising maps we obtain from the inequality (7.7)

$$t(f'(\bar{x})(x - \bar{x}) + y) + u(g(\bar{x}) + g'(\bar{x})(x - \bar{x}) + z_1) + v(h'(\bar{x})(x - \bar{x}))$$
$$\geq 0 \text{ for all } x \in \hat{S},\ y \in C_Y \text{ and } z_1 \in C_{Z_1}. \tag{7.8}$$

From the inequality (7.8) we get for $x = \bar{x}$

$$t(y) + u(g(\bar{x}) + z_1) \geq 0 \text{ for all } y \in C_Y \text{ and } z_1 \in C_{Z_1}. \tag{7.9}$$

For $z_1 = -g(\bar{x}) \in C_{Z_1}$ we conclude with the inequality (7.9)

$$t(y) \geq 0 \text{ for all } y \in C_Y$$

which implies $t \in C_{Y^*}$. For $y = 0_Y$ we obtain from the inequality (7.9)

$$u(g(\bar{x})) \geq -u(z_1) \text{ for all } z_1 \in C_{Z_1}. \tag{7.10}$$

From this inequality it follows immediately that

$$u(z_1) \geq 0 \text{ for all } z_1 \in C_{Z_1}$$

resulting in $u \in C_{Z_1^*}$. But the inequality (7.10) also implies $u(g(\bar{x})) \geq 0$. By assumption we have $g(\bar{x}) \in -C_{Z_1}$ so that we get $u(g(\bar{x})) \leq 0$. Consequently, the equality (7.6) is true. For the proof of the inequality (7.5) notice that for $y = 0_Y$ and $z_1 = -g(\bar{x})$ the inequality (7.8) leads to

$$t(f'(\bar{x})(x - \bar{x})) + u(g'(\bar{x})(x - \bar{x})) + v(h'(\bar{x})(x - \bar{x})) \geq 0 \text{ for all } x \in \hat{S}$$

or alternatively

$$(t \circ f'(\bar{x}) + u \circ g'(\bar{x}) + v \circ h'(\bar{x}))(x - \bar{x}) \geq 0 \text{ for all } x \in \hat{S}.$$

Finally, we investigate the case that, in addition to the given assumptions, there is an $\hat{x} \in \text{int}(\hat{S})$ with $g(\bar{x}) + g'(\bar{x})(\hat{x} - \bar{x}) \in -\text{int}(C_{Z_1})$ and $h'(\bar{x})(\hat{x} - \bar{x}) = 0_{Z_2}$ and the map $h'(\bar{x})$ is surjective. In this case the inequality (7.7) leads to

$$t(f'(\bar{x})(\hat{x} - \bar{x}) + y) > 0 \text{ for all } y \in \text{int}(C_Y)$$

which implies $t \neq 0_{Y^*}$. □

The necessary optimality conditions given in Theorem 7.4 generalize the well-known *Lagrange multiplier rule*. They also extend the so-called *F.-John conditions*. The additional assumption formulated in the second part of the preceding theorem under which the functional t is nonzero is called a *regularity assumption*. If $t \neq 0_{Y^*}$, then the necessary optimality conditions extend the so-called *Karush-Kuhn-Tucker-conditions*.

If the superset $\hat{S}$ of the constraint set S equals the whole space X, then the inequality (7.5) reduces to the equality

$$t \circ f'(\bar{x}) + u \circ g'(\bar{x}) + v \circ h'(\bar{x}) = 0_{X^*}.$$

The multiplier rule in Theorem 7.4 is formulated with a *real-valued Lagrangian* $t \circ f + u \circ g + v \circ h$. It will become obvious from the next theorem that this multiplier rule can also be formulated with a *vector-valued Lagrangian* $f + L_1 \circ g + L_2 \circ h$ where L_1 and L_2 are appropriate linear maps. There is no difference if we use a real-valued or a vector-valued Lagrangian.

7.1. Necessary Conditions for Minimal and Weakly Minimal Elements

Theorem 7.5. *Let the abstract optimization problem (7.2) be given under the assumption (7.1). For some $\bar{x} \in S$ assume that f, g and h are Fréchet differentiable at $\bar{x}$. Then the two statements (7.11) and (7.12) below are equivalent:*

$$\left.\begin{array}{l}\text{There are continuous linear functionals}\\ t \in C_{Y^*}\setminus\{0_{Y^*}\},\ u \in C_{Z_1^*}\ \text{and}\ v \in Z_2^*\ \text{with the properties}\\ (t \circ f'(\bar{x}) + u \circ g'(\bar{x}) + v \circ h'(\bar{x}))(x - \bar{x}) \geq 0\ \text{for all}\ x \in \hat{S}\\ \text{and}\ (u \circ g)(\bar{x}) = 0.\end{array}\right\} \quad (7.11)$$

$$\left.\begin{array}{l}\text{There are a continuous linear map}\ L_1 : Z_1 \to Y\\ \text{with}\ L_1(C_{Z_1}) \subset (\text{int}(C_Y) \cup \{0_Y\})\ \text{and a continuous}\\ \text{linear map}\ L_2 : Z_2 \to Y\ \text{with the properties}\\ (f'(\bar{x}) + L_1 \circ g'(\bar{x}) + L_2 \circ h'(\bar{x}))(x - \bar{x}) \notin -\text{int}(C_Y)\ \text{for all}\\ x \in \hat{S}\ \text{and}\ (L_1 \circ g)(\bar{x}) = 0_Y.\end{array}\right\} \quad (7.12)$$

Proof. First, we assume that the statement (7.11) is true. By Lemma 3.21, (c) there is a $\tilde{y} \in \text{int}(C_Y)$ with $t(\tilde{y}) = 1$. Then, following an idea due to Borwein [29, p. 62], we define the maps $L_1 : Z_1 \to Y$ and $L_2 : Z_2 \to Y$ by

$$L_1(z_1) = u(z_1)\tilde{y}\ \text{for all}\ z_1 \in Z_1 \quad (7.13)$$

and

$$L_2(z_2) = v(z_2)\tilde{y}\ \text{for all}\ z_2 \in Z_2.$$

Obviously, L_1 and L_2 are continuous linear maps, and we have

$$L_1(C_{Z_1}) \subset (\text{int}(C_Y) \cup \{0_Y\}).$$

Furthermore, we obtain $t \circ L_1 = u$ and $t \circ L_2 = v$. Consequently, the inequality in the statement (7.11) can be written as

$$(t \circ (f'(\bar{x}) + L_1 \circ g'(\bar{x}) + L_2 \circ h'(\bar{x})))(x - \bar{x}) \geq 0\ \text{for all}\ x \in \hat{S}.$$

Then we conclude with the scalarization result of Corollary 5.29

$$(f'(\bar{x}) + L_1 \circ g'(\bar{x}) + L_2 \circ h'(\bar{x}))(x - \bar{x}) \notin -\text{int}(C_Y)\ \text{for all}\ x \in \hat{S}.$$

Finally, with the equality (7.13) we get
$$(L_1 \circ g)(\bar{x}) = (u \circ g)(\bar{x})\tilde{y} = 0_Y.$$
Hence, the statement (7.12) is true.

For the second part of this proof we assume that the statement (7.12) is true. Then we have
$$(f'(\bar{x}) + L_1 \circ g'(\bar{x}) + L_2 \circ h'(\bar{x}))(x - \bar{x}) \notin -\text{int}(C_Y) \text{ for all } x \in \hat{S}.$$
By Corollary 5.29 and Lemma 3.15 there is a continuous linear functional $t \in C_{Y^*}\backslash\{0_{Y^*}\}$ with the property
$$(t \circ f'(\bar{x}) + t \circ L_1 \circ g'(\bar{x}) + t \circ L_2 \circ h'(\bar{x}))(x - \bar{x}) \geq 0 \text{ for all } x \in \hat{S}.$$
If we define $u := t \circ L_1$ and $v := t \circ L_2$, we obtain
$$(t \circ f'(\bar{x}) + u \circ g'(\bar{x}) + v \circ h'(\bar{x}))(x - \bar{x}) \geq 0 \text{ for all } x \in \hat{S}$$
and
$$(u \circ g)(\bar{x}) = (t \circ L_1 \circ g)(\bar{x}) = 0.$$
Furthermore, for every $z_1 \in C_{Z_1}$ it follows
$$u(z_1) = (t \circ L_1)(z_1) \geq 0$$
implying $u \in C_{Z_1^*}$. This completes the proof. $\square$

It is obvious from the previous proof that the image sets of the maps L_1 and L_2 are one-dimensional subspaces of Y.

For abstract optimization problems without explicit constraints the multiplier rule can also be used with g and h being the zero maps. But in this case a separate investigation leads to a much more general result.

Theorem 7.6. *Let S be a nonempty subset of a real linear space X, and let Y be a partially ordered linear space with an ordering cone $C_Y \neq Y$ which has a nonempty algebraic interior. Let $f : S \to Y$ be a given map. If $\bar{x} \in S$ is a weakly minimal solution of the abstract optimization problem*
$$\min_{x \in S} f(x) \tag{7.14}$$

7.1. Necessary Conditions for Minimal and Weakly Minimal Elements

and if f has a directional variation at x with respect to $-\mathrm{cor}(C_Y)$, then
$$f'(\bar{x})(x - \bar{x}) \notin -\mathrm{cor}(C_Y) \text{ for all } x \in S. \tag{7.15}$$

Proof. If the condition (7.15) is not true, i.e. for some $x \in S$
$$f'(\bar{x})(x - \bar{x}) \in -\mathrm{cor}(C_Y),$$
then by Definition 2.14 there is a $\bar{\lambda} > 0$ with $\hat{x} := \bar{x} + \bar{\lambda}(x - \bar{x}) \in S$ and $\frac{1}{\lambda}(f(\hat{x}) - f(\bar{x})) \in -\mathrm{cor}(C_Y)$. Consequently, we have
$$f(\hat{x}) \in (\{f(\bar{x})\} - \mathrm{cor}(C_Y)) \cap f(S)$$
which implies that $\bar{x}$ is no weakly minimal solution of the abstract optimization problem (7.14). $\square$

With the same argument as used in Theorem 7.5 the necessary optimality condition (7.15) in vector form is equivalent to an inequality, if the directional variation of f at $\bar{x}$ is convex-like.

Lemma 7.7. *Let S be a nonempty subset of a real linear space X, and let Y be a partially ordered linear space with an ordering cone $C_Y \neq Y$ which has a nonempty algebraic interior. Let $f : S \to Y$ be a map which has a directional variation at some $\bar{x} \in S$ with respect to $-\mathrm{cor}(C_Y)$. If there is a $t \in C_{Y'} \setminus \{0_{Y'}\}$ with*
$$(t \circ f'(\bar{x}))(x - \bar{x}) \geq 0 \text{ for all } x \in S, \tag{7.16}$$
then the condition (7.15) holds. If the map $f'(\bar{x})$ is convex-like, then the condition (7.15) implies the existence of a linear functional $t \in C_{Y'} \setminus \{0_{Y'}\}$ with the property (7.16).

Proof. If we assume that there is a $t \in C_{Y'} \setminus \{0_{Y'}\}$ with the property (7.16), then, by Theorem 5.28, we get immediately the condition (7.15).

Next we assume that the condition (7.15) holds. By Lemma 4.13, (b) we obtain
$$((f'(\bar{x})(S - \{\bar{x}\})) + C_Y) \cap (-\mathrm{cor}(C_Y)) = \emptyset.$$

Since $f'(\bar{x})$ is assumed to be convex-like, by Theorem 5.13 there is a linear functional $t \in C_{Y'} \backslash \{0_{Y'}\}$ so that the inequality (7.16) is satisfied. □

At the end of this section we turn our attention again to the generalized multiplier rule presented in Theorem 7.4. We specialize this result to a so-called *multiobjective optimization problem*, i.e., we consider the problem (7.2) in a finite dimensional setting.

Theorem 7.8. *Let $f : \mathbb{R}^n \to \mathbb{R}^m$, $g : \mathbb{R}^n \to \mathbb{R}^k$ and $h : \mathbb{R}^n \to \mathbb{R}^p$ be given vector functions, and let the constraint set S be given as*

$$S := \{x \in \mathbb{R}^n \mid g_i(x) \leq 0 \text{ for all } i \in \{1, \ldots, k\} \text{ and}$$
$$h_i(x) = 0 \text{ for all } i \in \{1, \ldots, p\}\}.$$

Let $\bar{x} \in S$ be a weakly minimal solution of the multiobjective optimization problem $\min_{x \in S} f(x)$ where the space $\mathbb{R}^m$ is assumed to be partially ordered in a natural way. Let f and g have partial derivatives at $\bar{x}$ and let h be continuously partially differentiable at $\bar{x}$. Moreover, let some $x \in \mathbb{R}^n$ exist with

$$\nabla g_i(\bar{x})^T (x - \bar{x}) < 0 \text{ for all } i \in I(\bar{x})$$

and

$$\nabla h_i(\bar{x})^T (x - \bar{x}) = 0 \text{ for all } i \in \{1, \ldots, p\}$$

where

$$I(\bar{x}) := \{i \in \{1, \ldots, k\} \mid g_i(\bar{x}) = 0\}$$

denotes the set of constraints being "active" at $\bar{x}$. Furthermore, let the gradients $\nabla h_1(\bar{x}), \ldots, \nabla h_p(\bar{x})$ be linearly independent. Then there are multipliers $t_i \geq 0$ (where at least one t_i, $i \in \{1, \ldots, m\}$, is nonzero), $u_i \geq 0$ ($i \in I(\bar{x})$) and $v_i \in \mathbb{R}$ ($i \in \{1, \ldots, p\}$) with the property

$$\sum_{i=1}^{m} t_i \nabla f_i(\bar{x}) + \sum_{i \in I(\bar{x})} u_i \nabla g_i(\bar{x}) + \sum_{i=1}^{p} v_i \nabla h_i(\bar{x}) = 0_{\mathbb{R}^n}.$$

7.1. Necessary Conditions for Minimal and Weakly Minimal Elements

Proof. We verify the assumptions in Theorem 7.4. Since the gradients $\nabla h_1(\bar{x}), \ldots, \nabla h_p(\bar{x})$ are linearly independent, the linear map $h'(\bar{x})$ is surjective. The ordering cone in Z_1 is given as $C_{Z_1} = \mathbb{R}_+^k$. Consequently, we have

$$\text{int}(C_{Z_1}) = \{x \in \mathbb{R}^k \mid x_i > 0 \text{ for all } i \in \{1,\ldots,k\}\},$$

and we get for a sufficiently small $\lambda > 0$ and $\hat{x} := \lambda x + (1-\lambda)\bar{x}$

$$g(\bar{x}) + g'(\bar{x})(\hat{x} - \bar{x}) = g(\bar{x}) + g'(\bar{x})(\lambda(x - \bar{x}))$$

$$= \begin{pmatrix} g_1(\bar{x}) + \lambda \nabla g_1(\bar{x})^T (x - \bar{x}) \\ \vdots \\ g_k(\bar{x}) + \lambda \nabla g_k(\bar{x})^T (x - \bar{x}) \end{pmatrix} \in -\text{int}(C_{Z_1})$$

and

$$h'(\bar{x})(\hat{x} - \bar{x}) = h'(\bar{x})(\lambda(x - \bar{x}))$$

$$= \lambda \begin{pmatrix} \nabla h_1(\bar{x})^T (x - \bar{x}) \\ \vdots \\ \nabla h_p(\bar{x})^T (x - \bar{x}) \end{pmatrix} = 0_{\mathbb{R}^p}.$$

Hence, the regularity assumption in Theorem 7.4 is fulfilled. Then there are multipliers $t_i \geq 0$ (where at least one t_i, $i \in \{1,\ldots,m\}$, is nonzero), $u_i \geq 0$ ($i \in I(\bar{x})$) and $v_i \in \mathbb{R}$ ($i \in \{1,\ldots,p\}$) with the property

$$\sum_{i=1}^m t_i \nabla f_i(\bar{x}) + \sum_{i=1}^k u_i \nabla g_i(\bar{x}) + \sum_{i=1}^p v_i \nabla h_i(\bar{x}) = 0_{\mathbb{R}^n} \tag{7.17}$$

and

$$\sum_{i=1}^m u_i g_i(\bar{x}) = 0. \tag{7.18}$$

Because of

$$g_i(\bar{x}) \leq 0 \text{ for all } i \in \{1,\ldots,k\},$$
$$u_i \geq 0 \text{ for all } i \in \{1,\ldots,k\}$$

and the equality (7.18) we conclude
$$u_i g_i(\bar{x}) = 0 \text{ for all } i \in \{1,\ldots,k\}.$$

For every $i \in \{1,\ldots,k\}\backslash I(\bar{x})$ we have $g_i(\bar{x}) < 0$ and, therefore, we get $u_i = 0$. Consequently, the equation (7.17) can be written as

$$\sum_{i=1}^m t_i \nabla f_i(\bar{x}) + \sum_{i \in I(\bar{x})} u_i \nabla g_i(\bar{x}) + \sum_{i=1}^p v_i \nabla h_i(\bar{x}) = 0_{\mathbb{R}^n}$$

which completes the proof. □

7.2 Sufficient Conditions for Minimal and Weakly Minimal Elements

In general, the necessary optimality conditions formulated in the previous section are not sufficient for minimal or weakly minimal solutions without additional assumptions. Therefore, in the first part of this section generalized quasiconvex maps are introduced. This generalized convexity concept is very useful for the proof of the sufficiency of the generalized multiplier rule which will be done in the second part of this section.

7.2.1 Generalized Quasiconvex Maps

In Section 2.1 we have already investigated convex maps and introduced one possible generalization. Another generalization of convex maps is presented in

Definition 7.9. Let S be a nonempty convex subset of a real linear space X, and let Y be a partially ordered linear space with an ordering cone C_Y. A map $f : S \to Y$ is called *quasiconvex* if

$$x_1, x_2 \in S \text{ with } f(x_1) - f(x_2) \in C_Y \qquad (7.19)$$

implies that

$$f(x_1) - f(\lambda x_1 + (1-\lambda)x_2) \in C_Y \text{ for all } \lambda \in [0,1]. \qquad (7.20)$$

7.2. Sufficient Conditions for Minimal and Weakly Minimal Elements

Every convex map $f : S \to Y$ is also quasiconvex, because the condition (7.19) implies

$$(1 - \lambda)(f(x_1) - f(x_2)) \in C_Y$$

and, therefore, we get (with (2.4))

$$f(x_1) - f(\lambda x_1 + (1 - \lambda)x_2) \in \{(1 - \lambda)(f(x_1) - f(x_2))\} + C_Y \subset C_Y.$$

A characterization of quasiconvex maps which is simple to prove is given in

Lemma 7.10. *Let S be a nonempty convex subset of a real linear space X, and let Y be a partially ordered linear space with an ordering cone C_Y. A map $f : S \to Y$ is quasiconvex if and only if for all $\bar{x} \in S$ the sets*

$$L_{\bar{x}} := \{x \in S \backslash \{\bar{x}\} \mid f(\bar{x}) - f(x) \in C_Y\} \tag{7.21}$$

contain $\{\lambda x + (1 - \lambda)\bar{x} \mid \lambda \in [0, 1]\}$ whenever $x \in L_{\bar{x}}$.

Next, we extend the class of quasiconvex maps considerably by the following definition.

Definition 7.11. Let S be a nonempty subset of a real linear space X, and let C be a nonempty subset of a real linear space Y. Let $\bar{x} \in S$ be a given element. A map $f : S \to Y$ is called *C-quasiconvex at $\bar{x}$* if the following holds: Whenever there is some $x \in S \backslash \{\bar{x}\}$ with $f(\bar{x}) - f(x) \in C$, then there is some $\tilde{x} \in S \backslash \{\bar{x}\}$ with

$$\left. \begin{array}{l} \lambda \tilde{x} + (1 - \lambda)\bar{x} \in S \text{ for all } \lambda \in (0, 1] \\ \text{and} \\ f(\bar{x}) - f(\lambda \tilde{x} + (1 - \lambda)\bar{x}) \in C \text{ for all } \lambda \in (0, 1]. \end{array} \right\} \tag{7.22}$$

Example 7.12.

(a) Every quasiconvex map $f : S \to Y$ is C_Y-quasiconvex at all $\bar{x} \in S$.

(b) Let the map $f : \mathbb{R} \to \mathbb{R}^2$ be given by
$$f(x) = (x, \sin x) \text{ for all } x \in \mathbb{R}$$
where the space $\mathbb{R}^2$ is partially ordered in the componentwise sense. The map f is $\mathbb{R}_+^2$-quasiconvex at 0 but it is not quasiconvex (at 0).

The following lemma shows that C-quasiconvexity of f at $\bar{x}$ can also be characterized by a property of the level set $L_{\bar{x}}$ in (7.21).

Lemma 7.13. *Let S be a nonempty subset of a real linear space X, and let C be a nonempty subset of a real linear space Y. Let $\bar{x} \in S$ be a given element. A map $f : S \to Y$ is C-quasiconvex at $\bar{x}$ if and only if the set*
$$L_{\bar{x}} := \{x \in S \setminus \{\bar{x}\} \mid f(\bar{x}) - f(x) \in C\}$$
is empty or it contains a half-open line segment starting at $\bar{x}$, excluding $\bar{x}$.

Proof. Rewrite the condition (7.22) as
$$\{\lambda \tilde{x} + (1 - \lambda)\bar{x} \mid \lambda \in [0, 1)\} \subset L_{\bar{x}}$$
and the statement of the lemma is clear. □

As it may be seen from Lemma 7.13 the relaxation of the requirement (7.20) to (7.22) by allowing $\tilde{x} \neq x_2$ extends the class of quasiconvex maps considerably.

If one asks for conditions under which local minima are also global minima, then it turns out that C-quasiconvexity characterizes this property.

Definition 7.14. Let S be a nonempty subset of a real linear space X, let Y be a partially ordered linear space with an ordering cone C_Y, and let $f : S \to Y$ be a given map. Consider the abstract optimization problem
$$\min_{x \in S} f(x). \tag{7.23}$$

7.2. Sufficient Conditions for Minimal and Weakly Minimal Elements

(a) An element $\bar{x} \in S$ is called a *local minimal solution* of the problem (7.23), if there is a set $U \subset X$ with $\bar{x} \in \text{cor}(U)$ so that $\bar{x}$ is a minimal solution of the problem (7.23) with S replaced by $S \cap \text{cor}(U)$.

(b) In addition, let the ordering cone have a nonempty algebraic interior. An element $\bar{x} \in S$ is called a *local weakly minimal solution* of the problem (7.23), if there is a set $U \subset X$ with $\bar{x} \in \text{cor}(U)$ so that $\bar{x}$ is a weakly minimal solution of the problem (7.23) with S replaced by $S \cap \text{cor}(U)$.

The following two theorems state a necessary and sufficient condition under which local minima are also global minima.

Theorem 7.15. *Let S be a nonempty subset of a real linear space X, let Y be a partially ordered linear space with an ordering cone $C_Y \neq \{0_Y\}$, and let $f : S \to Y$ be a given map. Let $\bar{x} \in S$ be a local minimal solution of the problem (7.23). Then $\bar{x}$ is a (global) minimal solution of the problem (7.23) if and only if the map f is $(C_Y \setminus (-C_Y))$-quasiconvex at $\bar{x}$.*

Proof. Suppose that $\bar{x} \in S$ is a local minimal solution of the problem (7.23). If $\bar{x}$ is not a minimal solution, then there is an $x \in S$ with $f(\bar{x}) - f(x) \in C_Y \setminus (-C_Y)$. Assume f is $(C_Y \setminus (-C_Y))$-quasiconvex at $\bar{x}$, then there is an $\tilde{x} \in S \setminus \{\bar{x}\}$ with

$$\lambda \tilde{x} + (1 - \lambda)\bar{x} \in S \text{ for all } \lambda \in (0, 1]$$

and

$$f(\bar{x}) - f(\lambda \tilde{x} + (1 - \lambda)\bar{x}) \in C_Y \setminus (-C_Y) \text{ for all } \lambda \in (0, 1]. \quad (7.24)$$

Since $\bar{x} \in \text{cor}(U)$ there is a $\bar{\lambda} \in (0, 1]$ with

$$\bar{\lambda} \tilde{x} + (1 - \bar{\lambda})\bar{x} \in S \cap \text{cor}(U)$$

and with (7.24) we get

$$f(\bar{x}) - f(\bar{\lambda} \tilde{x} + (1 - \bar{\lambda})\bar{x}) \in C_Y \setminus (-C_Y).$$

But this contradicts the assumption that $\bar{x}$ is a local minimal solution of the problem (7.23).

On the other hand if $\bar{x}$ is a minimal solution of the problem (7.23), then there is no $x \in S$ with $f(\bar{x}) - f(x) \in C_Y \backslash (-C_Y)$ and the $(C_Y \backslash (-C_Y))$-quasiconvexity of f at $\bar{x}$ holds trivially. □

The following theorem can be proved similarly.

Theorem 7.16. *Let S be a nonempty subset of a real linear space X, let Y be a partially ordered linear space with an ordering cone C_Y which has a nonempty algebraic interior, and let $f : S \to Y$ be a given map. Let $\bar{x} \in S$ be a local weakly minimal solution of the problem (7.23). Then $\bar{x}$ is a (global) weakly minimal solution of the problem (7.23) if and only if the map f is $\mathrm{cor}(C_Y)$-quasiconvex at $\bar{x}$.*

For the generalized multiplier rule we assume that the considered maps are, in a certain sense, differentiable. Therefore, it is reasonable to introduce an appropriate framework for differentiable C-quasiconvexity. In the next definition we use the notion of a directional variation introduced in Definition 2.14.

Definition 7.17. *Let S be a nonempty subset of a real linear space X, and let C_1 and $C_2 \subset C_3$ be nonempty subsets of a real linear space Y. Moreover, let $\bar{x} \in S$ be a given element and let a map $f : S \to Y$ have a directional variation at $\bar{x}$ with respect to C_3. The map f is called differentiably C_1-C_2-quasiconvex at $\bar{x}$ if the following holds: Whenever there is some $x \in S$ with*

$$x \neq \bar{x} \quad \text{and} \quad f(x) - f(\bar{x}) \in C_1, \tag{7.25}$$

then there is an $\tilde{x} \in S \backslash \{\bar{x}\}$ with

$$\left. \begin{array}{l} \lambda \tilde{x} + (1 - \lambda)\bar{x} \in S \text{ for all } \lambda \in (0, 1] \\ \text{and} \\ \quad f'(\bar{x})(\tilde{x} - \bar{x}) \in C_2. \end{array} \right\} \tag{7.26}$$

In the case of $C_1 = C_2 =: C$ the map f is simply called differentiably C-quasiconvex at $\bar{x}$.

7.2. Sufficient Conditions for Minimal and Weakly Minimal Elements

Example 7.18. Let S be a subset of a real normed space $(X, \|\cdot\|_X)$ which has a nonempty interior, and let $(Y, \|\cdot\|_Y)$ be a partially ordered normed space with an ordering cone C_Y. Moreover, let $f : S \to Y$ be a map which is Fréchet-differentiable at some $\bar{x} \in S$. Then the map f is called *pseudoconvex* at $\bar{x}$, if for all $x \in S$ the following holds:

$$f'(\bar{x})(x - \bar{x}) \in C_Y \implies f(x) - f(\bar{x}) \in C_Y.$$

This implication is equivalent to

$$f(x) - f(\bar{x}) \notin C_Y \implies f'(\bar{x})(x - \bar{x}) \notin C_Y.$$

Therefore, every map $f : S \to Y$ which is pseudoconvex at $\bar{x}$ is also differentiably $(Y \backslash C_Y)$-quasiconvex at $\bar{x}$. This shows that the class of pseudoconvex maps is contained in the larger class of differentiably C_1-C_2-quasiconvex maps.

With the next theorem we investigate some relations between C-quasiconvexity and differentiable C-quasiconvexity.

Theorem 7.19. *Let S be a nonempty subset of a real linear space X, and let $C \subset \hat{C}$ be nonempty subsets of a real linear space where $C \cup \{0_Y\}$ is assumed to be a cone. Moreover, let $\bar{x} \in S$ be a given element and let $f : S \to Y$ be a given map.*

(a) *If f is $(-C)$-quasiconvex at $\bar{x}$ and has a directional variation at $\bar{x}$ with respect to $\hat{C}$ and $Y \backslash C$, then f is differentiably C-quasiconvex at $\bar{x}$.*

(b) *If f is differentiably C-quasiconvex at $\bar{x}$ with a directional variation of f at $\bar{x}$ with respect to C, then f is $(-C)$-quasiconvex at $\bar{x}$.*

Proof.

(a) Let some $x \in S$ be given with (7.25). Since f is assumed to be $(-C)$-quasiconvex at $\bar{x}$, there is an $\tilde{x} \in S \backslash \{\bar{x}\}$ so that

$$\lambda \tilde{x} + (1 - \lambda)\bar{x} \in S \text{ for all } \lambda \in (0, 1]$$

and
$$f(\bar{x}) - f(\lambda \tilde{x} + (1-\lambda)\bar{x}) \in -C \text{ for all } \lambda \in (0,1]. \qquad (7.27)$$

Suppose that for all directional variations of f at $\bar{x}$ with respect to $\hat{C}$ and $Y\backslash C$ $f'(\bar{x})(\tilde{x} - \bar{x}) \notin C$. Then, from the definition of a directional variation with respect to $Y\backslash C$ there is a $\bar{\lambda} > 0$ with
$$\bar{x} + \lambda(\tilde{x} - \bar{x}) \in S \text{ for all } \lambda \in (0, \bar{\lambda}]$$
and
$$\frac{1}{\lambda}(f(\bar{x} + \lambda(\tilde{x} - \bar{x})) - f(\bar{x})) \notin C \text{ for all } \lambda \in (0, \bar{\lambda}].$$

By assumption $C \cup \{0_Y\}$ is a cone and, therefore, we conclude
$$f(\bar{x}) - f(\bar{x} + \lambda(\tilde{x} - \bar{x})) \notin -C \text{ for all } \lambda \in (0, \bar{\lambda}].$$

But this contradicts (7.27). Hence, for some directional variation of f at $\bar{x}$ with respect to $\hat{C}$ and $Y\backslash C$ we have $f'(\bar{x})(\tilde{x}-\bar{x}) \in C$ which shows that (7.25) implies (7.26) in Definition 7.17 with $C_1 = C_2 = C$ and $C_3 = \hat{C}$.

(b) Let some $x \in S$ be given with $x \neq \bar{x}$ and $f(x) - f(\bar{x}) \in C$. Then differentiable C-quasiconvexity of f at $\bar{x}$ implies that there is an $\tilde{x} \in S\backslash\{\bar{x}\}$ and a directional variation of f at $\bar{x}$ with respect to C with the property
$$\lambda \tilde{x} + (1-\lambda)\bar{x} \in S \text{ for all } \lambda \in [0,1]$$
and $f'(\bar{x})(\tilde{x} - \bar{x}) \in C$. Then by Definition 2.14 there is a $\bar{\lambda} > 0$ with
$$\bar{x} + \lambda(\tilde{x} - \bar{x}) \in S \text{ for all } \lambda \in (0, \bar{\lambda}]$$
and
$$\frac{1}{\lambda}(f(\bar{x} + \lambda(\tilde{x} - \bar{x})) - f(\bar{x})) \in C \text{ for all } \lambda \in (0, \bar{\lambda}].$$

Observing that $C \cup \{0_Y\}$ is a cone, we obtain
$$f(\bar{x}) - f(\bar{x} + \lambda(\tilde{x} - \bar{x})) \in -C \text{ for all } \lambda \in (0, \bar{\lambda}]$$
and the proof of the $(-C)$-quasiconvexity of f at $\bar{x}$ is complete. $\square$

7.2. Sufficient Conditions for Minimal and Weakly Minimal Elements

If one considers directional variations with respect to algebraically open sets, in the previous theorem under (a) and (b), one should assume that $\hat{C}$ and $Y \backslash C$ are algebraically open.

7.2.2 Sufficiency of the Generalized Multiplier Rule

The generalized multiplier rule introduced in the section 7.1 is now investigated again. We prove that this multiplier rule is a sufficient optimality condition for a substitute problem if and only if a certain composite map is generalized quasiconvex. Finally we discuss the results with respect to a multiobjective optimization problem.

Although we formulated the generalized multiplier rule for simplicity in a normed setting, we investigate this optimality condition now in a very general setting.

The standard assumption for the following results reads as follows:

$$\left.\begin{array}{l}\text{Let } \hat{S} \text{ be a nonempty subset of a real linear} \\ \text{space } X; \text{ let } Y, Z_1 \text{ and } Z_2 \text{ be partially ordered} \\ \text{linear spaces with the ordering cones } C_Y, C_{Z_1} \\ \text{and } C_{Z_2}, \text{ respectively; let } C_Y \text{ have a nonempty} \\ \text{algebraic interior and let } C_{Z_2} \text{ be pointed;} \\ \text{let } f: \hat{S} \to Y, \; g: \hat{S} \to Z_1 \text{ and } h: \hat{S} \to Z_2 \text{ be} \\ \text{given maps; let the constraint set} \\ S := \{x \in \hat{S} \mid g(x) \in -C_{Z_1} \text{ and } h(x) = 0_{Z_2}\} \text{ be} \\ \text{nonempty.}\end{array}\right\} \quad (7.28)$$

Under this assumption we investigate again the abstract optimization problem

$$\min_{x \in S} f(x). \qquad (7.29)$$

Theorem 7.20. *Let the abstract optimization problem (7.29) be given under the assumption (7.28), and suppose that for some $\bar{x} \in S$ there are nonempty sets G_0, G_1 and G_2 with $-cor(C_Y) \subset G_0 \subset Y$, $-C_{Z_1} + cone(\{g(\bar{x})\}) - cone(\{g(\bar{x})\}) \subset G_1 \subset Z_1$ and $0_{Z_2} \in G_2 \subset Z_2$ so that the maps f, g and h have directional variations at $\bar{x}$ with respect* to G_0, G_1 and G_2, respectively. Assume that there are some

$$t \in C_{Y'} \setminus \{0_{Y'}\}, \ u \in C_{Z_1'} \text{ and } v \in Z_2' \tag{7.30}$$

with

$$(t \circ f'(\bar{x}) + u \circ g'(\bar{x}) + v \circ h'(\bar{x}))(x - \bar{x}) \geq 0 \text{ for all } x \in \hat{S} \tag{7.31}$$

and

$$(u \circ g)(\bar{x}) = 0. \tag{7.32}$$

Then $\bar{x}$ is a weakly minimal solution of the problem (7.29) with S replaced by

$$\bar{S} := \{x \in \hat{S} \mid g(x) \in -C_{Z_1} + \text{cone}(\{g(\bar{x})\}) - \text{cone}(\{g(\bar{x})\}), \ h(x) = 0_{Z_2}\}$$

if and only if the composite map

$$(f, g, h) : \hat{S} \to Y \times Z_1 \times Z_2$$

is differentiably C-quasiconvex at $\bar{x}$ with

$$C := (-\text{cor}(C_Y)) \times (-C_{Z_1} + \text{cone}(\{g(\bar{x})\}) - \text{cone}(\{g(\bar{x})\})) \times \{0_{Z_2}\}. \tag{7.33}$$

Proof. Assume that the generalized multiplier rule (7.30) - (7.32) holds at some $\bar{x} \in S$. Then we assert that

$$(f'(\bar{x})(x - \bar{x}), \ g'(\bar{x})(x - \bar{x}), \ h'(\bar{x})(x - \bar{x})) \notin C \text{ for all } x \in \hat{S}. \tag{7.34}$$

For the proof of this assertion assume that there is an $x \in \hat{S}$ with

$$f'(\bar{x})(x - \bar{x}) \in -\text{cor}(C_Y),$$
$$g'(\bar{x})(x - \bar{x}) \in -C_{Z_1} + \text{cone}(\{g(\bar{x})\}) - \text{cone}(\{g(\bar{x})\}),$$
$$h'(\bar{x})(x - \bar{x}) = 0_{Z_2}.$$

With (7.30) and Lemma 1.26 we conclude for some $\alpha, \beta \geq 0$

$$\begin{aligned}(t \circ f'(\bar{x}) + u \circ g'(\bar{x}) + v \circ h'(\bar{x}))(x - \bar{x}) &< u(g'(\bar{x})(x - \bar{x})) \\ &\leq u(\alpha g(\bar{x})) - u(\beta g(\bar{x})) \\ &= (\alpha - \beta)u(g(\bar{x})).\end{aligned}$$

7.2. Sufficient Conditions for Minimal and Weakly Minimal Elements

But this inequality contradicts (7.31) and (7.32). Hence, the condition (7.34) holds.

If the composite map (f, g, h) is C-quasiconvex at $\bar{x}$, then it follows from (7.34)

$$(f(x) - f(\bar{x}), g(x) - g(\bar{x}), h(x) - h(\bar{x})) \notin C \text{ for all } x \in \hat{S}. \qquad (7.35)$$

The condition (7.35) means that there is no $x \in \hat{S}$ with

$$f(x) \in \{f(\bar{x})\} - \operatorname{cor}(C_Y),$$
$$g(x) \in \{g(\bar{x})\} - C_{Z_1} + \operatorname{cone}(\{g(\bar{x})\}) - \operatorname{cone}(\{g(\bar{x})\})$$
$$= -C_{Z_1} + \operatorname{cone}(\{g(\bar{x})\}) - \operatorname{cone}(\{g(\bar{x})\}),$$
$$h(x) = 0_{Z_2}.$$

If we notice that with

$$g(\bar{x}) \in -C_{Z_1} \subset -C_{Z_1} + \operatorname{cone}(\{g(\bar{x})\}) - \operatorname{cone}(\{g(\bar{x})\})$$

it also follows $\bar{x} \in \bar{S}$, then $\bar{x}$ is a weakly minimal solution of the abstract optimization problem

$$\min_{x \in \bar{S}} f(x). \qquad (7.36)$$

Now we assume in the converse case that $\bar{x}$ is a weakly minimal solution of the problem (7.36), then there is no $x \in \hat{S}$ with

$$f(x) \in \{f(\bar{x})\} - \operatorname{cor}(C_Y),$$
$$g(x) \in -C_{Z_1} + \operatorname{cone}(\{g(\bar{x})\}) - \operatorname{cone}(\{g(\bar{x})\})$$
$$= \{g(\bar{x})\} - C_{Z_1} + \operatorname{cone}(\{g(\bar{x})\}) - \operatorname{cone}(\{g(\bar{x})\}),$$
$$h(x) = 0_{Z_2},$$

i.e., the condition (7.35) is satisfied for all $x \in \hat{S}$. With the condition (7.34) we conclude that the map (f, g, h) is C-quasiconvex at $\bar{x}$. □

In the previous theorem we showed the equivalence of the generalized quasiconvexity with the sufficiency of the generalized multiplier rule of a substitute problem where S is replaced by $\bar{S}$. The set

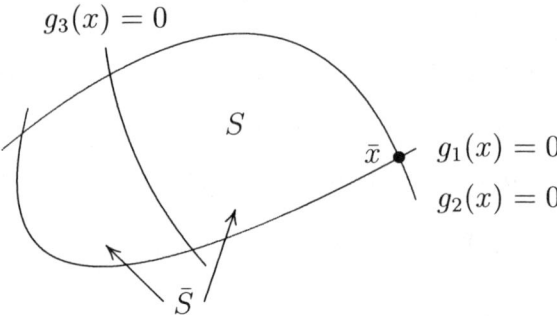

Figure 7.1: Illustration of the set $\bar{S}$.

$\mathrm{cone}(\{g(\bar{x})\}) - \mathrm{cone}(\{g(\bar{x})\})$ equals the onedimensional subspace of Z_1 spanned by $g(\bar{x})$. Figure 7.1 illustrates the modified constraint set $\bar{S}$.

For the original problem the following conclusion holds:

Corollary 7.21. *Let the assumptions of Theorem 7.20 be satisfied and let the map (f, g, h) be differentiably C-quasiconvex at $\bar{x} \in S$ with C given by (7.33). Then $\bar{x}$ is a weakly minimal solution of the problem (7.29).*

Proof. By Theorem 7.20 $\bar{x} \in S$ is a weakly minimal solution of the problem (7.36). For every $x \in S$ we have

$$\begin{aligned} g(x) &\in -C_{Z_1} \\ &\subset -C_{Z_1} + \mathrm{cone}(\{g(\bar{x})\}) - \mathrm{cone}(\{g(\bar{x})\}). \end{aligned}$$

Consequently we get $S \subset \bar{S}$ and, therefore, $\bar{x}$ is also a weakly minimal solution of the problem (7.29). □

If the generalized quasiconvexity assumption in Theorem 7.20 is strengthened, then a similar theorem holds for minimal solutions of the problem (7.29).

Theorem 7.22. *Let all the assumptions of Theorem 7.20 be sat-*

7.2. Sufficient Conditions for Minimal and Weakly Minimal Elements

isfied. Then $\bar{x} \in S$ is a minimal solution of the problem (7.36) if and only if the composite map (f, g, h) is differentiably C_1-C_2-quasiconvex at $\bar{x}$ with

$$C_1 := (-C_Y \setminus C_Y) \times (-C_{Z_1} + cone(\{g(\bar{x})\}) - cone(\{g(\bar{x})\})) \times \{0_{Z_2}\}$$

and

$$C_2 := (-cor(C_Y)) \times (-C_{Z_1} + cone(\{g(\bar{x})\}) - cone(\{g(\bar{x})\})) \times \{0_{Z_2}\}.$$

The proof of this theorem is almost identical to the one of Theorem 7.20 and, therefore, it is omitted. A result which is similar to that of Corollary 7.21 can also be obtained.

Finally, we investigate again the multiobjective optimization problem considered in Theorem 7.8. Recall that a real-valued function $f : \mathbb{R}^n \to \mathbb{R}$ which has partial derivatives at some $\bar{x} \in \mathbb{R}^n$ is called *pseudoconvex* at $\bar{x}$, if for every $x \in \mathbb{R}^n$

$$\nabla f(\bar{x})^T (x - \bar{x}) \geq 0 \implies f(x) \geq f(\bar{x})$$

(see Example 7.18), and it is *quasiconvex* at $\bar{x}$, if for every $x \in \mathbb{R}^n$

$$f(x) \leq f(\bar{x}) \implies \nabla f(\bar{x})^T (x - \bar{x}) \leq 0$$

(e.g., compare Mangasarian [220, ch. 9]).

Lemma 7.23. *Let $f : \mathbb{R}^n \to \mathbb{R}^m$, $g : \mathbb{R}^n \to \mathbb{R}^k$ and $h : \mathbb{R}^n \to \mathbb{R}^p$ be given vector functions. Let the constraint set S be given as*

$$S := \{x \in \mathbb{R}^n \mid g_i(x) \leq 0 \text{ for all } i \in \{1, \ldots, k\} \text{ and } \\ h_i(x) = 0 \text{ for all } i \in \{1, \ldots, p\}\}.$$

Let some $\bar{x} \in S$ be given and assume that the space $\mathbb{R}^m$ is partially ordered in a natural way. Let the vector functions f, g and h have partial derivatives at $\bar{x}$. If the functions $f_1, \ldots, f_m$ are pseudoconvex at $\bar{x}$ and the functions $h_1, \ldots, h_p, -h_1, \ldots, -h_p$ and g_i for all $i \in I(\bar{x})$ with

$$I(\bar{x}) := \{i \in \{1, \ldots, k\} \mid g_i(\bar{x}) = 0\}$$

are quasiconvex at $\bar{x}$, then the composite vector function (f, g, h) is differentiably C-quasiconvex at $\bar{x}$ with

$$C := (-int(\mathbb{R}_+^m)) \times (-\mathbb{R}_+^k + cone(\{g(\bar{x})\}) - cone(\{g(\bar{x})\})) \times \{0_{\mathbb{R}^p}\}.$$

Proof. Let some $x \in S$ be given with (7.25), i.e. $x \neq \bar{x}$ and

$$f_i(x) - f_i(\bar{x}) < 0 \text{ for all } i \in \{1, \ldots, m\},$$
$$g(x) - g(\bar{x}) \in -\mathbb{R}_+^k + cone(\{g(\bar{x})\}) - cone(\{g(\bar{x})\}) \quad (7.37)$$
$$h_i(x) - h_i(\bar{x}) = 0 \text{ for all } i \in \{1, \ldots, p\}.$$

The inequality (7.37) implies

$$g_i(x) - g_i(\bar{x}) \leq 0 \text{ for all } i \in I(\bar{x}).$$

Using the definition of pseudoconvex functions and the characterization of quasiconvex functions with partial derivatives the previous inequalities imply

$$f_i'(\bar{x})(x - \bar{x}) < 0 \text{ for all } i \in \{1, \ldots, m\},$$
$$g_i'(\bar{x})(x - \bar{x}) \leq 0 \text{ for all } i \in I(\bar{x}),$$
$$h_i'(\bar{x})(x - \bar{x}) = 0 \text{ for all } i \in \{1, \ldots, p\}.$$

Since $g_i(\bar{x}) < 0$ for all $i \in \{1, \ldots, k\} \setminus I(\bar{x})$, there are $\alpha, \beta \geq 0$ with

$$g_i'(\bar{x})(x - \bar{x}) \leq (\alpha - \beta) g_i(\bar{x}) \text{ for all } i \in \{1, \ldots, k\}.$$

Consequently, we get

$$(f, g, h)'(\bar{x})(x - \bar{x}) \in C$$

and we conclude that the condition (7.26) is fulfilled with $\tilde{x} := x$ and $C_2 := C$. □

Lemma 7.23 shows in particular that the convexity type conditions are only imposed on the active constraints. With Corollary 7.21 and Lemma 7.23 we immediately obtain a sufficient condition for a weakly minimal solution of a multiobjective optimization problem.

Corollary 7.24. *Let $f : \mathbb{R}^n \to \mathbb{R}^m$, $g : \mathbb{R}^n \to \mathbb{R}^k$ and $h : \mathbb{R}^n \to \mathbb{R}^p$ be given vector functions. Let the constraint set S be given as*

$$S := \{x \in \mathbb{R}^n \mid g_i(x) \leq 0 \text{ for all } i \in \{1, \ldots, k\} \text{ and } \\ h_i(x) = 0 \text{ for all } i \in \{1, \ldots, p\}\},$$

and let the space $\mathbb{R}^m$ be partially ordered in the natural way. Let some $\bar{x} \in S$ be given and assume that the vector functions f, g and h have partial derivatives at $\bar{x}$. Let the functions $f_1, \ldots, f_m$ be pseudoconvex at $\bar{x}$ and let the functions $h_1, \ldots, h_p, -h_1, \ldots, -h_p$ and g_i for all $i \in I(\bar{x})$ with

$$I(\bar{x}) := \{i \in \{1, \ldots, k\} \mid g_i(\bar{x}) = 0\}$$

be quasiconvex at $\bar{x}$. If there are multipliers $t_i \geq 0$ (where at least one t_i, $i \in \{1, \ldots, m\}$, is nonzero), $u_i \geq 0$ ($i \in I(\bar{x})$) and $v_i \in \mathbb{R}$ ($i \in \{1, \ldots, p\}$) with the property

$$\sum_{i=1}^{m} t_i \nabla f_i(\bar{x}) + \sum_{i \in I(\bar{x})} u_i \nabla g_i(\bar{x}) + \sum_{i=1}^{p} v_i \nabla h_i(\bar{x}) = 0_{\mathbb{R}^n},$$

then $\bar{x}$ is a weakly minimal solution of the multiobjective optimization problem $\min_{x \in S} f(x)$.

Notes

The investigation of necessary optimality conditions carried out in the section 7.1 for Banach spaces and Fréchet differentiable maps can be extended to much more general spaces and even to much more general differentiability notions. Kirsch-Warth-Werner [171] discuss these generalizations in their book in a profound way. The proof of the necessary condition presented in this book is based on similar work of Sachs [268], [269] and Kirsch-Warth-Werner [171]. The so-called F. John conditions were introduced by John [160] and the Karush-Kuhn-Tucker conditions became popular by the work of Kuhn-Tucker [187]. For a discussion of these necessary conditions for abstract optimization problems we also refer to Hurwicz [127], Borwein [29], Vogel [313], Penot [248], Hartwig [117], Oettli [242], Borwein [32], Craven [69] and

Minami [225], [226], [227], and others. In a paper of Jahn-Sachs [157] Theorem 7.5 can be found even in a non-topological setting. The necessary optimality condition in Theorem 7.6 given by Jahn-Sachs [156] extends a corresponding condition for scalar optimization problems (e.g., see Luenberger [217, p. 178]). In the case of a vector-valued objective map a similar condition is given by Sachs [268, p. 23], [269, p. 505] and Penot [248, p. 8].

The presentation of the section 7.2 is based on a paper of Jahn-Sachs [157]. The definition of quasiconvexity was first introduced by von Neumann [316, p. 307] and Nikaidô [239]. For abstract optimization problems this definition has been given by Hartwig [117] in a finite-dimensional setting and by Craven [69], Nehse [233] and Peemöller [246] for problems in infinite-dimensional spaces. Corollary 7.21 extends results of Vogel [313, p. 100], Hartwig [117, p. 313-314] (for another optimality notion) and Craven [69, p. 666-667].

Chapter 8

Duality

It is well-known from scalar optimization that, under appropriate assumptions, a maximization problem can be associated to a given minimization problem so that both problems have the same optimal values. Such a duality between a minimization and a maximization problem can also be formulated in vector optimization. In the first section we present a general duality principle for vector optimization problems. The following sections are devoted to a duality theory for abstract optimization problems. A generalization of the duality results known from linear programming is also given.

8.1 A General Duality Principle

The duality principle presented in this section is simple and it is based on a similar idea on which the duality theory for abstract optimization problems examined in the following section is based as well. This principle is designed in a way that, under appropriate assumptions, a minimal element of a subset of a partially ordered linear space is also a maximal element of an associated set.

Let P be a nonempty subset of a partially ordered linear space X with a pointed ordering cone $C \neq \{0_X\}$. Then we couple the *primal problem* of determining a minimal element of the set P with a *dual problem* of determining a maximal element of the complement set of

$P + (C\setminus\{0_X\})$. The set P is also called the *primal set*, and the set

$$D := X\setminus(P + (C\setminus\{0_X\})) \qquad (8.1)$$

is denoted as the *dual set* of our problem (see Fig. 8.1). The following

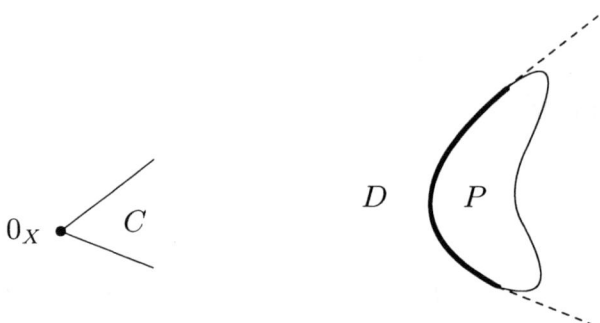

Figure 8.1: Illustration of the primal set P and the dual set D.

duality investigations are concentrated on the question: Under which assumption is a minimal element of the primal set P also a maximal element of the dual set D and vice versa? The following lemma is a key for the answer of this question.

Lemma 8.1. *Let P be a nonempty subset of a partially ordered linear space X with a pointed ordering cone $C \neq \{0_X\}$. If $\bar{x} \in P \cap D$ where D is defined in (8.1), then $\bar{x}$ is a minimal element of the set P and $\bar{x}$ is a maximal element of the set D.*

Proof. Since $\bar{x}$ is an element of the dual set D, it follows that $\bar{x} \notin P + (C\setminus\{0_X\})$ implying $(\{\bar{x}\} - (C\setminus\{0_X\})) \cap P = \emptyset$. But $\bar{x}$ also belongs to the primal set P and, therefore, $\bar{x}$ is a minimal element of the set P.

Since D is the complement set of $P + (C\setminus\{0_X\})$, we have $(P + (C\setminus\{0_X\})) \cap D = \emptyset$ and especially $(\{\bar{x}\} + (C\setminus\{0_X\})) \cap D = \emptyset$. If we notice that $\bar{x} \in D$, it is evident that $\bar{x}$ is a maximal element of the dual set D. □

8.1. A General Duality Principle

The following *duality theorem* is a consequence of the previous lemma.

Theorem 8.2. *Let P be a nonempty subset of a partially ordered linear space X with a pointed ordering cone $C \neq \{0_X\}$. Every minimal element of the primal set P is also a maximal element of the dual set D defined in (8.1).*

Proof. Let $\bar{x} \in P$ be a minimal element of the set P, and assume that $\bar{x} \notin D$. Then we have $\bar{x} \in P + (C \backslash \{0_X\})$ which is a contradiction to the minimality of $\bar{x}$. Consequently, $\bar{x}$ belongs to the dual set D, and Lemma 8.1 leads to the assertion. □

The next theorem is a so-called *converse duality theorem*.

Theorem 8.3. *Let P be a nonempty subset of a partially ordered linear space X with a pointed ordering cone $C \neq \{0_X\}$. If the complement set of $P + C$ is algebraically open, then every maximal element of the dual set D defined in (8.1) is also a minimal element of the set P.*

Proof. Let $\bar{x} \in D$ be a maximal element of the set D, and assume that $\bar{x} \notin P + C$. Since the set $X \backslash (P + C)$ is algebraically open, for every $h \in C \backslash \{0_X\}$ there is a $\bar{\lambda} > 0$ so that

$$\bar{x} + \lambda h \in X \backslash (P + C) \text{ for all } \lambda \in (0, \bar{\lambda}].$$

Then it follows $\bar{x} + \bar{\lambda} h \in D$ which contradicts the maximality of $\bar{x}$. Consequently, $\bar{x}$ is an element of the set $P + C$, and since $\bar{x}$ does not belong to $P + (C \backslash \{0_X\})$, we conclude $\bar{x} \in P$. Finally, Lemma 8.1 leads to the assertion. □

If the set $P + C$ is convex and algebraically closed, then the complement set of $P + C$ is algebraically open (compare also the proof of Lemma 1.22, (d)). But notice that, in general, the duality principle outlined in the two preceding theorems works even without any convexity assumptions on the set P or $P + C$.

8.2 Duality Theorems for Abstract Optimization Problems

In this section abstract optimization problems with inequality constraints are investigated and duality results for the minimality and weak minimality notion are presented. The following theory is based on the Lagrange formalism of chapter 7 (without differentiability assumptions) and on the duality principle of the previous section.

First, we list the standard assumption for the following theory:

$$\left.\begin{array}{l}\text{Let } \hat{S} \text{ be a nonempty convex subset of a real linear}\\ \text{space } X; \text{ let } Y \text{ and } Z \text{ be partially ordered}\\ \text{topological linear spaces with ordering cones}\\ C_Y \neq Y \text{ and } C_Z, \text{ respectively; let } f : \hat{S} \to Y \text{ and}\\ g : \hat{S} \to Z \text{ be convex maps; let the constraint set}\\ \quad S := \{x \in \hat{S} \mid g(x) \in -C_Z\}\\ \text{be nonempty.}\end{array}\right\} \quad (8.2)$$

Notice that under this assumption the set $f(S) + C_Y$ is convex (compare Theorem 2.11). Then we examine the abstract optimization problem

$$\min_{x \in S} f(x). \quad (8.3)$$

Instead of investigating the optimal solutions of the problem (8.3) we consider weakly minimal or almost properly minimal elements of the set $f(S) + C_Y$. If the ordering cone C_Y has a nonempty interior, we examine the problem:

$$\left.\begin{array}{l}\text{Determine a weakly minimal element of the set}\\ P_1 := f(S) + C_Y.\end{array}\right\} \quad (8.4)$$

If the quasi-interior $C_{Y^*}^{\#}$ of the dual ordering cone C_{Y^*} is nonempty, we formulate the problem:

$$\left.\begin{array}{l}\text{Determine an almost properly minimal element of the}\\ \text{set } P_2 := f(S).\end{array}\right\} \quad (8.5)$$

8.2. Duality Theorems for Abstract Optimization Problems

Next we assign *dual problems* to these two *primal problems*. If $\text{int}(C_Y) \neq \emptyset$, we define the problem which is dual to (8.4):

$$\left.\begin{array}{l}\text{Determine a weakly maximal element of the set}\\ D_1 := \{y \in Y \mid \text{ there are continuous linear functionals}\\ \quad t \in C_{Y^*}\setminus\{0_{Y^*}\} \text{ and } u \in C_{Z^*} \text{ with}\\ \quad (t \circ f + u \circ g)(x) \geq t(y) \text{ for all } x \in \hat{S}\}.\end{array}\right\} \quad (8.6)$$

For $C_{Y^*}^{\#} \neq \emptyset$ we formulate the problem which is dual to (8.5):

$$\left.\begin{array}{l}\text{Determine a maximal element of the set}\\ D_2 := \{y \in Y \mid \text{ there are continuous linear functionals}\\ \quad t \in C_{Y^*}^{\#} \text{ and } u \in C_{Z^*} \text{ with}\\ \quad (t \circ f + u \circ g)(x) \geq t(y) \text{ for all } x \in \hat{S}\}.\end{array}\right\} \quad (8.7)$$

Notice that the Krein-Rutman theorem (Theorem 3.38) gives a sufficient condition under which the set $C_{Y^*}^{\#}$ is nonempty. Moreover, if $C_{Y^*}^{\#}$ is nonempty, by Lemma 1.27, (b) the ordering cone C_Y is pointed and, therefore, the assumption $C_Y \neq Y$ is fulfilled. If $\text{int}(C_Y)$ is nonempty, by Lemma 3.21, (c) and the assumption $C_Y \neq Y$ the set $C_{Y^*}\setminus\{0_{Y^*}\}$ is nonempty.

With the next theorems we clarify in which sense the problems (8.4) and (8.6) and the problems (8.5) and (8.7) are dual to each other. First, we prove a *weak duality theorem*.

Theorem 8.4. *Let the assumption (8.2) be satisfied, and consider the problems (8.4) - (8.7).*

(a) *If* $\text{int}(C_Y) \neq \emptyset$, *then for every* $\bar{y} \in D_1$ *there is a* $t \in C_{Y^*}\setminus\{0_{Y^*}\}$ *with the property*

$$t(\bar{y}) \leq t(y) \text{ for all } y \in P_1.$$

(b) *If* $C_{Y^*}^{\#} \neq \emptyset$, *then for every* $\bar{y} \in D_2$ *there is a* $t \in C_{Y^*}^{\#}$ *with the property*

$$t(\bar{y}) \leq t(y) \text{ for all } y \in P_2.$$

Proof. We fix an arbitrary $\bar{y} \in D_1$ ($\bar{y} \in D_2$, respectively). Then there are continuous linear functionals $t \in C_{Y^*}\setminus\{0_{Y^*}\}$ ($t \in C_{Y^*}^\#$, respectively) and $u \in C_{Z^*}$ with

$$(t \circ f + u \circ g)(x) \geq t(\bar{y}) \text{ for all } x \in \hat{S}$$

which implies

$$(t \circ f)(x) \geq t(\bar{y}) \text{ for all } x \in S.$$

This inequality immediately leads to the assertions under (a) and (b). $\square$

The next lemma is useful for the proof of the following strong duality results. It can be compared with Lemma 8.1.

Lemma 8.5. *Let the assumption (8.2) be satisfied, and consider the problems (8.4) - (8.7).*

(a) Assume that $int(C_Y)$ is nonempty.

  (i) If $p \in P_1$ and $d \in D_1$, then $d - p \notin cor(C_Y)$.

  (ii) If $\bar{y} \in P_1 \cap D_1$, then $\bar{y}$ is a weakly minimal element of the set P_1 and $\bar{y}$ is a weakly maximal element of the set D_1.

(b) Assume that $C_{Y^*}^\#$ is nonempty.

  (i) If $p \in P_2$ and $d \in D_2$, then $d - p \notin C_Y\setminus\{0_Y\}$.

  (ii) If $\bar{y} \in P_2 \cap D_2$, then $\bar{y}$ is an almost properly minimal element of the set P_2 and $\bar{y}$ is a maximal element of the set D_2.

Proof.

(a) (i) Let $p \in P_1$ and $d \in D_1$ be arbitrarily given. If we assume that $d - p \in cor(C_Y)$, then we get with Lemma 3.21, (b)

$$t(d - p) > 0 \text{ for all } t \in C_{Y^*}\setminus\{0_{Y^*}\}$$

which contradicts Theorem 8.4, (a).

8.2. Duality Theorems for Abstract Optimization Problems

(ii) Let any $\bar{y} \in P_1 \cap D_1$ be given. Then we obtain with Lemma 8.5, (a), (i)
$$d \notin \{\bar{y}\} + \mathrm{cor}(C_Y) \text{ for all } d \in D_1$$
which implies that $\bar{y}$ is a weakly maximal element of the set D_1. Moreover, with Theorem 8.4, (a) and Theorem 5.28 $\bar{y}$ is also a weakly minimal element of the set P_1.

(b) (i) Let arbitrary elements $p \in P_2$ and $d \in D_2$ be given. If we assume that $d - p \in C_Y \setminus \{0_Y\}$, then we get
$$t(d-p) > 0 \text{ for all } t \in C_{Y*}^{\#}$$
which contradicts Theorem 8.4, (b).

(ii) We fix any $y \in P_2 \cap D_2$. Then we get with Lemma 8.5, (b), (i)
$$d \notin \{\bar{y}\} + (C_Y \setminus \{0_Y\}) \text{ for all } d \in D_2.$$
Consequently, $\bar{y}$ is a maximal element of the set D_2. Finally, with Theorem 8.4, (b) $\bar{y}$ is an almost properly minimal element of the set P_2 as well.

□

For the formulation of strong duality results we need the notions of normality and stability which are known from the scalar optimization theory (e.g., compare Ekeland-Temam [92, p. 51] or Rockafellar [261]). In this book we use the following

Definition 8.6. Let the assumption (8.2) be satisfied, and let $\varphi : \hat{S} \to \mathbb{R}$ be a convex functional.

(a) The scalar optimization problem
$$\inf_{x \in S} \varphi(x) \qquad (8.8)$$
is called *normal* if
$$\inf_{x \in S} \varphi(x) = \sup_{u \in C_{Z*}} \inf_{x \in \hat{S}} (\varphi + u \circ g)(x)$$
(where we do not assume that this number is finite).

(b) The scalar optimization problem (8.8) is called *stable*, if it is normal and if the problem

$$\sup_{u \in C_{Z^*}} \inf_{x \in \hat{S}} (\varphi + u \circ g)(x)$$

has at least one solution.

Theorem 8.7. *Let the assumption (8.2) be satisfied, and consider the problems (8.4) - (8.7).*

(a) *Let int(C_Y) be nonempty, and let $\bar{y}$ be any weakly minimal element of the set P_1. Let $t \in C_{Y^*} \setminus \{0_{Y^*}\}$ be a supporting functional to the set P_1 at $\bar{y}$ (the existence of t is ensured by Theorem 5.13 and Lemma 3.15), and let the scalar optimization problem*

$$\inf_{x \in S} (t \circ f)(x) \tag{8.9}$$

be stable. Then $\bar{y}$ is also a weakly maximal element of the set D_1.

(b) *Let $C_{Y^*}^\#$ be nonempty, and let $\bar{y}$ be an almost properly minimal element of the set P_2 with the continuous linear functional $t \in C_{Y^*}^\#$ given by Definition 5.23. Let the scalar optimization problem (8.9) be stable. Then $\bar{y}$ is also a maximal element of the set D_2.*

Proof. For simplicity we prove only part (a) of the assertion. The proof of the part (b) is similar. Let $\bar{y} \in P_1$ be any weakly minimal element of the set P_1 and let $t \in C_{Y^*} \setminus \{0_{Y^*}\}$ be a corresponding supporting functional, i.e. we have

$$t(\bar{y}) \leq t(y) \text{ for all } y \in P_1.$$

Consequently, there are an $\bar{x} \in S$ and a $\bar{c} \in C_Y$ with $\bar{y} = f(\bar{x}) + \bar{c}$ and

$$t(f(\bar{x}) + \bar{c}) \leq t(f(x) + c) \text{ for all } x \in S \text{ and all } c \in C_Y.$$

8.2. Duality Theorems for Abstract Optimization Problems

From this inequality we get $t(\bar{c}) = 0$ and

$$(t \circ f)(\bar{x}) \leq (t \circ f)(x) \text{ for all } x \in S.$$

By Lemma 2.7, (b) and Example 5.2, (a) the functional $t \circ f$ is convex. Hence, $\bar{x}$ is a solution of the convex optimization problem (8.9) which is assumed to be stable. Then there is a continuous linear functional $\bar{u} \in C_{Z^*}$ with

$$\inf_{x \in S} (t \circ f)(x) = \inf_{x \in \hat{S}} (t \circ f + \bar{u} \circ g)(x)$$

and

$$(t \circ f + \bar{u} \circ g)(x) \geq t(f(\bar{x})) \text{ for all } x \in \hat{S}.$$

Consequently, $\bar{y}$ belongs to the set $P_1 \cap D_1$ and an application of Lemma 8.5, (a), (ii) leads to the assertion. □

If the abstract optimization problem (8.3) satisfies the generalized Slater condition, i.e. there is an $x \in \hat{S}$ with $g(x) \in -\text{int}(C_Z)$, then the stability assumption of the previous theorem is satisfied (for a normed setting see, for instance, Krabs [184, p. 112–113]).

For the next duality result we need a technical lemma.

Lemma 8.8. *Let the assumption (8.2) be satisfied, and consider the problems (8.4) - (8.7). In addition, let Y be locally convex, and let the set P_1 be closed.*

(a) If the scalar optimization problem

$$\inf_{x \in S} (t \circ f)(x) \tag{8.10}$$

is normal for all $t \in C_{Y^*} \setminus \{0_{Y^*}\}$, then the complement set of P_1 is a subset of $\text{cor}(D_1)$.

(b) Let the sets $C_{Y^*}^{\#}$ and D_2 be nonempty. If the scalar optimization problem (8.10) is normal for all $t \in C_{Y^*}^{\#}$, then the complement set of $P_2 + C_Y \ (= P_1)$ is a subset of $\text{cor}(D_2)$.

Proof.

(a) Choose an arbitrary element $\bar{y} \in Y \backslash P_1$. Since the real linear space Y is locally convex and the set P_1 is convex and closed, by Theorem 3.18 there are a continuous linear functional $t \in Y^* \backslash \{0_{Y^*}\}$ and a real number α with

$$t(\bar{y}) < \alpha \leq t(y) \text{ for all } y \in P_1.$$

Obviously we have $t \in C_{Y^*} \backslash \{0_{Y^*}\}$. Moreover, we get

$$t(\bar{y}) < \inf_{y \in P_1} t(y) = \inf_{x \in S} (t \circ f)(x). \tag{8.11}$$

By assumption the scalar optimization problem

$$\inf_{x \in S} (t \circ f)(x)$$

is normal. Therefore, we conclude with (8.11) for some $u \in C_{Z^*}$

$$\inf_{x \in \hat{S}} (t \circ f + u \circ g)(x) > t(\bar{y}).$$

But this implies $\bar{y} \in \text{cor}(D_1)$.

(b) Fix any $\bar{y} \in Y \backslash (P_2 + C_Y)$. Again, by a separation theorem (Theorem 3.18) there are a continuous linear functional $t \in C_{Y^*} \backslash \{0_{Y^*}\}$ and a real number α with

$$t(\bar{y}) < \alpha \leq t(y) \text{ for all } y \in P_2 + C_Y.$$

Since the set D_2 is not empty, there is a $\tilde{y} \in D_2$ and with Theorem 8.4, (b) there is a continuous linear functional $\tilde{t} \in C_{Y^*}^{\#}$ with

$$\tilde{t}(\tilde{y}) \leq \tilde{t}(y) \text{ for all } y \in P_2.$$

Next, we define for every $\lambda \in (0, 1]$ a continuous linear functional

$$t_\lambda := \lambda \tilde{t} + (1 - \lambda) t$$

which belongs to $C_{Y^*}^{\#}$. Then we obtain with $\varepsilon := \alpha - t(\bar{y}) > 0$

$$\begin{aligned} t_\lambda(\bar{y}) &= t(\bar{y}) + \lambda(\tilde{t}(\bar{y}) - t(\bar{y})) \\ &= \alpha - \varepsilon + \lambda(\tilde{t}(\bar{y}) - \alpha + \varepsilon) \text{ for all } \lambda \in (0, 1] \end{aligned}$$

8.2. Duality Theorems for Abstract Optimization Problems

and

$$t_\lambda(y) \geq \alpha + \lambda(\tilde{t}(\tilde{y}) - \alpha) \text{ for all } \lambda \in (0,1] \text{ and all } y \in P_2 + C_Y.$$

For a sufficiently small $\bar{\lambda}$ we conclude

$$t_{\bar{\lambda}}(\bar{y}) < \alpha - \frac{\varepsilon}{2} \leq t_{\bar{\lambda}}(y) \text{ for all } y \in P_2 + C_Y$$

which implies

$$t_{\bar{\lambda}}(\bar{y}) < \inf_{y \in P_2 + C_Y} t_{\bar{\lambda}}(y) = \inf_{x \in S} (t_{\bar{\lambda}} \circ f)(x).$$

Because of the normality assumption we obtain for some $u \in C_{Z^*}$ with this inequality

$$\inf_{x \in \hat{S}} (t_{\bar{\lambda}} \circ f + u \circ g)(x) > t_{\bar{\lambda}}(\bar{y}).$$

Hence, $\bar{y}$ belongs to the algebraic interior of the set D_2.

$\square$

Again, if the abstract optimization problem (8.3) satisfies the generalized Slater condition, then the normality assumption in Lemma 8.8 is satisfied (for a normed setting see also Krabs [184, p. 103]).

Now we present a *strong converse duality theorem*.

Theorem 8.9. *Let the assumption (8.2) be satisfied, and consider the problems (8.4) - (8.7). In addition, let Y be locally convex, and let the set P_1 be closed.*

(a) *If the sets $\text{int}(C_Y)$ and D_1 are nonempty and if the scalar optimization problem (8.10) is normal for all $t \in C_{Y^*} \backslash \{0_{Y^*}\}$, then every weakly minimal element of the set D_1 is also a weakly minimal element of the set P_1.*

(b) *If the sets $C_{Y^*}^\#$ and D_2 are nonempty and if the scalar optimization problem (8.10) is normal for all $t \in C_{Y^*}^\#$, then every maximal element of the set D_2 is also an almost properly minimal element of the set P_2.*

Proof.

(a) Let $\bar{y}$ be any weakly maximal element of the set D_1. It is evident that $\bar{y} \notin \text{cor}(D_1)$ and, therefore, by Lemma 8.8, (a) $\bar{y} \in P_1$. Since $\bar{y} \in P_1 \cap D_1$, by Lemma 8.5, (a), (ii) $\bar{y}$ is also a weakly minimal element of the set P_1.

(b) Choose any maximal element $\bar{y}$ of the set D_2. Then we get $\bar{y} \notin \text{cor}(D_2)$ and with Lemma 8.8, (b) we conclude $\bar{y} \in P_2 + C_Y$. With Theorem 8.4, (b) we obtain even $\bar{y} \in P_2$, and an application of Lemma 8.5, (b), (ii) leads to the assertion.

$\square$

Summarizing the results of this section we have under appropriate assumptions that an element $\bar{y}$ is a weakly minimal element of the set P_1 if and only if $\bar{y}$ is a weakly maximal element of the set D_1. Every weakly minimal element of the set $f(S)$ is also a weakly minimal element of the set $P_1 = f(S) + C_Y$, but conversely, not every weakly minimal element of the set $f(S) + C_Y$ is a weakly minimal element of the set $f(S)$. Consequently, this duality theory is not completely applicable to the original abstract optimization problem (8.3). For the other duality theory which is related to the original problem (8.3) we have under suitable assumptions that an element $\bar{y}$ is an almost minimal element of the set P_2 if and only if $\bar{y}$ is a maximal element of the set D_2. The disadvantage of this theory is that it is not possible to get a corresponding result for minimal elements of the primal set P_2.

8.3 Specialization to Abstract Linear Optimization Problems

In this section we investigate special abstract optimization problems namely linear problems. It is the aim to transform the dual sets D_1 and D_2 (defined in (8.6) and (8.7)) in such a way that the relationship to the well-known dual problem in linear programming becomes more transparent.

8.3. Specialization to Abstract Linear Optimization Problems

The standard assumption which is needed now reads as follows:

$$\left.\begin{array}{l}\text{Let } X, Y \text{ and } Z \text{ be partially ordered separated}\\ \text{locally convex topological linear spaces with}\\ \text{ordering cones } C_X, C_Y \text{ and } C_Z, \text{ respectively;}\\ \text{let } C_Y \neq Y \text{ be nontrivial;}\\ \text{let } C: X \to Y \text{ and } A: X \to Z \text{ be continuous linear}\\ \text{maps;}\\ \text{let } b \in Z \text{ be a fixed vector;}\\ \text{let the constraint set } S := \{x \in C_X \mid A(x) - b \in C_Z\}\\ \text{be nonempty.}\end{array}\right\} \quad (8.12)$$

Under this assumption we consider the two primal problems (8.4) and (8.5) and formalize them as

$$\left.\begin{array}{l}\text{w-min } C(x) + y\\ \text{subject to the constraints}\\ A(x) - b \in C_Z\\ x \in C_X\\ y \in C_Y\end{array}\right\} \quad (8.13)$$

and

$$\left.\begin{array}{l}\text{a-p-min } C(x)\\ \text{subject to the constraints}\\ A(x) - b \in C_Z\\ x \in C_X,\end{array}\right\} \quad (8.14)$$

respectively. In this special case the two dual sets D_1 and D_2 of the problems (8.6) and (8.7) read

$$D_1 = \{y \in Y \mid \text{there are continuous linear functionals}$$
$$t \in C_{Y^*} \setminus \{0_{Y^*}\} \text{ and } u \in C_{Z^*} \text{ with}$$
$$t(C(x)) + u(-A(x) + b) \geq t(y) \text{ for all } x \in C_X\} \quad (8.15)$$

and

$$D_2 = \{y \in Y \mid \text{there are continuous linear functionals}$$
$$t \in C_{Y^*}^{\#} \text{ and } u \in C_{Z^*} \text{ with}$$
$$t(C(x)) + u(-A(x) + b) \geq t(y) \text{ for all } x \in C_X\}. \quad (8.16)$$

The next lemma gives a standard re-expression of the sets (8.15) and (8.16) without proof.

Lemma 8.10. *Let the assumption (8.12) be satisfied, and consider the sets in (8.15) and (8.16). Then:*

(a)
$$D_1 = \{y \in Y \mid \text{there are continuous linear functionals} \\ t \in C_{Y^*}\backslash\{0_{Y^*}\} \text{ and } u \in C_{Z^*} \text{ with} \\ C^*(t) - A^*(u) \in C_{X^*} \text{ and } t(y) \leq u(b)\}$$

(where C^* and A^* denote the adjoint maps of C and A, respectively).

(b)
$$D_2 = \{y \in Y \mid \text{there are continuous linear functionals} \\ t \in C_{Y^*}^{\#} \text{ and } u \in C_{Z^*} \text{ with} \\ C^*(t) - A^*(u) \in C_{X^*} \text{ and } t(y) \leq u(b)\}.$$

Another result which is simple to proof is given in

Lemma 8.11. *Let the assumption (8.12) be satisfied, and consider the sets in (8.15) and (8.16).*

(a) If $int(C_Y) \neq \emptyset$ and if $\bar{y}$ is a weakly maximal element of the set D_1 where $t \in C_{Y^*}\backslash\{0_{Y^*}\}$ and $u \in C_{Z^*}$ are given by definition, then it follows $t(\bar{y}) = u(b)$.

(b) If $C_{Y^*}^{\#} \neq \emptyset$ and if $\bar{y}$ is a maximal element of the set D_2 where $t \in C_{Y^*}^{\#}$ and $u \in C_{Z^*}$ are given by definition, then it follows $t(\bar{y}) = u(b)$.

Before we are able to prove the main result of this section we need an additional lemma. For simplicity we define the sets

$$\tilde{D}_1 = \{y \in Y \mid \text{there are a continuous linear functional} \\ t \in C_{Y^*}\backslash\{0_{Y^*}\} \text{ and a continuous linear}$$

8.3. Specialization to Abstract Linear Optimization Problems

$$\text{map } T : Z \to Y \text{ with } y = T(b),$$
$$T^*(t) \in C_{Z^*} \text{ and } (C - TA)^*(t) \in C_{X^*}\}$$
(8.17)

and

$$\tilde{D}_2 = \{y \in Y \mid \text{there are a continuous linear functional}$$
$$t \in C_{Y^*}^{\#} \text{ and a continuous linear map}$$
$$T : Z \to Y \text{ with } y = T(b), \ T^*(t) \in C_{Z^*} \text{ and}$$
$$(C - TA)^*(t) \in C_{X^*}\}.$$
(8.18)

Lemma 8.12. *Let the assumption (8.12) be satisfied, and consider the sets in (8.15) - (8.18). Then we have $\check{D}_1 \subset D_1$ and $\tilde{D}_2 \subset D_2$.*

Proof. We restrict ourselves to the proof of the inclusion $\tilde{D}_1 \subset D_1$. The case $\tilde{D}_1 = \emptyset$ is trivial. If $\tilde{D}_1$ is nonempty, choose any element $y \in \tilde{D}_1$. Then there are a continuous linear functional $t \in C_{Y^*}\setminus\{0_{Y^*}\}$ and a continuous linear map $T : Z \to Y$ with $y = T(b)$, $T^*(t) \in C_{Z^*}$ and $(C-TA)^*(t) \in C_{X^*}$. With Theorem 2.3, (a) we get for $u := T^*(t)$ the equality $t(y) = u(b)$. Since

$$(C - TA)^*(t) = C^*(t) - A^*(T^*(t))$$
$$= C^*(t) - A^*(u),$$

by Lemma 8.10, (a) we conclude $y \in D_1$. Hence, the inclusion $\tilde{D}_1 \subset D_1$ is true. □

Using the previous lemmas and Theorem 2.3 we obtain

Theorem 8.13. *Let the assumption (8.12) be satisfied, and consider the sets in (8.15) - (8.18).*

(a) *Assume that $int(C_Y)$ is nonempty.*

  (i) *Every weakly maximal element of the set $\tilde{D}_1$ is also a weakly maximal element of the set D_1.*

(ii) If $b \neq 0_Z$, then every weakly maximal element of the set D_1 is also a weakly maximal element of the set $\tilde{D}_1$.

(b) Assume that the set $C_{Y^*}^{\#}$ is nonempty.

(i) Every maximal element of the set $\tilde{D}_2$ is also a maximal element of the set D_2.

(ii) If $b \neq 0_Z$, then every maximal element of the set D_2 is also a maximal element of the set $\tilde{D}_2$.

Proof. For simplicity we prove only part (a) of the assertion. The proof of the other part is analogous.

(a) (i) First, we assume that $b \neq 0_Z$. The case $b = 0_Z$ will be treated later. Let $\bar{y}$ be any weakly maximal element of the set $\tilde{D}_1$ with a continuous linear functional $t \in C_{Y^*} \setminus \{0_{Y^*}\}$ and a continuous linear map $T : Z \to Y$ given by definition. Then we get with Lemma 8.12 that $\bar{y} = T(b) \in D_1$.

Assume that $\bar{y}$ is no weakly maximal element of the set D_1. Then there is some $\tilde{y} \in (\{\bar{y}\} + \text{cor}(C_Y)) \cap D_1$ with continuous linear functionals $\tilde{t} \in C_{Y^*} \setminus \{0_{Y^*}\}$ and $\tilde{u} \in C_{Z^*}$ given by definition. Without loss of generality the equality $\tilde{t}(\tilde{y}) = \tilde{u}(b)$ can be assumed (otherwise choose an appropriate $y \in \text{cor}(C_Y)$ with $\tilde{t}(\tilde{y} + y) = \tilde{u}(b)$). By Theorem 2.3, (b) there is a continuous linear map $\tilde{T} : Z \to Y$ with $\tilde{y} = \tilde{T}(b)$ and $\tilde{T}^*(\tilde{t}) = \tilde{u}$. Because of

$$C^*(\tilde{t}) - A^*(\tilde{u}) \in C_{X^*}$$

we get

$$(C - \tilde{T}A)^*(\tilde{t}) \in C_{X^*}$$

Then we obtain $\tilde{y} \in (\{\bar{y}\} + \text{cor}(C_Y)) \cap \tilde{D}_1$ which contradicts the assumption that $\bar{y}$ is a weakly maximal element of the set $\tilde{D}_1$. Hence, $\bar{y}$ is a weakly maximal element of the set D_1.

Finally, we assume that $b = 0_Z$. In this case we have $\tilde{D}_1 = \{0_Y\}$. By Lemma 8.12 we get $0_Y \in D_1$. If we

8.3. Specialization to Abstract Linear Optimization Problems

assume that 0_Y is not a weakly maximal element of the set D_1, then there is a $\tilde{y} \in \text{cor}(C_Y) \cap D_1$ with continuous linear functionals $\tilde{t} \in C_{Y^*} \setminus \{0_{Y^*}\}$ and $\tilde{u} \in C_{Z^*}$ given by definition. But then it follows $\tilde{t}(\tilde{y}) > 0$ which contradicts the inequality
$$\tilde{t}(\tilde{y}) \leq \tilde{u}(b) = 0.$$
Consequently, the zero element 0_Y is a weakly maximal element of the set D_1.

(ii) Let $\bar{y}$ be an arbitrary weakly maximal element of the set D_1. By Lemma 8.11, (a) it follows that $t(\bar{y}) = u(b)$ where the continuous linear functionals $t \in C_{Y^*} \setminus \{0_{Y^*}\}$ and $u \in C_{Z^*}$ are given by definition. With the same arguments as in part (a), (i) we obtain $\bar{y} \in \tilde{D}_1$ with $\bar{y}$ instead of $\tilde{y}$, t instead of $\tilde{t}$ and u instead of $\tilde{u}$. By Lemma 8.12 we conclude immediately that $\bar{y}$ is a weakly maximal element of the set $\tilde{D}_1$.

$\square$

It is the essential result of the previous theorem that under the assumption $b \neq 0_Z$ (which is needed in only one direction) the dual problems (8.6) and (8.7) with D_1 and D_2 given by (8.15) and (8.16), respectively, are equivalent to abstract optimization problems formalized as

$$\left.\begin{aligned}
&\text{w-max } T(b) \\
&\text{subject to the constraints} \\
&(C - TA)^*(t) \in C_{X^*} \\
&T^*(t) \in C_{Z^*} \\
&t \in C_{Y^*} \setminus \{0_{Y^*}\} \\
&T \in L(Z, Y)
\end{aligned}\right\} \quad (8.19)$$

and

$$\left.\begin{aligned}
&\max T(b) \\
&\text{subject to the constraints} \\
&(C - TA)^*(t) \in C_{X^*} \\
&T^*(t) \in C_{Z^*} \\
&t \in C_{Y^*}^{\#} \\
&T \in L(Z, Y)
\end{aligned}\right\} \quad (8.20)$$

where $L(Z,Y)$ denotes the linear space of continuous linear maps from Z to Y. Hence, the problem (8.19) is a possible dual problem to the primal problem (8.13) and (8.20) is a dual problem to the primal problem (8.14).

It is known from linear programming that the assumption $b \neq 0_Z$ is not needed. In fact, in the case of $Y = \mathbb{R}$ Theorem 8.13 can be proved without the assumption $b \neq 0_Z$. For $Y = \mathbb{R}$ we have namely $C_{Y^*}\setminus\{0_{Y^*}\} = C_{Y^*}^\# = \mathbb{R}_+\setminus\{0\}$ and, therefore, t is a positive real number. Consequently, the equation $t(\bar{y}) = u(b)$ leads to $\bar{y} = \frac{1}{t}u(b)$. Hence, Theorem 2.3 is not used for the proof of Theorem 8.13. But for abstract optimization problems the assumption $b \neq 0_Z$ is of importance. In the case of $b = 0_Z$ we have $\tilde{D}_1 = \tilde{D}_2 = \{0_Y\}$. Hence, 0_Y is the only weakly maximal element (and maximal element) of the set $\tilde{D}_1$ (and $\tilde{D}_2$, respectively). But one can present simple examples which show that the sets P_1 and P_2 have nonzero weakly minimal elements and nonzero almost properly minimal elements, respectively (e.g., see also Brumelle [48] and Gerstewitz-Göpfert-Lampe [104]). For instance, in the case of $Y := \mathbb{R}^2$ and $C_Y := \mathbb{R}_+^2$ the vector $(-1,1)$ is a weakly minimal element of the set $P_1 = f(S) + C_Y$ (and an almost properly minimal element of the set $P_2 = f(S)$) where

$$f(S) := \{(y_1, y_2) \in \mathbb{R}^2 \mid y_1 + y_2 \geq 0\}.$$

The vector $(-1,1)$ is also a weakly maximal element of the set D_1 (and a maximal element of the set D_2). But on the other hand we have $\tilde{D}_1 = \tilde{D}_2 = \{0_{\mathbb{R}^2}\}$.

It is simple to see that the two dual abstract optimization problems (8.19) and (8.20) generalize the known dual problem of linear programming. We remarked before that in the case of $Y = \mathbb{R}$ t is a positive real number. Therefore, after some elementary transformations, the dual problems (8.19) and (8.20) reduce to the scalar optimization problem

$$\max \; T(b)$$
subject to the constraints
$$C - A^*(T) \in C_{X^*}$$
$$T \in C_{Z^*}.$$

where now T and C are real-valued maps.

Notes

The duality approach of Section 8.2 is based on the duality theory of Schönfeld [279] which is generalized using the duality theory of Van Slyke-Wets [308] in the extended form of Krabs [184]. The duality theory for the almost proper minimality notion can also be found in a paper of Jahn [138]. The first duality results were obtained by Gale-Kuhn-Tucker [99] who investigated problems with a matrix-valued objective map. For abstract optimization problems in infinite-dimensional spaces there are only a few papers presenting such an approach. Breckner [44], Zowe [343], [344], Rosinger [267] and Gerstewitz-Göpfert-Lampe [104] generalized the Fenchel formalism to abstract optimization problems. Lehmann-Oettli [198] and Oettli [241] use the weak minimality notion for their investigations. Rosinger [266] examines dual problems in partially ordered sets. Nieuwenhuis [237] extends the duality theory of Van Slyke-Wets [308]. A comparison of the normality concept used in this book and the concept of Nieuwenhuis can be found in a paper of Borwein-Nieuwenhuis [39]. Lampe [196] carries out duality investigations using perturbation theory, and Corley [64] develops a saddle point theoretical approach. In the case of $Y = \mathbb{R}^n$ several nonlinear duality results are formulated by Schönfeld [279], di Guglielmo [79], Gros [112], Tanino-Sawaragi [298], Craven [68], Tanino-Sawaragi [299], Bitran [26], Brumelle [48], Nehse [234], Kawasaki [169], Nakayama [230], Tanino [297], and others. An overview on several duality concepts is given by Nakayama [231]. Theorem 8.7, (b) may also be found in a similar form in a paper of Borwein [29, p. 61, Thm. 3]. The results of Section 8.2 concerning the weak minimality and weak maximality notion are essentially included in a paper of Oettli [241].

Gale-Kuhn-Tucker [99] were also the first who investigated the duality between abstract linear optimization problems. The dual problem (8.7) with D_2 as in Lemma 8.10, (b) is the generalized dual problem of Gale-Kuhn-Tucker [99]. The dual problem (8.19) generalizes the dual problem of Isermann [133] formulated in a finite dimensional setting. In this special case Isermann [134] and Gerstewitz-Göpfert-Lampe [104] investigated the relationship between the sets D_2 and $\tilde{D}_2$ as well.

Part III
Mathematical Applications

The theory of vector optimization developed in the previous part of this book has many applications - not only in the applied sciences like engineering and economics but also in mathematical areas like approximation and games. As pointed out in Example 4.5 and Example 4.6 vector approximation problems and cooperative n person games are special abstract optimization problems. Therefore, many theorems from vector optimization can be applied to these special problems. In chapter 9 we discuss several results for vector approximation problems where we focus our attention mainly on necessary and sufficient optimality conditions. Cooperative n person differential games are the topic of chapter 10. The main part of this chapter is devoted to the study of a maximum principle for these games.

Chapter 9

Vector Approximation

Vector approximation problems are abstract approximation problems where a vectorial norm is used instead of a usual (real-valued) norm. Many important results known from approximation theory can be extended to this vector-valued case. After a short introduction we examine the relationship between vector approximation and simultaneous approximation, and we present the so-called generalized Kolmogorov condition. Moreover, we consider nonlinear and linear Chebyshev vector approximation problems and we formulate a generalized alternation theorem for these problems.

9.1 Introduction

In Example 4.5 we have already considered a vector approximation problem in a general form. For instance, if one wants to approximate not only a given function but also its derivative or its integral, then such a problem is a vector approximation problem. In the following we discuss a further example. We examine the free boundary Stefan problem (discussed by Reemtsen [255]):

$$u_{xx}(x,t) - u_t(x,t) = 0, \quad (x,t) \in D(s), \tag{9.1}$$
$$u_x(0,t) = g(t), \quad 0 < t \leq T, \tag{9.2}$$
$$u(s(t),t) = 0, \quad 0 < t \leq T, \tag{9.3}$$
$$u_x(s(t),t) = -\dot{s}(t), \quad 0 < t \leq T, \tag{9.4}$$
$$s(0) = 0 \tag{9.5}$$

where $g \in C([0,T])$ is a non-positive function with $g(0) < 0$ and
$$D(s) := \{(x,t) \in \mathbb{R}^2 \mid 0 < x < s(t),\ 0 < t \leq T\} \text{ for } s \in C([0,T]).$$
For the approximative solution of this problem one chooses the function
$$\bar{u}(x,t,a) = \sum_{i=0}^{l} a_i v_i(x,t)$$
with
$$v_i(x,t) = \sum_{k=0}^{[\frac{i}{2}]} \frac{i!}{(i-2k)!k!} x^{i-2k} t^k$$
($[\frac{i}{2}]$ denotes the largest integer number less than or equal to $\frac{i}{2}$) and
$$\bar{s}(t,b) = -g(0)t + \sum_{i=1}^{p} b_i t^{i+1}$$
(compare Reemtsen [255, p. 31–32]). For every $a \in \mathbb{R}^{l+1}$ $\bar{u}$ satisfies the partial differential equation (9.1) and for every $b \in \mathbb{R}^p$ $\bar{s}$ satisfies the equation (9.5). If we plug $\bar{u}$ and $\bar{s}$ in the equations (9.2), (9.3) and (9.4), then we obtain the error functions $\rho_1, \rho_2, \rho_3 \in C([0,T])$ with
$$\rho_1(t,a,b) := \bar{u}_x(0,t,a) - g(t) = \sum_{\substack{i=1 \\ i \text{ odd}}}^{l} a_i \frac{i!}{((i-1)/2)!} t^{(i-1)/2} - g(t),$$
$$\rho_2(t,a,b) := \bar{u}(\bar{s}(t,b),t,a) = \sum_{i=0}^{l} a_i v_i(\bar{s}(t,b),t)$$
and
$$\rho_3(t,a,b) := \bar{u}_x(\bar{s}(t,b),t,a) + \dot{\bar{s}}(t,b) = \sum_{i=1}^{l} a_i v_{ix}(\bar{s}(t,b),t) + \dot{\bar{s}}(t,b).$$
If $\|\cdot\|$ is any norm on $C([0,T])$, we formulate the following vector approximation problem for the approximative solution of the Stefan problem: Determine minimal or weakly minimal elements of the set
$$\{(\|\rho_1(\cdot,a,b)\|, \|\rho_2(\cdot,a,b)\|, \|\rho_3(\cdot,a,b)\|) \mid (a,b) \in \mathbb{R}^{l+1} \times \mathbb{R}^p\}$$

where the real linear space $\mathbb{R}^3$ is assumed to be partially ordered in the natural way.

For a more general type of vector approximation problems considered in this chapter we have the standard assumption:

$$\left.\begin{array}{l} \text{Let } S \text{ be a nonempty subset of a real linear space } X; \\ \text{let } Y \text{ be a partially ordered linear space with an} \\ \text{ordering cone } C_Y; \\ \text{let } \|\cdot\| : X \to Y \text{ be a vectorial norm (see} \\ \text{Definition 1.35);} \\ \text{let } \hat{x} \in X \text{ be a given element.} \end{array}\right\} \quad (9.6)$$

Then we consider the vector approximation problem formalized as

$$\min_{x \in S} \|x - \hat{x}\| \quad (9.7)$$

which means that we are looking for inverse images of minimal (or weakly minimal) elements of the set

$$V := \{\|x - \hat{x}\| \mid x \in S\}. \quad (9.8)$$

9.2 Simultaneous Approximation

In approximation theory a multiobjective approximation problem is often treated as a so-called *simultaneous approximation problem*:

$$\min_{x \in S} \| \|x - \hat{x}\| \|_Y \quad (9.9)$$

where we assume that the assumption (9.6) is satisfied and, in addition, $\|\cdot\|_Y$ is a (usual) norm on Y. The problem (9.9) is a scalar optimization problem. In the following we investigate the question: Are there any relationships between the solutions of the scalar optimization problem (9.9) and the inverse images of minimal (or weakly minimal) elements of the set V given in (9.8)? The answer of this question can be given immediately using certain scalarization results.

Theorem 9.1. *Let the assumption (9.6) be satisfied and, in addition, let the ordering cone C_Y be pointed and algebraically closed, and* let it have a nonempty algebraic interior. Moreover, let the set V be given as in (9.8) and assume that

$$\|x - \hat{x}\| \in \text{cor}(C_Y) \text{ for all } x \in S. \tag{9.10}$$

An element $\bar{y} \in V$ (with an inverse image $\bar{x} \in S$) is a minimal element of the set V if and only if there is a norm $\|\cdot\|_Y$ on Y which is monotonically increasing on C_Y with the property

$$\|\,\|\bar{x} - \hat{x}\|\,\|_Y < \|\,\|x - \hat{x}\|\,\|_Y \text{ for all } x \in S \text{ with } \bar{y} \neq \|x - \hat{x}\|.$$

Proof. This theorem follows immediately from Corollary 5.16, if we notice that the condition (9.10) is equivalent to the inclusion $V \subset \{0_Y\} + \text{cor}(C_Y)$. □

Theorem 9.2. *Let the assumption (9.6) be satisfied and, in addition, let the ordering cone C_Y have a nonempty algebraic interior. Moreover, let the set V be given as in (9.8) and assume that the condition (9.10) is fulfilled. An element $\bar{y} \in V$ (with an inverse image $\bar{x} \in S$) is a weakly minimal element of the set V if and only if there is a seminorm $\|\cdot\|_Y$ on Y which is strictly monotonically increasing on $\text{cor}(C_Y)$ with the property*

$$\|\,\|\bar{x} - \hat{x}\|\,\|_Y \leq \|\,\|x - \hat{x}\|\,\|_Y \text{ for all } x \in S.$$

Proof. Notice that the inclusion $V \subset \{0_Y\} + \text{cor}(C_Y)$ is satisfied and apply Corollary 5.26. □

The last two theorems show the strong connection between vector approximation and simultaneous approximation. After all, certain simultaneous approximation problems are scalarized vector approximation problems.

It is also possible to scalarize the vector approximation problem (9.7) by using linear functionals instead of norms.

Theorem 9.3. *Let the assumption (9.6) be satisfied, and let C_Y be pointed.*

9.2. Simultaneous Approximation

(a) If there are a linear functional $l \in C_{Y'}^{\#}$ and an element $\bar{x} \in S$ with the property
$$l(\|\bar{x} - \hat{x}\|) \leq l(\|x - \hat{x}\|) \text{ for all } x \in S,$$
then $\|\bar{x} - \hat{x}\|$ is a minimal element of the set V (given by (9.8)).

(b) In addition, let the set S be convex, and let the ordering cone C_Y be nontrivial. If the set $V + C_Y$ has a nonempty algebraic interior and $\bar{y}$ is a minimal element of the set V (with an inverse image $\bar{x} \in S$), then there is a linear functional $l \in C_{Y'} \setminus \{0_{Y'}\}$ with the property
$$l(\|\bar{x} - \hat{x}\|) \leq l(\|x - \hat{x}\|) \text{ for all } x \in S.$$

Proof. Part (a) of this theorem follows from Theorem 5.18, (b) and part (b) is a consequence of Theorem 5.4 and the fact that the set $V + C_Y$ is convex. □

Theorem 9.4. *Let the assumption (9.6) be satisfied, and let the ordering cone C_Y have a nonempty algebraic interior.*

(a) If there are a linear functional $l \in C_{Y'} \setminus \{0_{Y'}\}$ and an element $\bar{x} \in S$ with the property
$$l(\|\bar{x} - \hat{x}\|) \leq l(\|x - \hat{x}\|) \text{ for all } x \in S,$$
then $\|\bar{x} - \hat{x}\|$ is a weakly minimal element of the set V (given by (9.8)).

(b) In addition, let the set S be convex. If $\bar{y}$ is a weakly minimal element of the set V (with an inverse image $\bar{x} \in S$), then there is a linear functional $l \in C_{Y'} \setminus \{0_{Y'}\}$ with the property
$$l(\|\bar{x} - \hat{x}\|) \leq l(\|x - \hat{x}\|) \text{ for all } x \in S.$$

9.3 Generalized Kolmogorov Condition

In this section we discuss a generalization of the so-called Kolmogorov condition for the vector approximation problem (9.7). The Kolmogorov condition is a well-known optimality condition in approximation theory. As in the previous section we investigate the vector approximation problem (9.7), but now, for simplicity, we turn our attention only to the weak minimality notion. Results for the minimality and strong minimality notion are given by Oettli [243].

The following theorem presents the *generalized Kolmogorov condition* for weakly minimal elements of the set V.

Theorem 9.5. *Let the assumption (9.6) be satisfied. Let Y be order complete (i.e., every nonempty subset of Y which is bounded from below has an infimum) and pseudo-Daniell (i.e., every curve $\{y(\lambda) \in Y \mid \lambda \in (0,1]\}$ which decreases as $\lambda \downarrow 0$ and is bounded from below, if $\inf_{\lambda \in (0,1]} y(\lambda) \in -cor(C_Y)$, has the property: $y(\lambda) \in cor(C_Y)$ for sufficiently small $\lambda > 0$). Let C_Y have a nonempty abgebraic interior, let S be convex, and let $\bar{x} \in S$ be given. Then $\|\bar{x} - \hat{x}\|$ is a weakly minimal element of the set V (given by (9.8)) if and only if for every $x \in S$ there is a linear map $T_x : X \to Y$ so that*

$$T_x(x - \bar{x}) \notin -cor(C_Y), \tag{9.11}$$

$$T_x(\tilde{x}) \leq_{C_Y} \|\tilde{x}\| \text{ for all } \tilde{x} \in X \tag{9.12}$$

and

$$T_x(\bar{x} - \hat{x}) = \|\bar{x} - \hat{x}\|. \tag{9.13}$$

Proof.

(a) For every $x \in S$ let a linear map T_x be given satisfying the conditions (9.11), (9.12) and (9.13). Assume that $\bar{x}$ is no weakly minimal element of V. Then there is an $x \in S$ with

$$\|x - \hat{x}\| - \|\bar{x} - \hat{x}\| \in -\text{cor}(C_Y).$$

9.3. Generalized Kolmogorov Condition

With (9.12) for $\tilde{x} := x - \hat{x}$ and (9.13) we obtain

$$\begin{aligned} T_x(x - \bar{x}) &= T_x(x - \hat{x}) - T_x(\bar{x} - \hat{x}) \\ &\leq_{C_Y} \|x - \hat{x}\| - \|\bar{x} - \hat{x}\| \\ &\in -\text{cor}(C_Y) \end{aligned}$$

which contradicts the condition (9.11).

(b) Next, let $\bar{x} \in S$ be a weakly minimal element of V. Fix an arbitrary $x \in S$ and define $\bar{h} := x - \bar{x}$. In analogy to the proof of Lemma 2.24 one can show that the directional derivative $\|\bar{x} - \hat{x}\|'(\bar{h})$ of the vectorial norm exists (here we use a definition of the directional derivative given by $\inf_{\lambda \in (0,1]} \frac{1}{\lambda}(\|\bar{x}-\hat{x}+\lambda\bar{h}\|-\|\bar{x}-\hat{x}\|)$). Moreover, the map $\|\bar{x}-\hat{x}\|'(\cdot)$ is sublinear (see Zowe [344, Thm. 3.1.4]). By the vectorial version of the Hahn-Banach extension theorem given by Zowe [344, Thm. 2.1.1] there is a linear map $T_x : X \to Y$ with

$$T_x(h) \leq_{C_Y} \|\bar{x} - \hat{x}\|'(h) \text{ for all } h \in X \tag{9.14}$$

and

$$T_x(\bar{h}) = \|\bar{x} - \hat{x}\|'(\bar{h}). \tag{9.15}$$

The inequality (9.14) implies

$$T_x(h) \leq_{C_Y} \|\bar{x} - \hat{x} + h\| - \|\bar{x} - \hat{x}\| \text{ for all } h \in X. \tag{9.16}$$

For $h = \bar{x} - \hat{x}$ we get $T_x(\bar{x} - \hat{x}) \leq_{C_Y} \|\bar{x} - \hat{x}\|$ and for $h = -(\bar{x} - \hat{x})$ we obtain $T_x(-(\bar{x} - \hat{x})) \leq_{C_Y} -\|\bar{x} - \hat{x}\|$ implying $\|\bar{x} - \hat{x}\| \leq_{C_Y} T_x(\bar{x} - \hat{x})$. Since C_Y is assumed to be pointed, the equation (9.13) is shown. For the proof of the inequality (9.12) we conclude with (9.16)

$$T_x(\tilde{x}) \leq_{C_Y} \|\bar{x} - \hat{x} + \tilde{x}\| - \|\bar{x} - \hat{x}\| \leq_{C_Y} \|\tilde{x}\| \text{ for all } \tilde{x} \in X.$$

Finally, assume that the condition (9.11) does not hold, i.e., we have

$$T_x(\bar{h}) \in -\text{cor}(C_Y).$$

We then get with the equation (9.15) that $\|\bar{x}-\hat{x}\|'(\bar{h}) \in -\mathrm{cor}(C_Y)$ which implies, because Y is pseudo-Daniell,

$$\|\bar{x} - \hat{x} + \lambda\bar{h}\| - \|\bar{x} - \hat{x}\| \in -\mathrm{cor}(C_Y) \text{ for sufficiently small } \lambda > 0.$$

This is a contradiction to the weak minimality of $\|\bar{x} - \hat{x}\|$.

$\square$

Originally, the Kolmogorov condition was formulated in (scalar-valued) linear Chebyshev approximation. If $\|\cdot\|$ is a usual (scalar-valued) norm (i.e. $Y = \mathbb{R}$), then this condition reads

$$\max\left\{ x^*(x - \bar{x}) \mid x^*(\bar{x} - \hat{x}) = \|\bar{x} - \hat{x}\|_X \text{ and } \|x^*\|_{X^*} = 1 \right\} \geq 0$$

(e.g., see Jahn [149]). This inequality is equivalent to the conditions (9.11), (9.12) and (9.13) in this special case.

9.4 Nonlinear Chebyshev Vector Approximation

In this section we investigate the general vector approximation problem (9.7) in a special form. We assume that the vectorial norm is given componentwise as a Chebyshev norm. For this special type of problems we present an alternation theorem as a consequence of the generalized Lagrange multiplier rule.

Now we have the following standard assumption denoted by (9.17):

> Let S be a convex subset of $\mathbb{R}^n$ with a nonempty interior;
> let $\hat{S}$ be an open superset of S;
> let Ω be a compact Hausdorff space with at least $n + 2$ elements;
> let $C(\Omega)$ denote the real linear space of real-valued continuous functions on Ω equipped with the maximum norm $\|\cdot\|$ (see Example 1.49);
> let $f_1, \ldots, f_m : \hat{S} \to C(\Omega)$ be maps which are

9.4. Nonlinear Chebyshev Vector Approximation

Fréchet differentiable on $\hat{S}$;
let $z_1, \ldots, z_m \in C(\Omega)$ be given functions;
let the space $\mathbb{R}^m$ be partially ordered in the
natural way. (9.17)

The vectorial norm which is used implicitly is given as

$$\|(y_1, \ldots, y_m)\| := (\|y_1\|, \ldots, \|y_m\|) \text{ for all } y_1, \ldots, y_m \in C(\Omega).$$

Then the general vector approximation problem (9.7) reduces to

$$\min_{x \in S} \begin{pmatrix} \|f_1(x) - z_1\| \\ \vdots \\ \|f_m(x) - z_m\| \end{pmatrix} \quad (9.18)$$

and the set V is given as

$$V := \left\{ \begin{pmatrix} \|f_1(x) - z_1\| \\ \vdots \\ \|f_m(x) - z_m\| \end{pmatrix} \ \Big| \ x \in S \right\} \quad (9.19)$$

which is a subset of $\mathbb{R}^m$. For our investigations we consider the set $V + \mathbb{R}^m_+$ given as

$$V + \mathbb{R}^m_+ = \{y \in \mathbb{R}^m \mid \text{ there is an } x \in S \text{ so that }$$
$$\|f_k(x) - z_k\| \leq y_k \text{ for all } k \in \{1, \ldots, m\}\}.$$

This set is the image set of the objective map of the following abstract optimization problem formalized by

$$\left.\begin{array}{l} \min y \\ \text{subject to the constraints} \\ \left.\begin{array}{l} f_k(x)(t) - z_k(t) - y_k \leq 0 \\ -f_k(x)(t) + z_k(t) - y_k \leq 0 \end{array}\right\} \begin{array}{l} \text{for all } k \in \{1, \ldots, m\} \\ \text{and all } t \in \Omega \end{array} \\ x \in S, \ y \in \mathbb{R}^m. \end{array}\right\} \quad (9.20)$$

Roughly speaking, the problem (9.20) is obtained by introducing "slack variables" in the vector approximation problem (9.18). Such a transformation is advantageous because we can apply the generalized

Lagrange multiplier rule in order to get an optimality condition for the problem (9.18). With the following theorem we present, as a necessary optimality condition, an *alternation theorem for the nonlinear Chebyshev vector approximation problem* (9.18).

Theorem 9.6. *Let the assumption (9.17) be satisfied, and let the set V be given by (9.19). If $\bar{y}$ is a weakly minimal element of the set V (with an inverse image $\bar{x} \in S$) and if for every $k \in \{1, \ldots, m\}$ the Fréchet-derivative of f_k at $\bar{x}$ is given by*

$$f'_k(\bar{x})(x) = \sum_{i=1}^{n} x_i v_{ki} \text{ for all } x \in S \qquad (9.21)$$

with certain functions $v_{ki} \in C(\Omega)$, then there are non-negative numbers $\tau_1, \ldots, \tau_m$ where at least one τ_k is nonzero with the following property: For every $k \in \{1, \ldots, m\}$ with $\tau_k > 0$ there are p_k points $t_{k1}, \ldots, t_{kp_k} \in E_k(\bar{x})$ with

$$1 \leq p_k \leq \dim \text{ span } \{v_{k1}, \ldots, v_{kn}, e, f_k(\bar{x}) - z_k\} \leq n+2 \quad (e \equiv 1 \text{ on } \Omega),$$

$$E_k(\bar{x}) := \{t \in \Omega \mid |(f_k(\bar{x}) - z_k)(t)| = \|f_k(\bar{x}) - z_k\|\}$$

and there are real numbers $\lambda_{k1}, \ldots, \lambda_{kp_k}$ so that

$$\sum_{i=1}^{p_k} |\lambda_{ki}| = 1, \qquad (9.22)$$

$$\sum_{j=1}^{n}(x_j - \bar{x}_j) \sum_{\substack{k=1 \\ \tau_k > 0}}^{m} \tau_k \sum_{i=1}^{p_k} \lambda_{ki} v_{kj}(t_{ki}) \geq 0 \text{ for all } x \in S \qquad (9.23)$$

and

$$\lambda_{ki} \neq 0 \text{ for some } i \in \{1, \ldots, p_k\}$$
$$\implies (f_k(\bar{x}) - z_k)(t_{ki}) = \|f_k(x) - z_k\| \, sgn(\lambda_{ki}). \qquad (9.24)$$

Proof. Since $\bar{y}$ is assumed to be a weakly minimal element of the set V, by Lemma 4.13, (b) $\bar{y}$ is also a weakly minimal element

9.4. Nonlinear Chebyshev Vector Approximation

of the set $V + \mathbb{R}_+^m$. Hence, $(\bar{x}, \bar{y})$ is a weakly minimal solution of the transformed problem (9.20). Then, by Theorem 7.4 (notice that the regularity assumption is satisfied) there are non-negative numbers $\tau_1, \ldots, \tau_m$ where at least one τ_k is nonzero and certain continuous linear functionals $u_k, w_k \in C_{C(\Omega)^*}$, $k \in \{1, \ldots, m\}$, with

$$\tau_k = u_k(e) + w_k(e) \text{ for all } k \in \{1, \ldots, m\}, \tag{9.25}$$

$$\sum_{k=1}^m (u_k - w_k)(f_k'(\bar{x})(x - \bar{x})) \geq 0 \text{ for all } x \in S, \tag{9.26}$$

and

$$\left.\begin{array}{l} u_k(f_k(\bar{x}) - z_k - \bar{y}_k e) = 0 \\ w_k(-f_k(\bar{x}) + z_k - \bar{y}_k e) = 0 \end{array}\right\} \text{ for all } k \in \{1, \ldots, m\}. \tag{9.27}$$

It is clear that $C_{C(\Omega)^*}$ denotes the dual cone of the natural ordering cone in $C(\Omega)$. If $\tau_k = 0$ for some $k \in \{1, \ldots, m\}$, then it follows $u_k = w_k = 0_{C(\Omega)^*}$ and nothing needs to be shown. Otherwise define $\bar{u}_k = \frac{1}{\tau_k} u_k$, $\bar{w}_k = \frac{1}{\tau_k} w_k$ and a representation theorem for linear functionals on finite-dimensional subspaces of $C(\Omega)$ (see Krabs [184, IV 2.3–2.4]) gives the existence of q_k points $t_{ki}^+ \in \Omega$ and real numbers $\bar{\lambda}_{ki}^+ \geq 0$ for $i \in \{1, \ldots, q_k\}$ with

$$\bar{u}_k(f) = \sum_{i=1}^{q_k} \bar{\lambda}_{ki}^+ f(t_{ki}^+).$$

In a similar way there are r_k points $t_{ki}^- \in \Omega$ and real numbers $\bar{\lambda}_{ki}^- \geq 0$ for $i \in \{1, \ldots, r_k\}$ with

$$\bar{w}_k(f) = \sum_{j=1}^{r_k} \bar{\lambda}_{ki}^- f(t_{ki}^-).$$

If we define

$$\lambda_{ki} := \bar{\lambda}_{ki}^+ \text{ for all } i \in \{1, \ldots, q_k\}$$

and

$$\lambda_{k\ i+q_k} := -\bar{\lambda}_{ki}^- \text{ for all } i \in \{1, \ldots, r_k\},$$

and if we set $p_k := q_k + r_k$, then (9.25) is equivalent to (9.22), and (9.26) is equivalent to (9.23). The analogous application of a known result from optimization (e.g., compare Krabs [184, Thm. I.5.2]) leads to $p_k \leq \dim \text{span } \{v_{k1}, \ldots, v_{kn}, e, f_k(\bar{x}) - z_k\}$. For every $k \in \{1, \ldots, m\}$ the equations (9.27) can be written as

$$\sum_{i=1}^{q_k} \lambda_{ki}[(f_k(\bar{x}) - z_k)(t_{ki}^+) - \|f_k(\bar{x}) - z_k\|] = 0$$

and

$$\sum_{i=1}^{r_k} \lambda_{k\ i+q_k}[(f_k(\bar{x}) - z_k)(t_{ki}^-) + \|f_k(\bar{x}) - z_k\|] = 0$$

which is equivalent to the implication (9.24). □

The preceding proof shows the usefulness of the generalized multiplier rule. Theorem 9.6 gives a necessary optimality condition for the vector approximation problem (9.18). We know from Theorem 7.20 that the generalized multiplier rule is also a sufficient optimality condition if and only if a composite map is in a certain sense differentiably C-quasiconvex. The next theorem presents a so-called *representation condition* which implies the differentiable C-quasiconvexity of this composite map.

Theorem 9.7. *Let the assumption (9.17) be satisfied, and let the set V be given by (9.19). Moreover, let some $\bar{y} \in V$ (with an inverse image $\bar{x} \in S$) be given, and for every $k \in \{1, \ldots, m\}$ let the Fréchet-derivative of f_k at $\bar{x}$ be given by (9.21). Assume that there are nonnegative numbers $\tau_1, \ldots, \tau_m$ where at least one τ_k is nonzero with the following property: For every $k \in \{1, \ldots, m\}$ with $\tau_k > 0$ there are p_k points $\tau_{k1}, \ldots, t_{kp_k} \in E_k(\bar{x})$ with*

$$1 \leq p_k \leq \dim \text{span } \{v_{k1}, \ldots, v_{kn}, e, f_k(\bar{x}) - z_k\} \leq n+2 \quad (e \equiv 1 \text{ on } \Omega),$$

$$E_k(\bar{x}) := \{t \in \Omega \mid |(f_k(\bar{x}) - z_k)(t)| = \|f_k(\bar{x}) - z_k\|\}$$

and there are real numbers $\lambda_{k1}, \ldots, \lambda_{kp_k}$ so that the conditions (9.22), (9.23) and (9.24) are satisfied. Furthermore, let $f_1, \ldots, f_m$ satisfy the representation condition, i.e., for every $x \in S$ there are positive functions $\Psi_1(x, \bar{x}), \ldots, \Psi_m(x, \bar{x}) \in C(\Omega)$ and some $\tilde{x} \in S$ with

$$(f_k(x) - f_k(\bar{x}))(t) = \Psi_k(x, \bar{x})(t) \cdot (f_k'(\bar{x})(\tilde{x} - \bar{x}))(t)$$
$$\text{for all } t \in \Omega \text{ and all } k \in \{1, \ldots, m\}. \quad (9.28)$$

9.4. Nonlinear Chebyshev Vector Approximation

Then $\bar{y}$ is a weakly minimal element of the set V.

Proof. It is obvious from the proof of the previous theorem that the generalized multiplier rule is satisfied for the problem (9.20). In the following the objective map of this problem is denoted by f, that is
$$f(x,y) = y \text{ for all } (x,y) \in S \times \mathbb{R}^m.$$
The constraint map $g : S \times \mathbb{R}^m \to C(\Omega)^{2m}$ is denoted by

$$g(x,y) = \begin{pmatrix} f_1(x) - z_1 - y_1 e \\ -f_1(x) + z_1 - y_1 e \\ \vdots \\ f_m(x) - z_m - y_m e \\ -f_m(x) + z_m - y_m e \end{pmatrix} \text{ for all } (x,y) \in S \times \mathbb{R}^m.$$

The real linear space $C(\Omega)^{2m}$ is assumed to be partially ordered in the natural way (the ordering cone is denoted $C_{C(\Omega)^{2m}}$). If we show that the composite map (f, g) is differentiably C-quasiconvex at $(\bar{x}, \bar{y})$ with

$$C := (-\text{int}(\mathbb{R}^m_+)) \times (-C_{C(\Omega)^{2m}} + \text{cone}(\{g(\bar{x}, \bar{y})\}) - \text{cone}(\{g(\bar{x}, \bar{y})\})),$$

then, by Corollary 7.21, $(\bar{x}, \bar{y})$ is a weakly minimal solution of the problem (9.20). But this means that $\bar{y} \in V$ is a weakly minimal element of the set $V + \mathbb{R}^m_+$. But then we conclude with Lemma 4.13, (a) that $\bar{y}$ is also a weakly minimal element of the set V.

Hence, it remains to prove that the composite map (f, g) is differentiably C-quasiconvex at $(\bar{x}, \bar{y})$. Let $(x, y) \in S \times \mathbb{R}^m$ be arbitrarily given with the property
$$(f, g)(x, y) - (f, g)(\bar{x}, \bar{y}) \in C$$
which means for some $\alpha, \beta \geq 0$
$$y_k - \bar{y}_k < 0 \text{ for all } k \in \{1, \ldots, m\},$$

$$f_k(x) - y_k e - f_k(\bar{x}) + \bar{y}_k e$$
$$\leq \alpha(f_k(\bar{x}) - z_k - \bar{y}_k e) - \beta(f_k(\bar{x}) - z_k - \bar{y}_k e)$$
$$\text{for all } k \in \{1, \ldots, m\} \tag{9.29}$$

$$-f_k(x) - y_k e + f_k(\bar{x}) + \bar{y}_k e$$
$$\leq \alpha(-f_k(\bar{x}) + z_k - \bar{y}_k e) - \beta(-f_k(\bar{x}) + z_k - \bar{y}_k e)$$
$$\text{for all } k \in \{1, \ldots, m\} \tag{9.30}$$

(where $e \equiv 1$ on Ω). Then there are positive functions $\Psi_1(x, \bar{x}), \ldots, \Psi_m(x, \bar{x}) \in C(\Omega)$ and some $\tilde{x} \in S$ so that the equation (9.28) is satisfied. Furthermore there are positive real numbers $\alpha_1, \ldots, \alpha_m, \beta_1, \ldots, \beta_m$ with

$$0 < \alpha_k \leq \Psi_k(x, \bar{x})(t) \leq \beta_k \text{ for all } k \in \{1, \ldots, m\} \text{ and all } t \in \Omega,$$

and we define
$$\tilde{\alpha} := \frac{\alpha}{\max\{\beta_1, \ldots, \beta_m\}},$$

$$\tilde{\beta} := \frac{\beta}{\min\{\alpha_1, \ldots, \alpha_m\}}$$

and
$$\tilde{y}_k := \bar{y}_k + \frac{1}{\beta_k}(y_k - \bar{y}_k) < \bar{y}_k \text{ for all } k \in \{1, \ldots, m\}.$$

Then the inequality (9.29) implies with (9.28) and the feasibility of $(\bar{x}, \bar{y})$

$$f'_k(\bar{x})(\tilde{x} - \bar{x})(t) - (\tilde{y}_k - \bar{y}_k)$$
$$= \frac{1}{\Psi_k(x, \bar{x})(t)} \big(f_k(x) - f_k(\bar{x})\big)(t) - \frac{1}{\beta_k}\big(y_k - \bar{y}_k\big)$$
$$\leq \frac{1}{\Psi_k(x, \bar{x})(t)} \Big[y_k - \bar{y}_k + \alpha(f_k(\bar{x}) - z_k - \bar{y}_k e)(t)$$
$$\quad - \beta(f_k(\bar{x}) - z_k - \bar{y}_k e)(t)\Big] - \frac{1}{\beta_k}\big(y_k - \bar{y}_k\big)$$
$$\leq \frac{\alpha}{\beta_k}\big(f_k(\bar{x}) - z_k - \bar{y}_k e\big)(t) - \frac{\beta}{\alpha_k}\big(f_k(\bar{x}) - z_k - \bar{y}_k e\big)(t)$$
$$\quad + \Big(\frac{1}{\Psi_k(x, \bar{x})(t)} - \frac{1}{\beta_k}\Big)(y_k - \bar{y}_k)$$
$$\leq \tilde{\alpha}(f_k(\bar{x}) - z_k - \bar{y}_k e)(t) - \tilde{\beta}(f_k(\bar{x}) - z_k - \bar{y}_k e)(t)$$
$$\text{for all } k \in \{1, \ldots, m\} \text{ and all } t \in \Omega.$$

9.4. Nonlinear Chebyshev Vector Approximation

Similarly the inequality (9.30) implies

$$-f'_k(\bar{x})(\tilde{x}-\bar{x})(t) - (\tilde{y}_k - \bar{y}_k)$$
$$= \frac{1}{\Psi_k(x,\bar{x})(t)}\left(-f_k(x) + f_k(\bar{x})\right)(t) - \frac{1}{\beta_k}\left(y_k - \bar{y}_k\right)$$
$$\leq \frac{1}{\Psi_k(x,\bar{x})(t)}\Big[y_k - \bar{y}_k + \alpha(-f_k(\bar{x}) + z_k - \bar{y}_k e)(t)$$
$$-\beta(-f_k(\bar{x}) + z_k - \bar{y}_k e)(t)\Big] - \frac{1}{\beta_k}\left(y_k - \bar{y}_k\right)$$
$$\leq \frac{\alpha}{\beta_k}\left(-f_k(\bar{x}) + z_k - \bar{y}_k e\right)(t) - \frac{\beta}{\alpha_k}\left(-f_k(\bar{x}) + z_k - \bar{y}_k e\right)(t)$$
$$+\left(\frac{1}{\Psi_k(x,\bar{x})(t)} - \frac{1}{\beta_k}\right)(y_k - \bar{y}_k)$$
$$\leq \tilde{\alpha}(-f_k(\bar{x}) + z_k - \bar{y}_k e)(t) - \tilde{\beta}(-f_k(\bar{x}) + z_k - \bar{y}_k e)(t)$$

for all $k \in \{1,\ldots,m\}$ and all $t \in \Omega$.

Hence we get

$$(f,g)'(\bar{x},\bar{y})(\tilde{x}-\bar{x},\tilde{y}-\bar{y}) \in C.$$

This completes the proof. □

The representation condition in the previous theorem is satisfied for rational approximating families: Let functions $p_{ki} \in C(\Omega)$, $k \in \{1,\ldots,m\}$ and $i \in \{1,\ldots,n\}$, be given and define for some $n_k \in \{1,\ldots,n-1\}$, with $k \in \{1,\ldots,m\}$,

$$f_k(x)(t) = \frac{\sum_{i=1}^{n_k} x_i p_{ki}(t)}{\sum_{i=n_k+1}^{n} x_i p_{ki}(t)} \quad \text{for all } x \in \mathbb{R}^n \text{ and all } t \in \Omega$$

and

$$S := \left\{x \in \mathbb{R}^n \;\Big|\; \sum_{i=n_k+1}^{n} x_i p_{ki}(t) > 0 \text{ for all } t \in \Omega\right\}.$$

An easy computation shows that the equality (9.28) holds with

$$\Psi_k(x,\bar{x})(t) = \frac{\sum_{i=n_k+1}^{n} x_i p_{ki}(t)}{\sum_{i=n_k+1}^{n} \bar{x}_i p_{ki}(t)} \quad \text{for all } t \in \Omega$$

where $x = (x_1, \ldots, x_n)$ and $\bar{x} = (\bar{x}_1, \ldots, \bar{x}_n)$. For further discussion of these types of condition for the case $m = 1$ see Krabs [183].

9.5 Linear Chebyshev Vector Approximation

In the preceding section we investigated nonlinear Chebyshev vector approximation problems. Obviously, these results can also be applied for linear problems. It is the aim of this section to demonstrate the usefulness of the duality theory developed for abstract linear optimization problems. Using the dual problem we are able to formulate an alternation theorem which is comparable with the corresponding result of the previous section.

Specializing the assumption (9.17) we obtain our standard assumption as follows:

$$\left.\begin{array}{l}\text{Let } \Omega \text{ be a compact Hausdorff space with at least} \\ n+1 \text{ elements;} \\ \text{let } C(\Omega) \text{ denote the real linear space of real-} \\ \text{valued continuous functions on } \Omega \text{ equipped with} \\ \text{the maximum norm } \|\cdot\| \text{ (see Example 1.49);} \\ \text{for every } k \in \{1, \ldots, m\} \text{ let some functions} \\ v_{k1}, \ldots, v_{kn} \in C(\Omega) \text{ be given which are linearly} \\ \text{independent;} \\ \text{let } z_1, \ldots, z_m \in C(\Omega) \text{ be given functions;} \\ \text{let the space } \mathbb{R}^m \text{ be partially ordered in the} \\ \text{natural way.}\end{array}\right\} \quad (9.31)$$

Under this special assumption the set V (defined in (9.19)) reduces

9.5. Linear Chebyshev Vector Approximation

to

$$V := \left\{ \begin{pmatrix} \left\| \sum_{i=1}^{n} x_i v_{1i} - z_1 \right\| \\ \vdots \\ \left\| \sum_{i=1}^{n} x_i v_{mi} - z_m \right\| \end{pmatrix} \;\middle|\; x \in \mathbb{R}^n \right\}. \qquad (9.32)$$

A vector approximation problem for which the image set of the objective map equals V is also called a *linear Chebyshev vector approximation problem*. The following lemma shows that it makes sense to examine this vector optimization problem.

Lemma 9.8. *Let the assumption (9.31) be satisfied. The set V (given in (9.32)) has at least one almost properly minimal element and, therefore, also a minimal and weakly minimal element.*

Proof. Notice that $\sum_{k=1}^{m} \|\cdot\|$ is a norm on $C(\Omega)^m$. Then we obtain with a known existence theorem (compare Meinardus [223, p. 1]) that the scalar optimization problem

$$\min_{x \in \mathbb{R}^n} \sum_{k=1}^{m} \left\| \sum_{i=1}^{n} x_i v_{ki} - z_k \right\|$$

has at least one solution. Consequently, the set V has at least one almost properly minimal element which is also minimal by Theorem 5.18, (b) and weakly minimal by Theorem 5.28. □

9.5.1 Duality Results

In the following we formulate the dual problem of the linear Chebyshev vector approximation problem introduced previously. In Chapter 8 we presented duality results for two optimality notions. Since the primal problem (8.4) with $P_1 = V + \mathbb{R}^m_+$ (where V is given by (9.32)) is not equivalent to the vector optimization problem of determining weakly minimal elements of the set V, we turn our attention to the problem (8.5) with $P_2 = V$ which is formalized as follows:

$$\text{a-p-min}_{x \in \mathbb{R}^n} \begin{pmatrix} \left\| \sum_{i=1}^n x_i v_{1i} - z_1 \right\| \\ \vdots \\ \left\| \sum_{i=1}^n x_i v_{mi} - z_m \right\| \end{pmatrix}. \qquad (9.33)$$

This problem is equivalent to the abstract optimization problem

a-p-min y

subject to the constraints

$$\left. \begin{aligned} \sum_{i=1}^n x_i v_{ki}(t) + y_k &\geq z_k(t) \\ -\sum_{i=1}^n x_i v_{ki}(t) + y_k &\geq -z_k(t) \end{aligned} \right\} \quad \text{for all } t \in \Omega \text{ and all } k \in \{1, \ldots, m\} \qquad (9.34)$$

$x \in \mathbb{R}^n$, $y \in \mathbb{R}^m$.

The problem formalized in (9.34) can also be interpreted in the following way: Determine an almost properly minimal element of the set $V + \mathbb{R}_+^m$. Then the equivalence of the problems (9.33) and (9.34) is to be understood in the sense that an element of the set V is an almost properly minimal element of the set V if and only if it is an almost properly minimal element of the set $V + \mathbb{R}_+^m$.

Definition 9.9. Let the assumption (9.31) be satisfied. If

$$\bar{y} = \begin{pmatrix} \left\| \sum_{i=1}^n \bar{x}_i v_{1i} - z_1 \right\| \\ \vdots \\ \left\| \sum_{i=1}^n \bar{x}_i v_{mi} - z_m \right\| \end{pmatrix} \quad (\text{with } \bar{x} \in \mathbb{R}^n)$$

is an almost properly minimal element of the set V (given by (9.32)), then $\bar{x}$ is called an *almost properly minimal solution* of the linear Chebyshev vector approximation problem (9.33).

9.5. Linear Chebyshev Vector Approximation

The problem (9.34) is an abstract semi-infinite linear optimization problem. The dual problem reads as follows:

$$\max \begin{pmatrix} \sum_{k=1}^{m} (z_{1k}^{+*}(z_k) - z_{1k}^{-*}(z_k)) \\ \vdots \\ \sum_{k=1}^{m} (z_{mk}^{+*}(z_k) - z_{mk}^{-*}(z_k)) \end{pmatrix}$$

subject to the constraints

$$\sum_{k=1}^{m}\sum_{i=1}^{m} \tau_i(z_{ik}^{+*}(v_{kj}) - z_{ik}^{-*}(v_{kj})) = 0 \text{ for all } j \in \{1,\ldots,n\}$$

$$\sum_{i=1}^{m} \tau_i(z_{ik}^{+*}(e) + z_{ik}^{-*}(e)) = \tau_k \text{ for all } k \in \{1,\ldots,m\}$$

$$z_{ik}^{+*}, z_{ik}^{-*} \in C_{Z_k^*} \text{ for all } i,k \in \{1,\ldots,m\}$$

$$\tau_1,\ldots,\tau_m > 0.$$

(9.35)

By an easy computation one obtains this problem from the formulation in (8.20). Here, Z_k (with $k \in \{1,\ldots,m\}$) denotes the linear subspace of $C(\Omega)$ spanned by $v_{k1},\ldots,v_{kn}, e$ and z_k (again $e \equiv 1$ on Ω). $C_{Z_k^*}$ denotes the ordering cone of the dual space Z_k^*. The max-term in problem (9.35) means that we are looking for a maximal solution of this problem, i.e., we are interested in a feasible element whose image is a maximal element of the image set of the objective map.

Since $Z_1,\ldots,Z_m$ are finite-dimensional linear subspaces of $C(\Omega)$ and $e \in Z_k$ for all $k \in \{1,\ldots,m\}$, every functional $z_{ik}^{+*}, z_{ik}^{-*} \in C_{Z_k^*}$ (with $i,k \in \{1,\ldots,m\}$) can be represented in the form

$$\left.\begin{aligned} z_{ik}^{+*}(f) &= \sum_{l \in I_{ik}^+} \lambda_{ikl}^+ f(t_{ikl}^+) \\ z_{ik}^{-*}(f) &= \sum_{l \in I_{ik}^-} \lambda_{ikl}^- f(t_{ikl}^-) \end{aligned}\right\} \text{ for all } f \in Z_k$$

(e.g., compare Krabs [184, IV 2.3–2.4]) where $\lambda_{ikl}^+ \geq 0$, $\lambda_{ikl}^- \geq 0$, $t_{ikl}^+ \in \Omega$, $t_{ikl}^- \in \Omega$, and I_{ik}^+ and I_{ik}^- are finite index sets. Using this representation the problem (9.35) is equivalent to the following abstract optimization problem:

$$\max \begin{pmatrix} \sum_{k=1}^{m} \Big[\sum_{l \in I_{1k}^+} \lambda_{1kl}^+ z_k(t_{1kl}^+) - \sum_{l \in I_{1k}^-} \lambda_{1kl}^- z_k(t_{1kl}^-) \Big] \\ \vdots \\ \sum_{k=1}^{m} \Big[\sum_{l \in I_{mk}^+} \lambda_{mkl}^+ z_k(t_{mkl}^+) - \sum_{l \in I_{mk}^-} \lambda_{mkl}^- z_k(t_{mkl}^-) \Big] \end{pmatrix}$$

subject to the constraints

$$\sum_{k=1}^{m} \sum_{i=1}^{m} \tau_i \Big[\sum_{l \in I_{ik}^+} \lambda_{ikl}^+ v_{kj}(t_{ikl}^+) - \sum_{l \in I_{ik}^-} \lambda_{ikl}^- v_{kj}(t_{ikl}^-) \Big] = 0$$

$$\text{for all } j \in \{1, \ldots, n\}$$

$$\sum_{i=1}^{m} \tau_i \Big[\sum_{l \in I_{ik}^+} \lambda_{ikl}^+ + \sum_{l \in I_{ik}^-} \lambda_{ikl}^- \Big] = \tau_k \text{ for all } k \in \{1, \ldots, m\}$$

$$\Big(\lambda_{ikl}^+ \geq 0 \text{ for all } l \in I_{ik}^+, \ \lambda_{ikl}^- \geq 0 \text{ for all } l \in I_{ik}^- \Big)$$

$$\text{for all } i, k \in \{1, \ldots, m\}$$

$$\{ t_{ikl}^+ \mid l \in I_{ik}^+ \} \cup \{ t_{ikl}^- \mid l \in I_{ik}^- \} \subset \Omega \text{ for all } i, k \in \{1, \ldots, m\}$$

$$\tau_1, \ldots, \tau_m > 0$$

(9.36)

The problem (9.36) is formally the dual abstract optimization problem of the linear Chebyshev vector approximation problem (9.33). A first relationship between these two problems is given by

Theorem 9.10. *Let the assumption (9.31) be satisfied. For every almost properly minimal solution of the linear Chebyshev vector approximation problem (9.33) there is a maximal solution of the abstract optimization problem (9.36) so that the images of the objective maps are equal.*

Proof. The problem (9.34) which is equivalent to the linear Chebyshev vector approximation problem (9.33) satisfies the generalized Slater condition (for $x = 0_{\mathbb{R}^n}$ and $y_k > \|z_k\|$ for all $k \in$

9.5. Linear Chebyshev Vector Approximation

$\{1,\ldots,m\}$). Consequently, the strong duality theorem (Theorem 8.7, (b)) leads to the assertion. □

From Lemma 9.8 and Theorem 9.10 it follows immediately that the abstract optimization problem (9.36) has at least one maximal solution. In addition to Theorem 9.10 a strong converse duality theorem can be proved as well.

Theorem 9.11. *Let the assumption (9.31) be satisfied. For every maximal solution of the abstract optimization problem (9.36) there is an almost properly minimal solution of the linear Chebyshev vector approximation problem (9.33) so that the images of the objective maps are equal.*

Proof. The assertion follows from the strong converse duality theorem 8.9, (b), if we varify the assumption that the set $V + \mathbb{R}^m_+$ is closed.

Let $(y^l)_{l \in \mathbb{N}}$ be an arbitrary sequence in $V + \mathbb{R}^m_+$ converging to some $\bar{y} \in \mathbb{R}^m$. Then there are a sequence $(x^l)_{l \in \mathbb{N}}$ in $\mathbb{R}^n$ and a sequence $(z^l)_{l \in \mathbb{N}}$ in $\mathbb{R}^m_+$ with

$$y^l_k = \left\| \sum_{i=1}^m x^l_i v_{ki} - z_k \right\| + z^l_k \text{ for all } k \in \{1,\ldots,m\} \text{ and all } l \in \mathbb{N}.$$

Furthermore there is an $l' \in \mathbb{N}$ with

$$y^l_k \leq \bar{y}_k + 1 \text{ for all } k \in \{1,\ldots,m\} \text{ and all } l \in \mathbb{N} \text{ with } l \geq l'.$$

Consequently, we get for all $k \in \{1,\ldots,m\}$ and all $l \in \mathbb{N}$ with $l \geq l'$

$$\begin{aligned} \|z_k\| + \bar{y}_k + 1 &\geq \|z_k\| + \left\| \sum_{i=1}^n x^l_i v_{ki} - z_k \right\| + z^l_k \\ &\geq \left\| \sum_{i=1}^n x^l_i v_{ki} \right\| \\ &= \left\| \sum_{i=1}^n \frac{x^l_i}{\|x^l\|_{\mathbb{R}^n}} v_{ki} \right\| \|x^l\|_{\mathbb{R}^n}, \end{aligned} \quad (9.37)$$

if $x^l \neq 0_{\mathbb{R}^n}$. In the inequality (9.37) $\|\cdot\|_{\mathbb{R}^n}$ denotes the maximum norm on $\mathbb{R}^n$. Since the unit sphere

$$S := \{x \in \mathbb{R}^n \mid \|x\|_{\mathbb{R}^n} = 1\}$$

in $\mathbb{R}^n$ is compact, by a continuity argument for every $k \in \{1, \ldots, m\}$ there is an $\bar{x}^k \in S$ with

$$\gamma_k := \left\|\sum_{i=1}^n \bar{x}_i^k v_{ki}\right\| \leq \left\|\sum_{i=1}^n x_i v_{ki}\right\| \text{ for all } x \in S.$$

Since for every $k \in \{1, \ldots, m\}$ the functions $v_{k1}, \ldots, v_{kn}$ are linearly independent, we obtain $\gamma_k > 0$. Then we conclude with the inequality (9.37)

$$\|x^l\|_{\mathbb{R}^n} \leq \frac{1}{\gamma_k}\left(\|z_k\| + \bar{\gamma}_k + 1\right) \text{ for all } k \in \{1, \ldots, m\} \text{ and}$$
$$\text{all } l \in \mathbb{N} \text{ with } l \geq l'. \qquad (9.38)$$

This inequality is also valid for $x^l = 0_{\mathbb{R}^n}$ (for the inequality (9.38) compare also Collatz-Krabs [61, p. 184] and Reemtsen [255, p. 29]). Consequently, the sequence $(x^l)_{l \in \mathbb{N}}$ has a subsequence $(x^{l_j})_{j \in \mathbb{N}}$ converging to some $\bar{x} \in \mathbb{R}^n$. Then we get

$$\lim_{j \to \infty} z_k^{l_j} = \bar{y}_k - \left\|\sum_{i=1}^n \bar{x}_i v_{ki} - z_k\right\|.$$

Since $\mathbb{R}_+^m$ is closed, we also conclude

$$\bar{y}_k \geq \left\|\sum_{i=1}^n \bar{x}_i v_{ki} - z_k\right\| \text{ for all } k \in \{1, \ldots, m\}$$

implying $\bar{y} \in V + \mathbb{R}_+^m$. □

The dual problem (9.36) is a finite nonlinear optimization problem although the primal problem (9.33) is a semi-infinite linear optimization problem. The duality results are useful for the formulation of an alternation theorem.

9.5.2 An Alternation Theorem

In Section 9.4 we presented an alternation theorem for nonlinear Chebyshev vector approximation problems which is valid for linear Chebyshev vector approximation problems as well. But such an alternation theorem can also be obtained with the preceding duality results. In contrast to the theory in the scalar case we have the difficulty with linear Chebyshev vector approximation problems that the known complementary slackness theorem holds only in a weaker form. Moreover, the conditions given in Theorem 9.13 do not follow immediately from the constraints of the dual problem (9.36). In order to get a similar result as in Theorem 9.6 these constraints have to be transformed in an appropriate way.

Lemma 9.12. *Let the assumption (9.31) be satisfied. Moreover, let x be an almost properly minimal solution of the linear Chebyshev vector approximation problem (9.33), and let the tuple $(\lambda_{ikl}^+, \lambda_{ikl}^-, t_{ikl}^+, t_{ikl}^-, I_{ik}^+, I_{ik}^-, \tau)$ be a maximal solution of the vector optimization problem (9.36) so that the images of the objective maps are equal. Then it follows for all $i, k \in \{1, \ldots, m\}$:*

$\lambda_{ikl}^+ > 0$ *for some* $l \in I_{ik}^+$

$$\Longrightarrow \sum_{j=1}^n x_j v_{kj}(t_{ikl}^+) - z_k(t_{ikl}^+) = \left\| \sum_{j=1}^n x_j v_{kj} - z_k \right\|,$$

$\lambda_{ikl}^- > 0$ *for some* $l \in I_{ik}^-$

$$\Longrightarrow \sum_{j=1}^n x_j v_{kj}(t_{ikl}^-) - z_k(t_{ikl}^-) = -\left\| \sum_{j=1}^n x_j v_{kj} - z_k \right\|.$$

Proof. Let x be an almost properly minimal solution of the linear Chebyshev vector approximation problem (9.33), and let the tuple $(\lambda_{ikl}^+, \lambda_{ikl}^-, t_{ikl}^+, t_{ikl}^-, I_{ik}^+, I_{ik}^-, \tau)$ be a maximal solution of the problem (9.36) with

$$\left\|\sum_{j=1}^{n} x_j v_{ij} - z_i\right\| = \sum_{k=1}^{m}\left[\sum_{l\in I_{ik}^+}\lambda_{ikl}^+ z_k(t_{ikl}^+) - \sum_{l\in I_{ik}^-}\lambda_{ikl}^- z_k(t_{ikl}^-)\right]$$
$$\text{for all } i \in \{1,\ldots,m\}. \tag{9.39}$$

For the following transformations the equation (9.39) and the constraints of problem (9.36) are used:

$$\sum_{k=1}^{m}\sum_{i=1}^{m}\tau_i \sum_{l\in I_{ik}^+}\lambda_{ikl}^+\left[\sum_{j=1}^{n} x_j v_{kj}(t_{ikl}^+) - z_k(t_{ikl}^+) + \left\|\sum_{j=1}^{n} x_j v_{kj} - z_k\right\|\right]$$
$$+\sum_{k=1}^{m}\sum_{i=1}^{m}\tau_i \sum_{l\in I_{ik}^-}\lambda_{ikl}^-\left[-\sum_{j=1}^{n} x_j v_{kj}(t_{ikl}^-) + z_k(t_{ikl}^-) + \left\|\sum_{j=1}^{n} x_j v_{kj} - z_k\right\|\right]$$
$$= \sum_{k=1}^{m}\sum_{i=1}^{m}\tau_i\Bigg[\sum_{l\in I_{ik}^+}\lambda_{ikl}^+\left(-z_k(t_{ikl}^+) + \left\|\sum_{j=1}^{n} x_j v_{kj} - z_k\right\|\right)$$
$$+\sum_{l\in I_{ik}^-}\lambda_{ikl}^-\left(z_k(t_{ikl}^-) + \left\|\sum_{j=1}^{n} x_j v_{kj} - z_k\right\|\right)\Bigg]$$
$$= \sum_{i=1}^{m}\tau_i \sum_{k=1}^{m}\left[-\sum_{l\in I_{ik}^+}\lambda_{ikl}^+ z_k(t_{ikl}^+) + \sum_{l\in I_{ik}^-}\lambda_{ikl}^- z_k(t_{ikl}^-)\right]$$
$$+\sum_{k=1}^{m}\left\|\sum_{j=1}^{n} x_j v_{kj} - z_k\right\| \sum_{i=1}^{m}\tau_i\left[\sum_{l\in I_{ik}^+}\lambda_{ikl}^+ + \sum_{l\in I_{ik}^-}\lambda_{ikl}^-\right]$$
$$= -\sum_{i=1}^{m}\tau_i\left\|\sum_{j=1}^{n} x_j v_{ij} - z_i\right\| + \sum_{k=1}^{m}\tau_k\left\|\sum_{j=1}^{n} x_j v_{kj} - z_k\right\|$$
$$= 0.$$

If we notice that the coefficients λ_{ikl}^+, λ_{ikl}^- are non-negative and the coefficients τ_i are positive, from the previous equation it follows for all $i,k \in \{1,\ldots,m\}$

$$\sum_{l\in I_{ik}^+}\lambda_{ikl}^+\left[\sum_{j=1}^{n} x_j v_{kj}(t_{ikl}^+) - z_k(t_{ikl}^+) + \left\|\sum_{j=1}^{n} x_j v_{kj} - z_k\right\|\right] = 0$$

9.5. Linear Chebyshev Vector Approximation

and

$$\sum_{l\in I_{ik}^-}\lambda_{ikl}^-\left[-\sum_{j=1}^n x_j v_{kj}(t_{ikl}^-) + z_k(t_{ikl}^-) + \left\|\sum_{j=1}^n x_j v_{kj} - z_k\right\|\right] = 0$$

Because of the nonnegativity of the coefficients $\lambda_{ikl}^+, \lambda_{ikl}^-$ and the terms in brackets we immediately obtain the assertion. □

With Lemma 9.12 we are now able to formulate the announced *alternation theorem for the linear Chebyshev vector approximation problem (9.33)*.

Theorem 9.13. *Let the assumption (9.31) be satisfied. An element $\bar{x} \in \mathbb{R}^n$ is an almost properly minimal solution of the linear Chebyshev vector approximation problem (9.33) if and only if for every $k \in \{1,\ldots,m\}$ there are $p_k \leq n+1$ elements $\bar{t}_{k1},\ldots,\bar{t}_{kp_k}$ in the set*

$$E_k(\bar{x}) := \left\{ t \in \Omega \,\bigg|\, \left|\sum_{i=1}^n \bar{x}_i v_{ki}(t) - z_k(t)\right| = \left\|\sum_{i=1}^n \bar{x}_i v_{ki} - z_k\right\|\right\}$$

and real numbers $\bar{\lambda}_{k1},\ldots,\bar{\lambda}_{kp_k}$ as well as a positive real number $\bar{\tau}_k$ so that

$$\sum_{i=1}^{p_k} |\bar{\lambda}_{ki}| = 1 \text{ for all } k \in \{1,\ldots,m\}, \tag{9.40}$$

$$\sum_{k=1}^m \bar{\tau}_k \sum_{i=1}^{p_k} \bar{\lambda}_{ki} v_{kj}(\bar{t}_{ki}) = 0 \text{ for all } j \in \{1,\ldots,n\} \tag{9.41}$$

and

$$\bar{\lambda}_{ki} \neq 0 \text{ for some } k \in \{1,\ldots,m\} \text{ and some } i \in \{1,\ldots,p_k\}$$
$$\implies \sum_{j=1}^n \bar{x}_j v_{kj}(\bar{t}_{ki}) - z_k(\bar{t}_{ki}) = \left\|\sum_{j=1}^n \bar{x}_j v_{kj} - z_k\right\| \text{sgn}\,(\bar{\lambda}_{ki}). \tag{9.42}$$

Proof. First, we prove the sufficiency of the above conditions for an almost properly minimal solution of the problem (9.33). Therefore, we assume that for some $\bar{x} \in \mathbb{R}^n$ and all $k \in \{1,\ldots,m\}$ there are arbitrarily given elements $\bar{t}_{ki} \in E_k(\bar{x})$ and real numbers $\bar{\lambda}_{ki}$ as well as a positive real number $\bar{\tau}_k$. Moreover, we assume that the equations (9.40), (9.41) and the implication (9.42) are satisfied. Then we have:

$$\sum_{k=1}^{m} \bar{\tau}_k \left\| \sum_{j=1}^{n} \bar{x}_j v_{kj} - z_k \right\|$$

$$= \sum_{k=1}^{m} \bar{\tau}_k \sum_{i=1}^{p_k} |\bar{\lambda}_{ki}| \left\| \sum_{j=1}^{n} \bar{x}_j v_{kj} - z_k \right\| \quad \text{(by (9.40))}$$

$$= \sum_{k=1}^{m} \bar{\tau}_k \sum_{i=1}^{p_k} \text{sgn}(\bar{\lambda}_{ki}) |\bar{\lambda}_{ki}| \left(\sum_{j=1}^{n} \bar{x}_j v_{kj}(\bar{t}_{ki}) - z_k(\bar{t}_{ki}) \right) \quad \text{(by (9.42))}$$

$$= \sum_{k=1}^{m} \bar{\tau}_k \sum_{i=1}^{p_k} \bar{\lambda}_{ki} \sum_{j=1}^{n} \bar{x}_j v_{kj}(\bar{t}_{ki}) - \sum_{k=1}^{m} \bar{\tau}_k \sum_{i=1}^{p_k} \bar{\lambda}_{ki} z_k(\bar{t}_{ki})$$

$$= -\sum_{k=1}^{m} \bar{\tau}_k \sum_{i=1}^{p_k} \bar{\lambda}_{ki} z_k(\bar{t}_{ki}) \quad \text{(by (9.41))}$$

$$= \sum_{k=1}^{m} \bar{\tau}_k \sum_{i=1}^{p_k} \bar{\lambda}_{ki} \sum_{j=1}^{n} x_j v_{kj}(\bar{t}_{ki}) - \sum_{k=1}^{m} \bar{\tau}_k \sum_{i=1}^{p_k} \bar{\lambda}_{ki} z_k(\bar{t}_{ki}) \quad \text{(by (9.41))}$$

$$= \sum_{k=1}^{m} \bar{\tau}_k \sum_{i=1}^{p_k} \bar{\lambda}_{ki} \left(\sum_{j=1}^{n} x_j v_{kj}(\bar{t}_{ki}) - z_k(\bar{t}_{ki}) \right)$$

$$\leq \sum_{k=1}^{m} \bar{\tau}_k \sum_{i=1}^{p_k} |\bar{\lambda}_{ki}| \left| \sum_{j=1}^{n} x_j v_{kj}(\bar{t}_{ki}) - z_k(\bar{t}_{ki}) \right|$$

$$\leq \sum_{k=1}^{m} \bar{\tau}_k \sum_{i=1}^{p_k} |\bar{\lambda}_{ki}| \left\| \sum_{j=1}^{n} x_j v_{kj} - z_k \right\|$$

$$= \sum_{k=1}^{m} \bar{\tau}_k \left\| \sum_{j=1}^{n} x_j v_{kj} - z_k \right\| \text{ for all } x \in \mathbb{R}^n \quad \text{(by (9.40))}.$$

Consequently, $\bar{x}$ is an almost properly minimal solution of the linear Chebyshev vector approximation problem (9.33).

Next, we prove the necessity of the conditions given in this theorem for an arbitrary almost properly minimal solution $\bar{x}$ of the problem (9.33).

9.5. Linear Chebyshev Vector Approximation

By Theorem 9.10 there is a maximal solution $(\lambda_{ikl}^+, \lambda_{ikl}^-, t_{ikl}^+, t_{ikl}^-, I_{ik}^+, I_{ik}^-, \tau)$ of the dual problem (9.36) so that the images of the objective maps are equal. In particular, the following equations are satisfied:

$$\sum_{k=1}^{m}\sum_{i=1}^{m} \tau_i \left[\sum_{l \in I_{ik}^+} \lambda_{ikl}^+ v_{kj}(t_{ikl}^+) - \sum_{l \in I_{ik}^-} \lambda_{ikl}^- v_{kj}(t_{ikl}^-) \right] = 0 \text{ for all } j \in \{1, \ldots, n\},$$
(9.43)

$$\sum_{i=1}^{m} \tau_i \left[\sum_{l \in I_{ik}^+} \lambda_{ikl}^+ + \sum_{l \in I_{ik}^-} \lambda_{ikl}^- \right] = \tau_k \text{ for all } k \in \{1, \ldots, m\}. \quad (9.44)$$

If we introduce new index sets I_k^+, I_k^- and new variables λ_{kl}^+, λ_{kl}^-, t_{kl}^+, t_{kl}^- the terms

$$\sum_{i=1}^{m}\sum_{l \in I_{ik}^+} \tau_i \lambda_{ikl}^+ v_{kj}(t_{ikl}^+), \quad \sum_{i=1}^{m}\sum_{l \in I_{ik}^-} \tau_i \lambda_{ikl}^- v_{kj}(t_{ikl}^-),$$

$$\sum_{i=1}^{m}\sum_{l \in I_{ik}^+} \tau_i \lambda_{ikl}^+ \quad \text{and} \quad \sum_{i=1}^{m}\sum_{l \in I_{ik}^-} \tau_i \lambda_{ikl}^-$$

appearing in the equations (9.43) and (9.44) can also be written in the following way:

$$\sum_{l \in I_k^+} \lambda_{kl}^+ v_{kj}(t_{kl}^+), \quad \sum_{l \in I_k^-} \lambda_{kl}^- v_{kj}(t_{kl}^-),$$

$$\sum_{l \in I_k^+} \lambda_{kl}^+ \quad \text{and} \quad \sum_{l \in I_k^-} \lambda_{kl}^-.$$

Then the equations (9.43) and (9.44) are equivalent to

$$\sum_{k=1}^{m} \left[\sum_{l \in I_k^+} \lambda_{kl}^+ v_{kj}(t_{kl}^+) - \sum_{l \in I_k^-} \lambda_{kl}^- v_{kj}(t_{kl}^-) \right] = 0 \text{ for all } j \in \{1, \ldots, n\},$$
(9.45)

$$\sum_{l \in I_k^+} \lambda_{kl}^+ + \sum_{l \in I_k^-} \lambda_{kl}^- = \tau_k \text{ for all } k \in \{1, \ldots, m\}. \quad (9.46)$$

Now we replace the numbers λ_{kl}^+ and λ_{kl}^- by $\tau_k \tilde{\lambda}_{kl}^+$ and $\tau_k \tilde{\lambda}_{kl}^-$, respectively, and we obtain equations which are equivalent to (9.45) and (9.46):

$$\sum_{k=1}^{m} \tau_k \Big[\sum_{l \in I_k^+} \tilde{\lambda}_{kl}^+ v_{kj}(t_{kl}^+) - \sum_{l \in I_k^-} \tilde{\lambda}_{kl}^- v_{kj}(t_{kl}^-) \Big] = 0 \text{ for all } j \in \{1,\ldots,n\},$$
(9.47)

$$\sum_{l \in I_k^+} \tilde{\lambda}_{kl}^+ + \sum_{l \in I_k^-} \tilde{\lambda}_{kl}^- = 1 \text{ for all } k \in \{1,\ldots,m\}. \quad (9.48)$$

If we notice that the implications in Lemma 9.12 are true, then because of the positivity of the numbers $\tau_1,\ldots,\tau_m$ the following implications are also true for the new variables ($k \in \{1,\ldots,m\}$):

$$\tilde{\lambda}_{kl}^+ > 0 \text{ for some } l \in I_k^+$$
$$\implies \sum_{j=1}^{n} \bar{x}_j v_{kj}(t_{kl}^+) - z_k(t_{kl}^+) = \Big\| \sum_{j=1}^{n} \bar{x}_j v_{kj} - z_k \Big\|,$$

$$\tilde{\lambda}_{kl}^- > 0 \text{ for some } l \in I_k^-$$
$$\implies \sum_{j=1}^{n} \bar{x}_j v_{kj}(t_{kl}^-) - z_k(t_{kl}^-) = -\Big\| \sum_{j=1}^{n} \bar{x}_j v_{kj} - z_k \Big\|.$$

If we define for every $k \in \{1,\ldots,m\}$ and every $j \in \{1,\ldots,n\}$ the real number

$$\beta_{kj} := \sum_{l \in I_k^+} \tilde{\lambda}_{kl}^+ v_{kj}(t_{kl}^+) - \sum_{l \in I_k^-} \tilde{\lambda}_{kl}^- v_{kj}(t_{kl}^-), \quad (9.49)$$

then we have for all $k \in \{1,\ldots,m\}$

$$\sum_{l \in I_k^+} \tilde{\lambda}_{kl}^+ v_{kj}(t_{kl}^+) - \sum_{l \in I_k^-} \tilde{\lambda}_{kl}^- v_{kj}(t_{kl}^-) = \beta_{kj} \text{ for all } j \in \{1,\ldots,n\}.$$

By an analogous application of a known result from optimization (for instance, see Krabs [184, Thm. I.5.2]) for every $k \in \{1,\ldots,m\}$ there are index sets $\hat{I}_k^+ \subset I_k^+$ and $\hat{I}_k^- \subset I_k^-$ where the magnitude of the set

9.5. Linear Chebyshev Vector Approximation

$\hat{I}_k^+ \cup \hat{I}_k^-$ is not larger than $n+1$ and non-negative real numbers $\hat{\lambda}_{kl}^+$ and $\hat{\lambda}_{kl}^-$ so that:

$$\sum_{l \in \hat{I}_k^+} \hat{\lambda}_{kl}^+ v_{kj}(t_{kl}^+) - \sum_{l \in \hat{I}_k^-} \hat{\lambda}_{kl}^- v_{kj}(t_{kl}^-) = \beta_{kj} \text{ for all } j \in \{1, \ldots, n\}, \quad (9.50)$$

$$\sum_{l \in \hat{I}_k^+} \hat{\lambda}_{kl}^+ + \sum_{l \in \hat{I}_k^-} \hat{\lambda}_{kl}^- = 1, \quad (9.51)$$

$\hat{\lambda}_{kl}^+ > 0$ for some $l \in \hat{I}_k^+$

$$\implies \sum_{j=1}^n \bar{x}_j v_{kj}(t_{kl}^+) - z_k(t_{kl}^+) = \left\| \sum_{j=1}^n \bar{x}_j v_{kj} - z_k \right\|, \quad (9.52)$$

$\hat{\lambda}_{kl}^- > 0$ for some $l \in \hat{I}_k^-$

$$\implies \sum_{j=1}^n \bar{x}_j v_{kj}(t_{kl}^-) - z_k(t_{kl}^-) = -\left\| \sum_{j=1}^n \bar{x}_j v_{kj} - z_k \right\|. \quad (9.53)$$

With (9.47), (9.49) and (9.50) we get

$$\sum_{k=1}^m \tau_k \left[\sum_{l \in \hat{I}_k^+} \hat{\lambda}_{kl}^+ v_{kj}(t_{kl}^+) - \sum_{l \in \hat{I}_k^-} \hat{\lambda}_{kl}^- v_{kj}(t_{kl}^-) \right] = 0 \text{ for all } j \in \{1, \ldots, n\}. \quad (9.54)$$

Finally, we define again some new variables $\bar{\lambda}_{kl}^+ := \hat{\lambda}_{kl}^+$ and $\bar{\lambda}_{kl}^- := -\hat{\lambda}_{kl}^-$. Then we conclude with (9.54) and (9.51):

$$\sum_{k=1}^m \tau_k \left[\sum_{l \in \hat{I}_k^+} \bar{\lambda}_{kl}^+ v_{kj}(t_{kl}^+) + \sum_{l \in \hat{I}_k^-} \bar{\lambda}_{kl}^- v_{kj}(t_{kl}^-) \right] = 0 \text{ for all } j \in \{1, \ldots, n\},$$

$$\sum_{l \in \hat{I}_k^+} |\bar{\lambda}_{kl}^+| + \sum_{l \in \hat{I}_k^-} |\bar{\lambda}_{kl}^-| = 1 \text{ for all } k \in \{1, \ldots, m\},$$

and from (9.52) and (9.53) we get the implications

$\bar{\lambda}_{kl}^+ \neq 0$ for some $k \in \{1, \ldots, m\}$ and some $l \in \hat{I}_k^+$

$$\implies \sum_{j=1}^n \bar{x}_j v_{kj}(t_{kl}^+) - z_k(t_{kl}^+) = \left\| \sum_{j=1}^n \bar{x}_j v_{kj} - z_k \right\| \text{sgn}(\bar{\lambda}_{kl}^+),$$

$\bar{\lambda}_{kl}^- > 0$ for some $k \in \{1,\ldots,m\}$ and some $l \in \hat{I}_k^-$

$$\Longrightarrow \sum_{j=1}^n \bar{x}_j v_{kj}(t_{\overline{kl}}^-) - z_k(t_{\overline{kl}}^-) = \left\| \sum_{j=1}^n \bar{x}_j v_{kj} - z_k \right\| \operatorname{sgn}(\bar{\lambda}_{kl}^-).$$

This leads immediately to the assertion, if we notice that for every $k \in \{1,\ldots,m\}$ the set $\hat{I}_k^+ \cup \hat{I}_k^-$ consists of at most $n+1$ indices. □

This alternation theorem gives necessary and sufficient conditions for an almost properly minimal solution of the linear Chebyshev vector approximation problem (9.33). This result generalizes a known theorem of linear Chebyshev approximation.

Example 9.14. We investigate the following linear Chebyshev vector approximation problem

$$\underset{x \in \mathbb{R}}{\text{a-p-min}} \begin{pmatrix} \|xv - \sinh\| \\ \vdots \\ \|xv' - \cosh\| \end{pmatrix}. \tag{9.55}$$

In our standard assumption (9.31) we have now $\Omega = [0,2]$, $m = 2$, $n = 1$, $z_1 = \sinh$, $z_2 = \cosh$, $v_{11} = v$ (identity on $[0,2]$) and $v_{21} = v'$ ($\equiv 1$). With Theorem 9.13 the necessary and sufficient conditions for an almost properly minimal solution $\bar{x}$ of the linear Chebyshev vector approximation problem (9.55) are given as:

$|\bar{\lambda}_{11}| + |\bar{\lambda}_{12}| = 1,$
$|\bar{\lambda}_{21}| + |\bar{\lambda}_{22}| = 1,$
$\bar{\tau}_1 \bar{\lambda}_{11} \bar{t}_{11} + \bar{\tau}_1 \bar{\lambda}_{12} \bar{t}_{12} + \bar{\tau}_2 \bar{\lambda}_{21} + \bar{\tau}_2 \bar{\lambda}_{22} = 0,$
$\bar{\lambda}_{11} \neq 0 \implies \bar{x} \bar{t}_{11} - \sinh \bar{t}_{11} = \|\bar{x} v - \sinh\| \operatorname{sgn}(\bar{\lambda}_{11}),$
$\bar{\lambda}_{12} \neq 0 \implies \bar{x} \bar{t}_{12} - \sinh \bar{t}_{12} = \|\bar{x} v - \sinh\| \operatorname{sgn}(\bar{\lambda}_{12}),$
$\bar{\lambda}_{21} \neq 0 \implies \bar{x} - \cosh \bar{t}_{21} = \|\bar{x} v' - \cosh\| \operatorname{sgn}(\bar{\lambda}_{21}),$
$\bar{\lambda}_{22} \neq 0 \implies \bar{x} - \cosh \bar{t}_{22} = \|\bar{x} v' - \cosh\| \operatorname{sgn}(\bar{\lambda}_{22}),$
$\bar{t}_{11}, \bar{t}_{12} \in E_1(\bar{x}),$
$\bar{t}_{21}, \bar{t}_{22} \in E_2(\bar{x}),$
$\bar{\lambda}_{11}, \bar{\lambda}_{12}, \bar{\lambda}_{21}, \bar{\lambda}_{22} \in \mathbb{R},$
$\bar{\tau}_1, \bar{\tau}_2 > 0.$

From these conditions one obtains after some calculations that $\bar{x}$ is an almost properly minimal solution of the linear Chebyshev vector approximation problem (9.55) if and only if $\bar{x} \in [\bar{x}_1, \bar{x}_2]$ with $\bar{x}_1 \approx 1.600233$ and $\bar{x}_2 \approx 2.381098$. Figure 9.1 illustrates the approximation of the functions sinh and cosh, if we choose the almost properly minimal solution $\bar{x} = 2$.

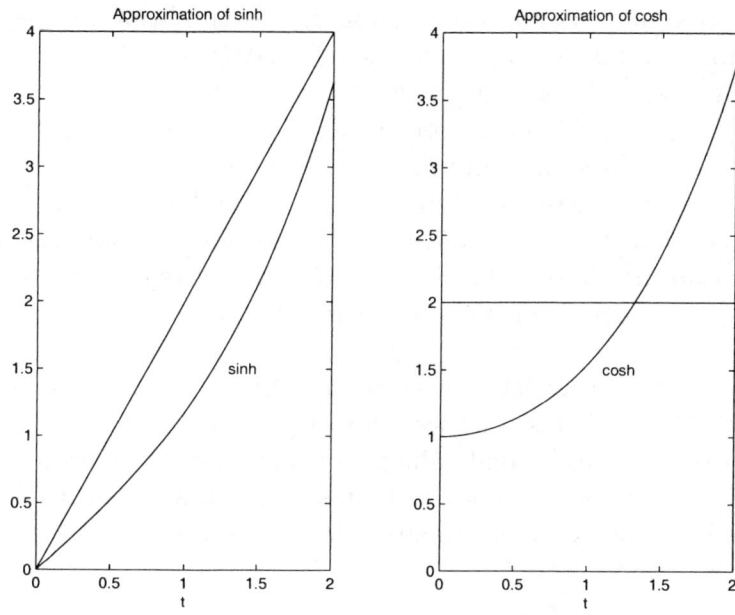

Figure 9.1: Illustration of the approximation of sinh and cosh using $\bar{x} = 2$.

Notes

Several authors investigated vector approximation problems very early, for instance, Bacopoulos [12], Gearhart [100] and others. The example presented in the introduction of this chapter is discussed in detail by Reemtsen [255].

The results concerning the simultaneous approximation in section 9.2 can essentially be found in a paper of Jahn [141]. Theorem 9.1 and Theorem 9.2 extend some results of Bacopoulos-Godini-Singer [15] (see also Bacopoulos-Singer [16], [17] and Bacopoulos-Godini-Singer [13], [14]).

The generalized Kolmogorov condition generalizes an optimality condition of Kolmogorov [177] which was introduced for linear Chebyshev approximation. For a discussion of this condition in the case of scalar approximation the reader is referred to the book of Kirsch-Warth-Werner [171] and a paper of Krabs [185]. Theorem 9.5 was given by Oettli [243] in a more general form; he considers convex maps instead of vectorial norms. Oettli's paper generalizes results of Wanka [318] who extended Theorem 9.5 in the book [145].

The section 9.4 on nonlinear Chebyshev vector approximation is based on an article of Jahn-Sachs [157]. In the real-valued case the representation condition mentioned in Theorem 9.7 was introduced by Krabs [183].

The results of the last section can also be found in a paper of Jahn [139]. Problems of this type were also investigated by Behringer [21] and Censor [50]. In the real-valued case a similar alternation theorem can also be found in the book of Krabs [184, Thm. I.5.6]. Censor [50] examined the necessity of the alternation result as well.

Chapter 10

Cooperative n Player Differential Games

In contrast to the theory of cooperative games introduced by John von Neumann, this chapter is devoted to deterministic differential games with n players behaving exclusively cooperatively. Such games can be described as vector optimization problems. After some basic remarks on the cooperation concept we present necessary and sufficient conditions for optimal and weakly optimal controls concerning a system of ordinary differential equations. In the last section we discuss a special cooperative differential game with a linear differential equation in a Hilbert space.

10.1 Basic Remarks on the Cooperation Concept

Cooperative n player differential games are especially qualified to be formulated as vector optimization problems. The concept used in this book differs from that of John von Neumann because the game is assumed to be exclusively cooperative (e.g., compare also the book of Burger [49, p. 29 and p. 129]). For our investigations we have the following standard assumption denoted by (10.1):

> We assume that n players (individuals or groups) take part in the game (let $n \in \mathbb{N}$ be a fixed number). Let E be a real linear

space and let $Y_1, \ldots, Y_n$ be given partially ordered linear spaces with the ordering cones $C_{Y_1}, \ldots, C_{Y_n}$. In E let a nonempty subset be given - the so-called set of all *playable* $(n+2)$-tuples $(x, u_1, \ldots, u_n, \hat{t})$ where u_i denotes a feasible control of player i, x is a resulting state and $\hat{t}$ represents the terminal time of the control problem. In the following sections the set S will be described in detail. The goal of the i-th player ($i \in \{1, \ldots, n\}$) is formulated by an objective map $v_i : S \to Y_i$.

Under the assumption (10.1) the cooperative game reads as follows: Determine a playable $(n+2)$-tuple $(\bar{x}, \bar{u}_1, \ldots, \bar{u}_n, \bar{t}) \in S$ which is "preferred" by all players because of their cooperation.

For the mathematical description of the cooperation concept it is reasonable to consider the product space $Y := \prod_{i=1}^{n} Y_i$ with the product partial ordering induced by the product ordering cone $C_Y := \prod_{i=1}^{n} C_{Y_i}$, and we define the objective map $v : S \to Y$ by

$$v(x, u_1, \ldots, u_n, \hat{t}) = \begin{pmatrix} v_1(x, u_1, \ldots, u_n, \hat{t}) \\ \vdots \\ v_n(x, u_1, \ldots, u_n, \hat{t}) \end{pmatrix}$$

for all $(x, u_1, \ldots, u_n, \hat{t}) \in S$.

With this map we introduce the following

Definition 10.1. Let the assumption (10.1) be satisfied.

(a) A playable $(n+2)$-tuple $(\bar{x}, \bar{u}_1, \ldots, \bar{u}_n, \bar{t}) \in S$ is called *optimal*, if $v(\bar{x}, \bar{u}_1, \ldots, \bar{u}_n, \bar{t})$ is a minimal element of the set $\prod_{i=1}^{n} v_i(S)$. In this case the control $\bar{u}_i$ ($i \in \{1, \ldots, n\}$) is said to be an *optimal control* of the i-th player.

(b) In addition, let the product ordering cone C_Y have a nonempty algebraic interior. A playable $(n+2)$-tuple $(\bar{x}, \bar{u}_1, \ldots, \bar{u}_n, \bar{t}) \in S$ is called *weakly optimal*, if $v(\bar{x}, \bar{u}_1, \ldots, \bar{u}_n, \bar{t})$ is a weakly minimal element of the set $\prod_{i=1}^{n} v_i(S)$. In this case the control $\bar{u}_i$ ($i \in \{1,\ldots,n\}$) is said to be a *weakly optimal control* of the i-th player.

Using the product ordering we get an adequate description of the cooperation, since playable $(n+2)$-tuples are "preferred" if and only if they are "preferred" by each player. In the case of only one player, i.e. for $n = 1$, this cooperative game reduces to a usual control problem with a vector-valued objective map.

Finally, we note without proof how to obtain the dual ordering cone and its quasi-interior with respect to the product space Y.

Lemma 10.2. *Let the assumption (10.1) be satisfied, and let Y and C_Y be given as $Y := \prod_{i=1}^{n} Y_i$ and $C_Y := \prod_{i=1}^{n} C_{Y_i}$, respectively. Then the dual ordering cone $C_{Y'}$ equals $\prod_{i=1}^{n} C_{Y'_i}$ and its quasi-interior $C_{Y'}^{\#}$ equals $\prod_{i=1}^{n} C_{Y'_i}^{\#}$. If $Y_1, \ldots, Y_n$ are topological linear spaces, then $C_{Y^*} := \prod_{i=1}^{n} C_{Y_i^*}$ and $C_{Y^*}^{\#} := \prod_{i=1}^{n} C_{Y_i^*}^{\#}$.*

10.2 A Maximum Principle

In this section we investigate a cooperative differential game where the state equation is a nonlinear differential equation and the values of the controls are restricted to a certain set. For this game we derive a maximum principle as a necessary optimality condition which extends the well-known maximum principle given by Pontryagin-Boltyanskii-Gamkrelidze-Mishchenko [253] and Hestenes [120] to these games. The question under which assumption this maximum principle is also a sufficient optimality condition is investigated in the second part of this section. The Hamilton-Jacobi-Bellmann equations are also discussed in this context.

In the following we consider a cooperative n player differential game formulated in

Problem 10.3. On a fixed time interval $[t_0, t_1]$ (with $t_0 < t_1$) let the state equation be given as

$$\dot{x}(t) = f(x(t), u_1(t), \ldots, u_n(t)) \text{ almost everywhere on } [t_0, t_1]. \quad (10.2)$$

We assume that $u_1 \in L_\infty^{s_1}([t_0, t_1]), \ldots, u_n \in L_\infty^{s_n}([t_0, t_1])$ are the controls of the n players $(s_1, \ldots, s_n \in \mathbb{N})$ satisfying the condition

$$u_i(t) \in \Omega_i \text{ almost everywhere on } [t_0, t_1].$$

$\Omega_1, \ldots, \Omega_n$ are assumed to be nonempty subsets of the real linear spaces $\mathbb{R}^{s_1}, \ldots, \mathbb{R}^{s_n}$. Let $f : \mathbb{R}^m \times \mathbb{R}^{s_1} \times \cdots \times \mathbb{R}^{s_n} \to \mathbb{R}^m$ (with $m \in \mathbb{N}$) be a vector function which is Lipschitz continuous with respect to $x, u_1, \ldots, u_n$. The solutions x of the differential equation (10.2) are defined as absolutely continuous vector functions in the sense of Carathéodory (e.g., see Curtain-Pritchard [75, p. 122]), i.e.

$$x \in W_{1,\infty}^m([t_0, t_1]) := \{y : [t_0, t_1] \to \mathbb{R}^m \mid y \text{ is} \\ \text{absolutely continuous on } [t_0, t_1] \\ \text{and } \dot{y} \in L_\infty^m([t_0, t_1])\}.$$

Let

$$x(t_0) = x_0 \quad (10.3)$$

be the initial condition with some fixed $x_0 \in \mathbb{R}^m$; the terminal condition reads as

$$x(t_1) \in Q \quad (10.4)$$

where the target set Q is assumed to be a nonempty subset of $\mathbb{R}^m$. In this case the set S of all playable $(n+2)$-tuples $(x, u_1, \ldots, u_n, \hat{t})$ is defined as follows

$$S := \{(x, u_1, \ldots, u_n, \hat{t}) \mid \hat{t} \in [t_0, t_1]; \text{ for all } i \in \{1, \ldots, n\} \\ \text{we have } u_i \in L_\infty^{s_i}([t_0, t_1]) \text{ and } u_i(t) \in \Omega_i \\ \text{almost everywhere on } [t_0, t_1]; \ x \in W_{1,\infty}^m([t_0, t_1]); \\ u_1, \ldots, u_n \text{ and } x \text{ satisfy the equations (10.2) and} \\ (10.3) \text{ and the condition (10.4)}\}.$$

10.2. A Maximum Principle

Let $(Y_1, \|\cdot\|_{Y_1}), \ldots, (Y_n, \|\cdot\|_{Y_n})$ be real Banach spaces. Then the objective map $v_i : S \to Y_i$ of the i-th player is assumed to be given as

$$v_i(x, u_1, \ldots, u_n, \hat{t}) = h_i(x(\hat{t})) + \int_{t_0}^{\hat{t}} f_i^0(x(t), u_1(t), \ldots, u_n(t))\, dt$$

$$\text{for all } (x, u_1, \ldots, u_n, \hat{t}) \in S \qquad (10.5)$$

where $h_i : \mathbb{R}^m \to Y_i$ and $f_i^0 : \mathbb{R}^m \times \mathbb{R}^{s_1} \times \cdots \times \mathbb{R}^{s_n} \to Y_i$ are given maps. For every $(x, u_1, \ldots, u_n, \hat{t}) \in S$ the composition $f_i^0 \circ (x, u_1, \ldots, u_n)$ is assumed to be Bochner integrable (e.g., compare Curtain-Pritchard [75, p. 88]). Therefore, the integral appearing in (10.5) is a Bochner integral.

10.2.1 Necessary Conditions for Optimal and Weakly Optimal Controls

In this subsection we aim at an optimality condition for the cooperative differential game outlined in Problem 10.3. For simplicity we restrict ourselves to the special case that the terminal time $\hat{t} = t_1$ is fixed. In order to be able to prove our main result we need

Lemma 10.4. *Let A be a matrix function on $[t_0, t_1]$ with real coefficients. If Φ is the unique solution of the equations*

$$\left.\begin{array}{l} \dot{\Phi}(t) = A(t)\Phi(t) \text{ almost everywhere on } [t_0, t_1], \\ \Phi(t_0) = I \text{ (identity)}, \end{array}\right\} \qquad (10.6)$$

then for an arbitrary $y \in W_{1,\infty}^m([t_0, t_1])$ the function

$$x(\cdot) = y(\cdot) + \Phi(\cdot) \int_{t_0}^{\cdot} \Phi^{-1}(s) A(s) y(s)\, ds \qquad (10.7)$$

satisfies the integral equation

$$x(\cdot) - \int_{t_0}^{\cdot} A(s) x(s)\, ds = y(\cdot). \qquad (10.8)$$

Proof. For an arbitrary $y \in W_{1,\infty}^m([t_0, t_1])$ we get from (10.7) using integration by parts that

$$x(\cdot) - \int_{t_0}^{\cdot} A(s)x(s)\,ds$$

$$= y(\cdot) + \Phi(\cdot) \int_{t_0}^{\cdot} \Phi^{-1}(s)A(s)y(s)\,ds$$

$$- \int_{t_0}^{\cdot} A(s)\left[y(s) + \Phi(s)\int_{t_0}^{s} \Phi^{-1}(\sigma)A(\sigma)y(\sigma)\,d\sigma\right] ds$$

$$= y(\cdot) + \Phi(\cdot) \int_{t_0}^{\cdot} \Phi^{-1}(s)A(s)y(s)\,ds - \int_{t_0}^{\cdot} A(s)y(s)\,ds$$

$$- \int_{t_0}^{\cdot} \dot{\Phi}(s) \int_{t_0}^{s} \Phi^{-1}(\sigma)A(\sigma)y(\sigma)\,d\sigma\,ds$$

$$= y(\cdot) + \Phi(\cdot) \int_{t_0}^{\cdot} \Phi^{-1}(s)A(s)y(s)\,ds - \int_{t_0}^{\cdot} A(s)y(s)\,ds$$

$$- \Phi(\cdot) \int_{t_0}^{\cdot} \Phi^{-1}(s)A(s)y(s)\,ds + \int_{t_0}^{\cdot} \Phi(s)\Phi^{-1}(s)A(s)y(s)\,ds$$

$$= y(\cdot).$$

Hence, the equation (10.8) is satisfied. $\square$

Now we are able to formulate a *Pontryagin maximum principle* for the cooperative n player differential game introduced in Problem 10.3.

Theorem 10.5. *Let the cooperative n player differential game formulated in Problem 10.3 be given with a fixed terminal time $\hat{t} = t_1$ and the target set*

$$Q := \{\tilde{x} \in \mathbb{R}^m \mid g(\tilde{x}) = 0_{\mathbb{R}^r}\}$$

10.2. A Maximum Principle

where $g : \mathbb{R}^m \to \mathbb{R}^r$ (with $r \in \mathbb{N}$) is a continuously differentiable vector function. The maps $h_1, \ldots, h_n$ and $f_1^0, \ldots, f_n^0$ are assumed to be continuously Fréchet differentiable. Let f be continuously partially differentiable. Moreover, for every $i \in \{1, \ldots, n\}$ let the sets Ω_i be convex and let it have a nonempty interior. Let the ordering cones $C_{Y_1}, \ldots, C_{Y_n}$ have a nonempty algebraic interior. Let $\bar{u}_1, \ldots, \bar{u}_n$ be weakly optimal controls of the n players, and let $\bar{x} \in W_{1,\infty}^m([t_0, t_1])$ be the resulting state. Furthermore, let the matrix $\frac{\partial g}{\partial x}(\bar{x}(t_1))$ have maximal rank. Then there are continuous linear functionals $l_i \in C_{Y_i^*}$ (for all $i \in \{1, \ldots, n\}$), a vector function $w \in W_{1,\infty}^m([t_0, t_1])$ and a vector $a \in \mathbb{R}^r$ so that $(l_1, \ldots, l_n, w) \neq (0_{Y_1^*}, \ldots, 0_{Y_n^*}, 0_{W_{1,\infty}^m([t_0,t_1])})$ and

(a)

$$-\dot{w}(t)^T = w(t)^T \frac{\partial f}{\partial x}(\bar{x}(t), \bar{u}_1(t), \ldots, \bar{u}_n(t))$$
$$- \sum_{i=1}^{n} \left(l_i \circ \frac{\partial f_i^0}{\partial x} \right)(\bar{x}(t), \bar{u}_1(t), \ldots, \bar{u}_n(t))$$

almost everywhere on $[t_0, t_1]$, \hfill (10.9)

(b)

$$-w(t_1)^T = a^T \frac{\partial g}{\partial x}(\bar{x}(t_1)) + \sum_{i=1}^{n} \left(l_i \circ \frac{\partial h_i}{\partial x} \right)(\bar{x}(t_1)), \quad (10.10)$$

(c) for every $k \in \{1, \ldots, n\}$ and every $u_k \in L_\infty^{s_k}([t_0, t_1])$ with $u_k(t) \in \Omega_k$ almost everywhere on $[t_0, t_1]$ we have

$$\left[w(t)^T \frac{\partial f}{\partial u_k}(\bar{x}(t), \bar{u}_1(t), \ldots, \bar{u}_n(t)) \right.$$
$$\left. - \sum_{i=1}^{n} \left(l_i \circ \frac{\partial f_i^0}{\partial u_k} \right)(\bar{x}(t), \bar{u}_1(t), \ldots, \bar{u}_n(t)) \right] (u_k(t) - \bar{u}_k(t)) \leq 0$$

almost everywhere on $[t_0, t_1]$. \hfill (10.11)

Proof. Let $\bar{u}_1, \ldots, \bar{u}_n$ be any weakly optimal controls of the n players with a resulting state $\bar{x}$. For a better formulation of the considered cooperative game as an abstract optimization problem we introduce the product space

$$L := W_{1,\infty}^m([t_0, t_1]) \times L_\infty^{s_1}([t_0, t_1]) \times \cdots \times L_\infty^{s_n}([t_0, t_1]).$$

Instead of writing $(x, u_1, \ldots, u_n)$ for an arbitrary element of L we use the abbreviation (x,u). The objective map $F: L \to Y_1 \times \cdots \times Y_n$ is defined by

$$F(x, u) = \begin{pmatrix} h_1(x(t_1)) + \int_{t_0}^{t_1} f_1^0(x(t), u(t))\, dt \\ \vdots \\ h_n(x(t_1)) + \int_{t_0}^{t_1} f_n^0(x(t), u(t))\, dt \end{pmatrix} \quad \text{for all } (x, u) \in L,$$

and the constraint map $G: L \to W_{1,\infty}^m([t_0, t_1]) \times \mathbb{R}^r$ is given by

$$G(x, u) = \begin{pmatrix} x(\cdot) - x_0 - \int_{t_0}^{\cdot} f(x(s), u(s))\, ds \\ g(x(t_1)) \end{pmatrix} \quad \text{for all } (x, u) \in L.$$

Then the considered cooperative n player differential game can be formulated as

$$\left. \begin{array}{l} \min F(x, u) \\ \text{subject to the constraints} \\ (x, u) \in \hat{S} := \{(x, u) \in L \mid u_i(t) \in \Omega_i \text{ almost everywhere} \\ \qquad\qquad\qquad\qquad\qquad\qquad \text{on } [t_0, t_1]\ (i \in \{1, \ldots, n\})\} \\ G(x, u) = 0_{W_{1,\infty}^m([t_0, t_1]) \times \mathbb{R}^r}. \end{array} \right\} \quad (10.12)$$

By assumption $(\bar{x}, \bar{u})$ is a weakly minimal solution of this abstract optimization problem. It is our aim to apply the generalized Lagrange multiplier rule (Theorem 7.4) to this special problem. But first, we briefly check the required assumptions. By an extensive computation

10.2. A Maximum Principle

one can see that F is Fréchet differentiable at $(\bar{x}, \bar{u})$ and that the Fréchet derivative of F at $(\bar{x}, \bar{u})$ is given as

$$F'(\bar{x}, \bar{u})(x, u) =$$

$$\begin{pmatrix} \frac{\partial h_1}{\partial x}(\bar{x}(t_1))x(t_1) + \int_{t_0}^{t_1} \left[\frac{\partial f_1^0}{\partial x}(\bar{x}(s), \bar{u}(s))x(s) + \sum_{j=1}^{n} \frac{\partial f_1^0}{\partial u_j}(\bar{x}(s), \bar{u}(s))u_j(s) \right] ds \\ \vdots \\ \frac{\partial h_n}{\partial x}(\bar{x}(t_1))x(t_1) + \int_{t_0}^{t_1} \left[\frac{\partial f_n^0}{\partial x}(\bar{x}(s), \bar{u}(s))x(s) + \sum_{j=1}^{n} \frac{\partial f_n^0}{\partial u_j}(\bar{x}(s), \bar{u}(s))u_j(s) \right] ds \end{pmatrix}$$

for all $(x, u) \in L$

(for a proof notice that for every Bochner integrable function y with values in a real Banach space $(Y, \|\cdot\|_Y)$ one has

$$\left\| \int y(s) \, ds \right\|_Y \leq \int \|y(s)\|_Y \, ds$$

where the integral on the left side of this inequality is a Bochner integral and the integral on the right side is a Lebesgue integral). Moreover, the map G is continuously Fréchet differentiable at $(\bar{x}, \bar{u})$ and its Fréchet derivative is given by

$$G'(\bar{x}, \bar{u})(x, u) =$$

$$\begin{pmatrix} x(\cdot) - \int_{t_0}^{\cdot} \left[\frac{\partial f}{\partial x}(\bar{x}(s), \bar{u}(s))x(s) + \sum_{j=1}^{n} \frac{\partial f}{\partial u_j}(\bar{x}(s), \bar{u}(s))u_j(s) \right] ds \\ \frac{\partial g}{\partial x}(\bar{x}(t_1))x(t_1) \end{pmatrix}$$

for all $(x, u) \in L$.

Furthermore, since for every $i \in \{1, \ldots, n\}$ the sets Ω_i are assumed to be convex with a nonempty interior, the superset $\hat{S}$ defined in problem (10.12) is also convex and it has a nonempty interior. Then, by Theorem 7.4, there are linear functionals $l_1 \in C_{Y_1^*}, \ldots, l_n \in C_{Y_n^*}$ (compare also Lemma 10.2) and $l \in W_{1,\infty}^m([t_0, t_1])'$ and a vector $a \in \mathbb{R}^r$ with $(l_1, \ldots, l_n, l, a) \neq (0_{Y_1^*}, \ldots, 0_{Y_n^*}, 0_{W_{1,\infty}^m([t_0,t_1])'}, 0_{\mathbb{R}^r})$ and

$((l_1, \ldots, l_n) \circ F'(\bar{x}, \bar{u}) + (l, a) \circ G'(\bar{x}, \bar{u}))(x - \bar{x}, u - \bar{u}) \geq 0$ for all $(x, u) \in \hat{S}$

(since we do not prove that $G'(\bar{x}, \bar{u})(L)$ is closed, we cannot assert that the linear functional l is continuous). This inequality implies

$$\sum_{i=1}^{n} l_i \left[\frac{\partial h_i}{\partial x}(\bar{x}(t_1))(x(t_1) - \bar{x}(t_1)) + \int_{t_0}^{t_1} \left[\frac{\partial f_i^0}{\partial x}(\bar{x}(s), \bar{u}(s))(x(s) - \bar{x}(s)) \right. \right.$$

$$\left. + \sum_{j=1}^{n} \frac{\partial f_i^0}{\partial u_j}(\bar{x}(s), \bar{u}(s))(u_j(s) - \bar{u}_j(s)) \right] ds \right] + l\left[x(\cdot) - \bar{x}(\cdot)\right.$$

$$- \int_{t_0}^{\cdot} \left[\frac{\partial f}{\partial x}(\bar{x}(s), \bar{u}(s))(x(s) - \bar{x}(s)) \right.$$

$$\left. + \sum_{j=1}^{n} \frac{\partial f}{\partial u_j}(\bar{x}(s), \bar{u}(s))(u_j(s) - \bar{u}_j(s)) \right] ds \right]$$

$$+ a^T \frac{\partial g}{\partial x}(\bar{x}(t_1))(x(t_1) - \bar{x}(t_1)) \geq 0 \text{ for all } (x, u) \in \hat{S}. \quad (10.13)$$

If we plug $u = (\bar{u}_1, \ldots, \bar{u}_n)$ into the inequality (10.13) we get

$$\sum_{i=1}^{n} l_i \left[\frac{\partial h_i}{\partial x}(\bar{x}(t_1))(x(t_1) - \bar{x}(t_1)) \right.$$

$$\left. + \int_{t_0}^{t_1} \frac{\partial f_i^0}{\partial x}(\bar{x}(s), \bar{u}(s))(x(s) - \bar{x}(s)) \, ds \right]$$

$$+ l\left[x(\cdot) - \bar{x}(\cdot) - \int_{t_0}^{\cdot} \frac{\partial f}{\partial x}(\bar{x}(s), \bar{u}(s))(x(s) - \bar{x}(s)) \, ds \right]$$

$$+ a^T \frac{\partial g}{\partial x}(\bar{x}(t_1))(x(t_1) - \bar{x}(t_1)) \geq 0 \text{ for all } x \in W_{1,\infty}^m([t_0, t_1])$$

resulting in

$$\sum_{i=1}^{n} l_i \left[\frac{\partial h_i}{\partial x}(\bar{x}(t_1))x(t_1) + \int_{t_0}^{t_1} \frac{\partial f_i^0}{\partial x}(\bar{x}(s), \bar{u}(s))x(s) \, ds \right]$$

$$+ l\left[x(\cdot) - \int_{t_0}^{\cdot} \frac{\partial f}{\partial x}(\bar{x}(s), \bar{u}(s))x(s) \, ds \right]$$

10.2. A Maximum Principle

$$+a^T \frac{\partial g}{\partial x}(\bar{x}(t_1))x(t_1) = 0 \text{ for all } x \in W_{1,\infty}^m([t_0,t_1]). \quad (10.14)$$

For $x = \bar{x}$ it follows from the inequality (10.13)

$$\sum_{i=1}^n l_i \left[\int_{t_0}^{t_1} \sum_{j=1}^n \frac{\partial f_i^0}{\partial u_j}(\bar{x}(s),\bar{u}(s))(u_j(s) - \bar{u}_j(s))\,ds \right]$$

$$+l\left[-\int_{t_0}^{\cdot} \sum_{j=1}^n \frac{\partial f}{\partial u_j}(\bar{x}(s),\bar{u}(s))(u_j(s) - \bar{u}_j(s))\,ds \right] \geq 0$$

for all $(u_1,\ldots,u_n) \in L_\infty^{s_1}([t_0,t_1]) \times \cdots \times L_\infty^{s_n}([t_0,t_1])$
with $u_i(t) \in \Omega_i$ almost everywhere on $[t_0,t_1]$ ($i \in \{1,\ldots,n\}$).

For every $k \in \{1,\ldots,n\}$ we obtain with $u_j = \bar{u}_j$ for $j \in \{1,\ldots,n\}\setminus\{k\}$

$$\sum_{i=1}^n l_i \left[\int_{t_0}^{t_1} \frac{\partial f_i^0}{\partial u_k}(\bar{x}(s),\bar{u}(s))(u_k(s) - \bar{u}_k(s))\,ds \right]$$

$$-l\left[\int_{t_0}^{\cdot} \frac{\partial f}{\partial u_k}(\bar{x}(s),\bar{u}(s))(u_k(s) - \bar{u}_k(s))\,ds \right] \geq 0$$

for all $u_k \in L_\infty^{s_k}([t_0,t_1])$ with $u_k(t) \in \Omega_k$
almost everywhere on $[t_0,t_1]$. $\quad (10.15)$

Next, we investigate the equation (10.14) and we try to characterize the linear functional l. For every $y \in W_{1,\infty}^m([t_0,t_1])$ we obtain with (10.14) and Lemma 10.4 where $A(t) := \frac{\partial f}{\partial x}(\bar{x}(t),\bar{u}(t))$ and Φ is the unique solution of (10.6)

$$\begin{aligned}l(y) &= -\left(\sum_{i=1}^n \left(l_i \circ \frac{\partial h_i}{\partial x}\right)(\bar{x}(t_1)) + a^T \frac{\partial g}{\partial x}(\bar{x}(t_1))\right)x(t_1)\\ &\quad - \sum_{i=1}^n l_i\left(\int_{t_0}^{t_1} \frac{\partial f_i^0}{\partial x}(\bar{x}(s),\bar{u}(s))x(s)\,ds\right)\\ &= -\left(\sum_{i=1}^n \left(l_i \circ \frac{\partial h_i}{\partial x}\right)(\bar{x}(t_1)) + a^T \frac{\partial g}{\partial x}(\bar{x}(t_1))\right)\left(y(t_1)\right)\end{aligned}$$

$$+\Phi(t_1)\int_{t_0}^{t_1} \Phi^{-1}(s)\frac{\partial f}{\partial x}(\bar{x}(s),\bar{u}(s))y(s)\,ds\Bigg)$$

$$-\int_{t_0}^{t_1}\sum_{i=1}^{n}\left(l_i \circ \frac{\partial f_i^0}{\partial x}\right)(\bar{x}(s),\bar{u}(s))\Big(y(s)$$

$$+\Phi(s)\int_{t_0}^{s}\Phi^{-1}(\sigma)\frac{\partial f}{\partial x}(\bar{x}(\sigma),\bar{u}(\sigma))y(\sigma)\,d\sigma\Big)ds.$$

Notice that the l_i ($i \in \{1,\ldots,n\}$) are written behind the integral sign; this is possible because every l_i is a continuous linear functional (compare Hille-Phillips [121, p. 83–84] or Warga [319, p. 82]). In the following we use the abbreviations

$$b := \sum_{i=1}^{n}\left(l_i \circ \frac{\partial h_i}{\partial x}\right)(\bar{x}(t_1)) + a^T\frac{\partial g}{\partial x}(\bar{x}(t_1)),$$

$$c(s) := \sum_{i=1}^{n}\left(l_i \circ \frac{\partial f_i^0}{\partial x}\right)(\bar{x}(s),\bar{u}(s))$$

and

$$d(s) := \frac{\partial f}{\partial x}(\bar{x}(s),\bar{u}(s)).$$

Using integration by parts we get for every $y \in W_{1,\infty}^m([t_0,t_1])$

$$l(y) = -b\Big(y(t_1) + \Phi(t_1)\int_{t_0}^{t_1}\Phi^{-1}(s)d(s)y(s)\,ds\Big)$$

$$-\int_{t_0}^{t_1}c(s)\Big(y(s)+\Phi(s)\int_{t_0}^{s}\Phi^{-1}(\sigma)d(\sigma)y(\sigma)\,d\sigma\Big)ds$$

$$= -b\Big(y(t_1)+\Phi(t_1)\int_{t_0}^{t_1}\Phi^{-1}(s)d(s)y(s)\,ds\Big) - \int_{t_0}^{t_1}c(s)y(s)\,ds$$

$$-\int_{t_0}^{t_1}c(s)\Phi(s)\,ds\int_{t_0}^{t}\Phi^{-1}(s)d(s)y(s)\,ds\Big|_{t_0}^{t_1}$$

10.2. A Maximum Principle

$$+ \int_{t_0}^{t_1} \int_{t_0}^{t} c(s)\Phi(s)\,ds\ \Phi^{-1}(t)d(t)y(t)\,dt$$

$$= -b\left(y(t_1) + \Phi(t_1)\int_{t_0}^{t_1} \Phi^{-1}(s)d(s)y(s)\,ds\right) - \int_{t_0}^{t_1} c(s)y(s)\,ds$$

$$- \int_{t_0}^{t_1} c(s)\Phi(s)\,ds \int_{t_0}^{t_1} \Phi^{-1}(t)d(t)y(t)\,dt$$

$$+ \int_{t_0}^{t_1} \int_{t_0}^{t} c(s)\Phi(s)\,ds\ \Phi^{-1}(t)d(t)y(t)\,dt$$

$$= -b\,y(t_1) + \int_{t_0}^{t_1} \Big[-b\,\Phi(t_1)\Phi^{-1}(t)\,d(t) - c(t)$$

$$- \int_{t}^{t_1} c(s)\Phi(s)\,ds\ \Phi^{-1}(t)d(t)\Big] y(t)\,dt.$$

For the expression in brackets we introduce the abbreviation $p(t)^T$, that is

$$p(t)^T = -b\,\Phi(t_1)\Phi^{-1}(t)\,d(t) - c(t)$$

$$- \int_{t}^{t_1} c(s)\Phi(s)\,ds\ \Phi^{-1}(t)d(t) \text{ almost everywhere on } [t_0, t_1].$$

With the differential equation in (10.6) we obtain

$$\dot{\Phi^{-1}}(t) = -\Phi^{-1}(t)\dot{\Phi}(t)\Phi^{-1}(t)$$
$$= -\Phi^{-1}(t)d(t)\Phi(t)\Phi^{-1}(t)$$
$$= -\Phi^{-1}(t)d(t) \text{ almost everywhere on } [t_0, t_1]$$

and, therefore, we conclude

$$p(t)^T = b\,\Phi(t_1)\dot{\Phi^{-1}}(t) - c(t) + \int_{t}^{t_1} c(s)\Phi(s)\,ds\ \dot{\Phi^{-1}}(t)$$

almost everywhere on $[t_0, t_1]$.

For
$$w(t)^T := -b\,\Phi(t_1)\Phi^{-1}(t) - \int_t^{t_1} c(s)\Phi(s)\,ds\;\Phi^{-1}(t) \text{ for all } t \in [t_0, t_1]$$

it follows
$$\dot{w}(t) = -p(t) \text{ almost everywhere on } [t_0, t_1].$$

Then we get
$$w(t_1) = -b$$

and

$$w(t)^T d(t) - c(t)$$
$$= -b\,\Phi(t_1)\Phi^{-1}(t)d(t) - \int_t^{t_1} c(s)\Phi(s)\,ds\;\Phi^{-1}(t)d(t) - c(t)$$
$$= p(t)^T = -\dot{w}(t)^T \text{ almost everywhere on } [t_0, t_1].$$

Consequently, w satisfies the differential equation

$$-\dot{w}(t)^T = w(t)^T \frac{\partial f}{\partial x}(\bar{x}(t), \bar{u}(t)) - \sum_{i=1}^n \left(l_i \circ \frac{\partial f_i^0}{\partial x}\right)(\bar{x}(t), \bar{u}(t))$$
almost everywhere on $[t_0, t_1]$

and the terminal condition

$$-w(t_1)^T = \sum_{i=1}^n \left(l_i \circ \frac{\partial h_i}{\partial x}\right)(\bar{x}(t_1)) + a^T \frac{\partial g}{\partial x}(\bar{x}(t_1)). \qquad (10.16)$$

Hence, the conditions (10.9) and (10.10) are satisfied. Then the linear functional l can be represented as

$$l(y) = w(t_1)^T y(t_1) - \int_{t_0}^{t_1} \dot{w}(t)^T y(t)\,dt \text{ for all } y \in W_{1,\infty}^m([t_0, t_1])$$

and, therefore, the linear functional l is continuous.

10.2. A Maximum Principle

Next, we assert that $(l_1, \ldots, l_n, w) \neq (0_{Y_1^*}, \ldots, 0_{Y_n^*}, 0_{W_{1,\infty}^m([t_0,t_1])})$. If we assume that the $(n+1)$-tuple $(l_1, \ldots, l_n, w)$ is zero, then we conclude $l = 0_{W_{1,\infty}^m([t_0,t_1])^*}$, and with the equality (10.16) and the assumption that $\frac{\partial g}{\partial x}(\bar{x}(t_1))$ has maximal rank it follows $a = 0_{\mathbb{R}^r}$ as well. But this contradicts the fact that the $(n+2)$-tuple $(l_1, \ldots, l_n, l, a)$ is nonzero. Hence, the $(n+1)$-tuple $(l_1, \ldots, l_n, w)$ is nonzero. Finally, we turn our attention to the inequality (10.15). Using integration by parts we obtain for every $k \in \{1, \ldots, n\}$

$$0 \leq \int_{t_0}^{t_1} \sum_{i=1}^{n} \left(l_i \circ \frac{\partial f_i^0}{\partial u_k}\right)(\bar{x}(s), \bar{u}(s))(u_k(s) - \bar{u}_k(s))\, ds$$

$$-l\left(\int_{t_0}^{\cdot} \frac{\partial f}{\partial u_k}(\bar{x}(s), \bar{u}(s))(u_k(s) - \bar{u}_k(s))\, ds\right)$$

$$= \int_{t_0}^{t_1} \sum_{i=1}^{n} \left(l_i \circ \frac{\partial f_i^0}{\partial u_k}\right)(\bar{x}(s), \bar{u}(s))(u_k(s) - \bar{u}_k(s))\, ds$$

$$- w(t_1)^T \int_{t_0}^{t_0} \frac{\partial f}{\partial u_k}(\bar{x}(s), \bar{u}(s))(u_k(s) - \bar{u}_k(s))\, ds$$

$$+ \int_{t_0}^{t_1} \dot{w}(t)^T \int_{t_0}^{t} \frac{\partial f}{\partial u_k}(\bar{x}(s), \bar{u}(s))(u_k(s) - \bar{u}_k(s))\, ds\, dt$$

$$= \int_{t_0}^{t_1} \sum_{i=1}^{n} \left(l_i \circ \frac{\partial f_i^0}{\partial u_k}\right)(\bar{x}(s), \bar{u}(s))(u_k(s) - \bar{u}_k(s))\, ds$$

$$- w(t_1)^T \int_{t_0}^{t_1} \frac{\partial f}{\partial u_k}(\bar{x}(s), \bar{u}(s))(u_k(s) - \bar{u}_k(s))\, ds$$

$$+ w(t)^T \int_{t_0}^{t} \frac{\partial f}{\partial u_k}(\bar{x}(s), \bar{u}(s))(u_k(s) - \bar{u}_k(s))\, ds \Big|_{t_0}^{t_1}$$

$$-\int_{t_0}^{t_1} w(t)^T \frac{\partial f}{\partial u_k}(\bar{x}(t), \bar{u}(t))(u_k(t) - \bar{u}_k(t))\, dt$$

$$= \int_{t_0}^{t_1} \Big[\sum_{i=1}^{n} \Big(l_i \circ \frac{\partial f_i^0}{\partial u_k}\Big)(\bar{x}(t), \bar{u}(t)) - w(t)^T \frac{\partial f}{\partial u_k}(\bar{x}(t), \bar{u}(t))\Big]$$

$(u_k(t) - \bar{u}_k(t))\, dt$ for all $u_k \in L_\infty^{s_k}([t_0, t_1])$
with $u_k(t) \in \Omega_k$ almost everywhere on $[t_0, t_1]$.

Then for every $k \in \{1, \ldots, n\}$ and every $u_k \in L_\infty^{s_k}([t_0, t_1])$ with $u_k(t) \in \Omega_k$ almost everywhere on $[t_0, t_1]$ we conclude

$$\Big[-\sum_{i=1}^{n}\Big(l_i \circ \frac{\partial f_i^0}{\partial u_k}\Big)(\bar{x}(t), \bar{u}(t))$$

$$-w(t)^T \frac{\partial f}{\partial u_k}(\bar{x}(t), \bar{u}(t))\Big](u_k(t) - \bar{u}_k(t)) \leq 0$$

almost everywhere on $[t_0, t_1]$.

Consequently, the inequality (10.11) is satisfied which completes the proof of this theorem. □

The maximum principle of the preceding theorem consists mainly of three types. The differential equation (10.9) is also called the *adjoint equation*, the terminal condition (10.10) is called *transversality condition* and for every $k \in \{1, \ldots, n\}$ the inequality (10.11) is said to be the *local Pontryagin maximum principle* (compare also the book of Pontryagin-Boltyanskii-Gamkrelidze-Mishchenko [253]). If we define the so-called *Hamiltonian map*

$$H : W_{1,\infty}^m([t_0,t_1]) \times L_\infty^{s_1}([t_0,t_1]) \times \cdots \times L_\infty^{s_n}([t_0,t_1])$$
$$\times W_{1,\infty}^m([t_0,t_1]) \times Y_1^* \times \cdots \times Y_n^* \longrightarrow W_{1,\infty}^m([t_0,t_1])$$

by

$$H(x, u_1, \ldots, u_n, w, y_1^*, \ldots, y_n^*)(t)$$
$$:= w(t)^T f(x(t), u_1(t), \ldots, u_n(t)) - \sum_{i=1}^{n}(y_i^* \circ f_i^0)(x(t), u_1(t), \ldots, u_n(t))$$

almost everywhere on $[t_0, t_1]$,

10.2. A Maximum Principle

then the adjoint equation (10.9) can be written as

$$-\dot{w}(t)^T = \frac{\partial H}{\partial x}(\bar{x}, \bar{u}_1, \ldots, \bar{u}_n, w, l_1, \ldots, l_n)(t)$$

almost everywhere on $[t_0, t_1]$.

Moreover, in this case for every $k \in \{1, \ldots, n\}$ the local Pontryagin maximum principle (10.11) can also be formulated as follows: For every $u_k \in L_\infty^{s_k}([t_0, t_1])$ with $u_k(t) \in \Omega_k$ almost everywhere on $[t_0, t_1]$ we have

$$\frac{\partial H}{\partial u_k}(\bar{x}, \bar{u}_1, \ldots \bar{u}_n, w, l_1, \ldots, l_n)(t)(u_k(t) - \bar{u}_k(t)) \leq 0$$

almost everywhere on $[t_0, t_1]$.

The maximum principle of Theorem 10.5 is in fact an extended F. John condition. In order to get a necessary optimality condition of the Karush-Kuhn-Tucker type we need an additional regularity assumption. Under this assumption the n-tuple $(l_1, \ldots, l_n)$ is even nonzero. It can be shown that this regularity assumption is fulfilled, if, in addition to the assumptions of Theorem 10.5, the adjoint equation (10.9) is *completely controllable* (for this notion see, for instance, Girsanov [106, p. 65]). The proof of this assertion can be done as the proof of a similar result presented in the book of Girsanov [106, p. 64–68].

Using Lemma 4.14 we finally present a maximum principle for optimal controls of the n players.

Theorem 10.6. *If the ordering cones $C_{Y_1}, \ldots, C_{Y_n}$ are pointed, then Theorem 10.5 remains valid if $\bar{u}_1, \ldots, \bar{u}_n$ are optimal controls of the n players.*

10.2.2 Sufficient Conditions for Optimal and Weakly Optimal Controls

The maximum principle which is derived as a necessary optimality condition is now investigated again. We present conditions under which the maximum principle is even sufficient for optimal and weakly optimal controls. Another sufficient condition is given using the Hamilton-Jacobi-Bellmann equations.

First, we consider again Problem 10.3 with a fixed terminal time $\hat{t} = t_1$. In this case we get the following maximum principle as a sufficient optimality condition.

Theorem 10.7. *Let the cooperative n player differential game formulated in Problem 10.3 be given with a fixed terminal time $\hat{t} = t_1$ and the target set*

$$Q := \{\tilde{x} \in \mathbb{R}^m \mid g(\tilde{x}) = 0_{\mathbb{R}^r}\}$$

where $g : \mathbb{R}^m \to \mathbb{R}^r$ (with $r \in \mathbb{N}$) is a given vector function. Let any $(\bar{x}, \bar{u}_1, \ldots, \bar{u}_n, t_1) \in S$ be given. Let the vector function g be differentiable at $\bar{x}(t_1)$; for every $i \in \{1, \ldots, n\}$ let the map h_i be convex at $\bar{x}(t_1)$ and Fréchet differentiable at $\bar{x}(t_1)$; for every $i \in \{1, \ldots, n\}$ let the map f_i^0 be convex and Fréchet differentiable; let the map f be partially differentiable with respect to $x, u_1, \ldots, u_n$. For every $i \in \{1, \ldots, n\}$ let a continuous linear functional $l_i \in C_{Y_i^*}^{\#}$ be given. Moreover, assume that there are a function $w \in W_{1,\infty}^m([t_0, t_1])$ and a vector $a \in \mathbb{R}^r$ so that for every $(x, u_1, \ldots, u_n, t_1) \in S$ the following conditions are satisfied:

(a)

$$-\dot{w}(t)^T = w(t)^T \frac{\partial f}{\partial x}(\bar{x}(t), \bar{u}_1(t), \ldots, \bar{u}_n(t))$$
$$- \sum_{i=1}^n \left(l_i \circ \frac{\partial f_i^0}{\partial x}\right)(\bar{x}(t), \bar{u}_1(t), \ldots, \bar{u}_n(t))$$

almost everywhere on $[t_0, t_1]$; (10.17)

(b)

$$-w(t_1)^T = a^T \frac{\partial g}{\partial x}(\bar{x}(t_1)) + \sum_{i=1}^n \left(l_i \circ \frac{\partial h_i}{\partial x}\right)(\bar{x}(t_1)); \quad (10.18)$$

(c) *for every $k \in \{1, \ldots, n\}$ and every $u_k \in L_\infty^{s_k}([t_0, t_1])$ with $u_k(t) \in \Omega_k$ almost everywhere on $[t_0, t_1]$ we have*

$$\left[w(t)^T \frac{\partial f}{\partial u_k}(\bar{x}(t), \bar{u}_1(t), \ldots, \bar{u}_n(t))\right.$$

10.2. A Maximum Principle

$$-\sum_{i=1}^{n}\left(l_i \circ \frac{\partial f_i^0}{\partial u_k}\right)(\bar{x}(t), \bar{u}_1(t), \ldots, \bar{u}_n(t))\right](u_k(t) - \bar{u}_k(t)) \leq 0$$

almost everywhere on $[t_0, t_1]$; (10.19)

(d) *let* $a^T g(\cdot)$ *be quasiconvex at* $\bar{x}(t_1)$ *(in a componentwise sense as outlined on page 185), and almost everywhere on* $[t_0, t_1]$ *let the functional defined by* $-w(t)^T f(x(t), u_1(t), \ldots, u_n(t))$ *be convex at* $(\bar{x}(t), \bar{u}_1(t), \ldots, \bar{u}_n(t))$.

Then $\bar{u}_1, \ldots, \bar{u}_n$ are optimal controls of the n players.

Proof. Let any $(n+2)$-tuple $(x, u_1, \ldots, u_n, t_1) \in S$ be given. With the differential equations (10.17) and (10.2) we obtain

$$-\frac{d}{dt}(w(t)^T(x(t) - \bar{x}(t)))$$
$$= -\dot{w}(t)^T(x(t) - \bar{x}(t)) - w(t)^T(\dot{x}(t) - \dot{\bar{x}}(t))$$
$$= \left[w(t)^T \frac{\partial f}{\partial x}(\bar{x}(t), \bar{u}_1(t), \ldots, \bar{u}_n(t))\right.$$
$$\left. - \sum_{i=1}^{n}\left(l_i \circ \frac{\partial f_i^0}{\partial x}\right)(\bar{x}(t), \bar{u}_1(t), \ldots, \bar{u}_n(t))\right](x(t) - \bar{x}(t))$$
$$- w(t)^T[f(x(t), u_1(t), \ldots, u_n(t)) - f(\bar{x}(t), \bar{u}_1(t), \ldots, \bar{u}_n(t))]$$
almost everywhere on $[t_0, t_1]$.

Then we get

$$\sum_{i=1}^{n} l_i(f_i^0(x(t), u_1(t), \ldots, u_n(t)) - f_i^0(\bar{x}(t), \bar{u}_1(t), \ldots, \bar{u}_n(t)))$$
$$-\frac{d}{dt}(w(t)^T(x(t) - \bar{x}(t)))$$
$$= \sum_{i=1}^{n} l_i\bigg(f_i^0(x(t), u_1(t), \ldots, u_n(t)) - f_i^0(\bar{x}(t), \bar{u}_1(t), \ldots, \bar{u}_n(t))$$
$$- \frac{\partial f_i^0}{\partial x}(\bar{x}(t), \bar{u}_1(t), \ldots, \bar{u}_n(t))(x(t) - \bar{x}(t))\bigg)$$
$$- w(t)^T\bigg(f(x(t), u_1(t), \ldots, u_n(t)) - f(\bar{x}(t), \bar{u}_1(t), \ldots, \bar{u}_n(t))$$

$$-\frac{\partial f}{\partial x}(\bar{x}(t),\bar{u}_1(t),\ldots,\bar{u}_n(t))(x(t)-\bar{x}(t))\Big)$$
almost everywhere on $[t_0,t_1]$. \hfill (10.20)

Since every f_i^0 ($i \in \{1,\ldots,n\}$) is a convex map and Fréchet differentiable and every $l_i \in C_{Y_i^*}^{\#}$ ($i \in \{1,\ldots,n\}$) is a continuous and monotonically increasing linear functional, the functional $l_i \circ f_i^0$ is also convex (compare Lemma 2.7,(b)) and it is even Fréchet differentiable. By assumption the vector function f is partially differentiable with respect to $x, u_1, \ldots, u_n$; and almost everywhere on $[t_0, t_1]$ the functional defined by $-w(t)^T f(\tilde{x}(t), \tilde{u}_1(t), \ldots, \tilde{u}_n(t))$ (for any $(\tilde{x},\tilde{u}_1,\ldots,\tilde{u}_n,t_1) \in S$) is convex at $(\bar{x}(t),\bar{u}_1(t),\ldots,\bar{u}_n(t))$. Consequently we conclude using (10.20) and (10.19):

$$\sum_{i=1}^n l_i(f_i^0(x(t),u_1(t),\ldots,u_n(t)) - f_i^0(\bar{x}(t),\bar{u}_1(t),\ldots,\bar{u}_n(t)))$$
$$-\frac{d}{dt}(w(t)^T(x(t)-\bar{x}(t))) \geq 0 \text{ almost everywhere on } [t_0,t_1].$$

If we notice that $x(t_0) = \bar{x}(t_0) = x_0$, then integration leads to the inequality

$$\sum_{i=1}^n l_i\Big(\int_{t_0}^{t_1}[f_i^0(x(t),u_1(t),\ldots,u_n(t)) - f_i^0(\bar{x}(t),\bar{u}_1(t),\ldots,\bar{u}_n(t))]\,dt\Big)$$
$$-w(t_1)^T(x(t_1)-\bar{x}(t_1)) \geq 0. \hfill (10.21)$$

The inequality (10.21) is valid because for every $i \in \{1,\ldots,n\}$ l_i is a continuous linear functional and the map $f_i^0 \circ (\tilde{x},\tilde{u}_1,\ldots,\tilde{u}_n)$ is Bochner integrable for all $(\tilde{x},\tilde{u}_1,\ldots,\tilde{u}_n,t_1) \in S$ (see Hille-Phillips [121, p. 83–84] or Warga [319, p. 82]). Obviously, the integral appearing in the inequality (10.21) is a Bochner integral. With the equation (10.18) and the Fréchet differentiability and convexity of the maps $h_1,\ldots,h_n$ at $\bar{x}(t_1)$ we conclude:

$$-w(t_1)^T(x(t_1)-\bar{x}(t_1))$$
$$= a^T\frac{\partial g}{\partial x}(\bar{x}(t_1))(x(t_1)-\bar{x}(t_1))$$

10.2. A Maximum Principle

$$+ \sum_{i=1}^{n} \left(l_i \circ \frac{\partial h_i}{\partial x} \right)(\bar{x}(t_1))(x(t_1) - \bar{x}(t_1))$$

$$\leq a^T \frac{\partial g}{\partial x}(\bar{x}(t_1))(x(t_1) - \bar{x}(t_1))$$

$$+ \sum_{i=1}^{n} l_i(h_i(x(t_1)) - h_i(\bar{x}(t_1))). \quad (10.22)$$

Because of the differentiability and quasiconvexity of $a^T g(\cdot)$ at $\bar{x}(t_1)$ (compare page 185) the equation

$$0 = a^T g(x(t_1)) - a^T g(\bar{x}(t_1))$$

implies the inequality

$$0 \geq a^T \frac{\partial g}{\partial x}(\bar{x}(t_1))(x(t_1) - \bar{x}(t_1)). \quad (10.23)$$

Then the inequalities (10.22) and (10.23) lead to the inequality

$$-w(t_1)^T (x(t_1) - \bar{x}(t_1)) \leq \sum_{i=1}^{n} l_i(h_i(x(t_1)) - h_i(\bar{x}(t_1)))$$

which implies, with (10.21),

$$\sum_{i=1}^{n} l_i \left(h_i(x(t_1)) + \int_{t_0}^{t_1} f_i^0(x(t), u_1(t), \ldots, u_n(t)) \, dt \right)$$

$$\geq \sum_{i=1}^{n} l_i \left(h_i(\bar{x}(t_1)) + \int_{t_0}^{t_1} f_i^0(\bar{x}(t), \bar{u}_1(t), \ldots, \bar{u}_n(t)) \, dt \right)$$

resulting in

$$\sum_{i=1}^{n} l_i(v_i(x, u_1, \ldots, u_n, t_1)) \geq \sum_{i=1}^{n} l_i(v_i(\bar{x}, \bar{u}_1, \ldots, \bar{u}_n, t_1)).$$

Finally, an application of Theorem 5.18, (b) and Lemma 10.2 leads to the assertion. □

A similar sufficient condition can also be formulated for weakly optimal controls.

Theorem 10.8. *Let the ordering cones $C_{Y_1}, \ldots, C_{Y_n}$ have a nonempty algebraic interior. If the sets $C_{Y_i^*}^{\#}$ ($i \in \{1, \ldots, n\}$) are replaced by the dual ordering cones $C_{Y_i^*}$ where for at least one $i' \in \{1, \ldots, n\}$ $C_{Y_{i'}^*} \neq \{0_{Y_{i'}^*}\}$, then Theorem 10.7 remains valid for weakly optimal controls, i.e. in this case $\bar{u}_1, \ldots, \bar{u}_n$ are weakly optimal controls of the n players.*

Next, we consider an example which shows how the maximum principle can be used for the determination of optimal controls

Example 10.9. Two divisions of a conglomerate company are in competition because a certain product is produced by both divisions. For a fixed planing period $[0, t_1]$ the rate of demand at time t_1 and the profits of both divisions should be maximized by advertising for the product.

In the following $x_1(t)$ and $x_2(t)$ describe the rate of demand for both factories at time t. $u_1(t)$ and $u_2(t)$ denote the rate of expenditure for advertising for each division. Based on market observations it is assumed that the change of the rate of demand depends on the rate of demand and the rate of expenditure for advertising as follows:

$$\left. \begin{array}{l} \dot{x}_1(t) = 12u_1(t) - 2u_1(t)^2 - x_1(t) - u_2(t) \\ \dot{x}_2(t) = 12u_2(t) - 2u_2(t)^2 - x_2(t) - u_1(t) \end{array} \right\} \begin{array}{l} \text{almost} \\ \text{everywhere} \\ \text{on } [0, t_1]. \end{array}$$

Moreover assume that

$$x_1(0) = x_{1_0} \quad \text{and} \quad x_2(0) = x_{2_0}$$

where x_{1_0} and x_{2_0} are given initial rates of demand. For feasible advertising intensities we require

$$u_1(t) \in [0, \hat{u}_1] \text{ and } u_2(t) \in [0, \hat{u}_2] \text{ almost everywhere on } [0, t_1]$$

where $\hat{u}_1$ and $\hat{u}_2$ are positive real numbers. We assume that the profits

10.2. A Maximum Principle

of both divisions are given as

$$\int_0^{t_1} \left(\frac{1}{3}x_1(t) - u_1(t)\right) dt$$

and

$$\int_0^{t_1} \left(\frac{1}{3}x_2(t) - u_2(t)\right) dt,$$

respectively. Consequently, the objective maps v_1 and v_2 which have to be minimized read as

$$v_1(x, u_1, u_2, t_1) = \begin{pmatrix} -x_1(t_1) \\ \int_0^{t_1} (u_1(t) - \frac{1}{3}x_1(t))\, dt \end{pmatrix}$$

and

$$v_2(x, u_1, u_2, t_1) = \begin{pmatrix} -x_2(t_1) \\ \int_0^{t_1} (u_2(t) - \frac{1}{3}x_2(t))\, dt \end{pmatrix}.$$

Finally, we assume that each division (player) can give a convex cone $C_{\mathbb{R}_1^2}$ and $C_{\mathbb{R}_2^2}$, respectively, for which $C_{\mathbb{R}_1^2}^{\#} \neq \emptyset$ and $C_{\mathbb{R}_2^2}^{\#} \neq \emptyset$ (we restrict ourselves to the determination of optimal controls). For arbitrary vectors $\begin{pmatrix} \alpha_1 \\ \beta_1 \end{pmatrix} \in C_{\mathbb{R}_1^2}^{\#}$ and $\begin{pmatrix} \alpha_2 \\ \beta_2 \end{pmatrix} \in C_{\mathbb{R}_2^2}^{\#}$ the function w with

$$w(t) = \begin{pmatrix} \left(\alpha_1 - \frac{\beta_1}{3}\right) e^{t-t_1} + \frac{\beta_1}{3} \\ \left(\alpha_2 - \frac{\beta_2}{3}\right) e^{t-t_1} + \frac{\beta_2}{3} \end{pmatrix} \quad \text{for all } t \in [0, t_1]$$

satisfies the adjoint equation (10.17) and the transversality condition (10.18) (in this case we have $g \equiv 0_{\mathbb{R}^r}$).

In order to get concrete results we choose, for simplicity, $C_{\mathbb{R}_1^2} = C_{\mathbb{R}_2^2} = \mathbb{R}_+^2$. The weights are chosen as $\alpha_1 = \beta_1 = 1$, $\alpha_2 = 2$ and $\beta_2 = 3$. Moreover, we assume $t_1 = 1$ and $\hat{u}_1 = \hat{u}_2 = 4$. Then the vector function

w is componentwise non-negative on $[0, 1]$. The vector function f is concave with respect to x, u_1 and u_2. Consequently, the assumption (d) of Theorem 10.7 is satisfied. The controls $\bar{u}_1$ and $\bar{u}_2$ given by

$$\bar{u}_1(t) = \frac{6 + 21\,e^{t-1}}{4 + 8\,e^{t-1}} \quad \text{for all } t \in [0, 1]$$

and

$$\bar{u}_2(t) = \frac{13 + 17\,e^{t-1}}{6 + 6\,e^{t-1}} \quad \text{for all } t \in [0, 1]$$

satisfy the local Pontryagin maximum principle (10.19). In fact, $\bar{u}_1$ and $\bar{u}_2$ fulfill all assumptions of Theorem 10.7. Consequently, $\bar{u}_1$ and $\bar{u}_2$ are optimal controls for both divisions (see Fig. 10.1).

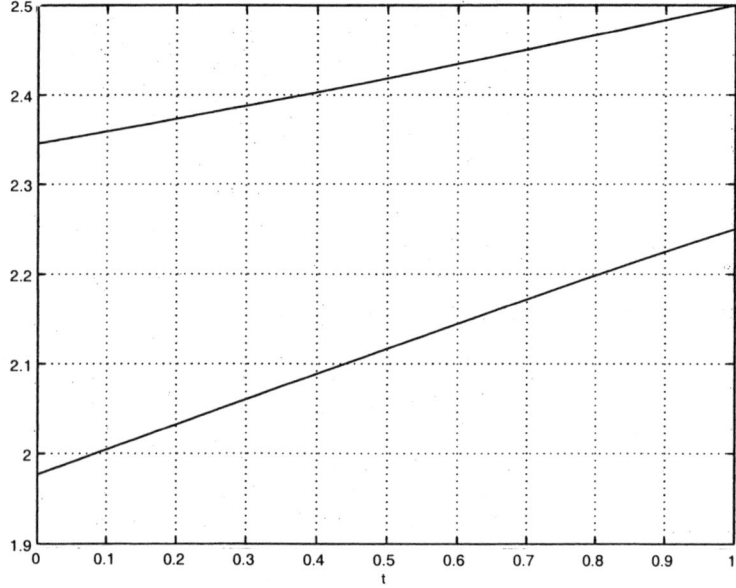

Figure 10.1: Illustration of the optimal controls $\bar{u}_1$ (lower curve) and $\bar{u}_2$ (upper curve).

Using the Hamilton-Jacobi-Bellmann equations it is also possible to formulate sufficient conditions for optimal controls. We present

10.2. A Maximum Principle

conditions for cooperative differential games with a free terminal time $\hat{t}$.

Theorem 10.10. *Let the cooperative n player differential game formulated in Problem 10.3 be given. Let $(\bar{x}, \bar{u}_1, \ldots, \bar{u}_n, \bar{t}) \in S$ be a given $(n+2)$-tuple. For every $i \in \{1, \ldots, n\}$ let a continuous linear functional $l_i \in C_{Y_i^*}^\#$ be given. Moreover, assume that there is a Lipschitz continuous function $w : \mathbb{R}^m \to \mathbb{R}$ with a componentwise weak derivative $\frac{\partial w}{\partial y} \in L_\infty^m([t_0, t_1])$ so that for all $(x, u_1, \ldots, u_n, \hat{t}) \in S$ the following holds:*

(a)
$$w(x(\hat{t})) = \sum_{i=1}^n l_i(h_i(x(\hat{t}))), \qquad (10.24)$$

(b)
$$\int_{t_0}^{\bar{t}} \left[\frac{\partial w}{\partial y}(\bar{x}(t)) f(\bar{x}(t), \bar{u}_1(t), \ldots, \bar{u}_n(t)) \right. $$
$$\left. + \sum_{i=1}^n l_i(f_i^0(\bar{x}(t), \bar{u}_1(t), \ldots, \bar{u}_n(t))) \right] dt = 0, \qquad (10.25)$$

(c)
$$\frac{\partial w}{\partial y}(x(t)) f(x(t), u_1(t), \ldots, u_n(t))$$
$$+ \sum_{i=1}^n l_i(f_i^0(x(t), u_1(t), \ldots, u_n(t))) \geq 0$$

almost everywhere on $[t_0, \hat{t}]$. (10.26)

Then $\bar{u}_1, \ldots, \bar{u}_n$ are optimal controls of the n players.

Proof. Let $(x, u_1, \ldots, u_n, \hat{t}) \in S$ be any playable $(n+2)$-tuple. Then we obtain with (10.2), (10.24) and (10.3)

$$\int_{t_0}^{\hat{t}} \frac{\partial w}{\partial y}(x(t)) f(x(t), u_1(t), \ldots, u_n(t)) \, dt$$

$$= \int_{t_0}^{\hat{t}} \frac{\partial w}{\partial y}(x(t))\dot{x}(t)\, dt$$

$$= w(x(\hat{t})) - w(x(t_0))$$

$$= \sum_{i=1}^{n} l_i(h_i(x(\hat{t}))) - w(x_0). \tag{10.27}$$

With (10.25), (10.2), (10.24) and (10.3) we analogously get

$$\int_{t_0}^{\bar{t}} \left[\sum_{i=1}^{n} l_i(f_i^0(\bar{x}(t), \bar{u}_1(t), \ldots, \bar{u}_n(t)))\right] dt$$

$$= -\int_{t_0}^{\bar{t}} \frac{\partial w}{\partial y}(\bar{x}(t)) f(\bar{x}(t), \bar{u}_1(t), \ldots, \bar{u}_n(t))\, dt$$

$$= -\int_{t_0}^{\bar{t}} \frac{\partial w}{\partial y}(\bar{x}(t))\dot{\bar{x}}(t)\, dt$$

$$= -\sum_{i=1}^{n} l_i(h_i(\bar{x}(\bar{t}))) - w(x_0).$$

With the equation (10.27) it follows

$$\int_{t_0}^{\bar{t}} \left[\sum_{i=1}^{n} l_i(f_i^0(\bar{x}(t), \bar{u}_1(t), \ldots, \bar{u}_n(t)))\right] dt$$

$$= \sum_{i=1}^{n} l_i(h_i(x(\hat{t})) - h_i(\bar{x}(\bar{t})))$$

$$- \int_{t_0}^{\hat{t}} \frac{\partial w}{\partial y}(x(t)) f(x(t), u_1(t), \ldots, u_n(t))\, dt.$$

Then we get

$$\sum_{i=1}^{n} l_i(v_i(x, u_1, \ldots, u_n, \hat{t})) - \sum_{i=1}^{n} l_i(v_i(\bar{x}, \bar{u}_1, \ldots, \bar{u}_n, \bar{t}))$$

10.2. A Maximum Principle

$$= \sum_{i=1}^{n} \Big[l_i(h_i(x(\hat{t})) - h_i(\bar{x}(\bar{t})))$$

$$+ l_i \Big(\int_{t_0}^{\hat{t}} f_i^0(x(t), u_1(t), \ldots, u_n(t)) \, dt$$

$$- \int_{t_0}^{\bar{t}} f_i^0(\bar{x}(t), \bar{u}_1(t), \ldots, \bar{u}_n(t)) \, dt \Big]$$

$$= \int_{t_0}^{\hat{t}} \Big[\frac{\partial w}{\partial y}(x(t)) f(x(t), u_1(t), \ldots, u_n(t))$$

$$+ \sum_{i=1}^{n} l_i(f_i^0(x(t), u_1(t), \ldots, u_n(t))) \Big] dt$$

which is non-negative because of the inequality (10.26). For the last conclusion we use the fact that every l_i ($i \in \{1, \ldots, n\}$) is a continuous linear functional (compare Hille-Phillips [121, p. 83–84] or Warga [319, p. 82]). Finally, Theorem 5.18, (b) and Lemma 10.2 lead to the assertion. □

The assumption $\frac{\partial w}{\partial y} \in L_\infty^m([t_0, t_1])$ given in the previous theorem can be replaced by the weaker assumption $\frac{\partial w}{\partial y} \in L_1^m([t_0, t_1])$, if for all $(x, u_1, \ldots, u_n, \hat{t}) \in S$ and all $j \in \{1, \ldots, m\}$ the condition

$$f_j(x(t), u_1(t), \ldots, u_n(t)) > 0 \text{ almost everywhere on } [t_0, \hat{t}]$$

is satisfied (compare Warga [319, p. 98]).

An example which shows the applicability of Theorem 10.10 can be found in the book of Leitmann [201, p. 32–36] (see also Leitmann-Liu [204]).

The next theorem presents a similar result as Theorem 10.10 for weakly optimal controls.

Theorem 10.11. *Let the ordering cones $C_{Y_1}, \ldots, C_{Y_n}$ have a non-empty algebraic interior. If the sets $C_{Y_i^*}^{\#}$ ($i \in \{1, \ldots, n\}$) are replaced* by the dual ordering cones $C_{Y_{i'}^*}$ where for at least one $i' \in \{1,\ldots,n\}$ $C_{Y_{i'}^*} \neq \{0_{Y_{i'}^*}\}$, then Theorem 10.10 remains valid for weakly optimal controls, i.e. in this case $\bar{u}_1,\ldots,\bar{u}_n$ are weakly optimal controls of the n players.

10.3 A Special Cooperative n Player Differential Game

In this last section of this chapter a special cooperative differential game is investigated which extends the so-called least squares problem known from control theory.

Problem 10.12. We turn our attention to an infinite-dimensional autonomous linear system

$$\dot{x}(t) = Ax(t) + \sum_{i=1}^{n} B_i u_i(t) \text{ for all } t \in (0, t_1) \tag{10.28}$$

with the initial condition

$$x(0) = x_0 \tag{10.29}$$

where t_1 is a given positive terminal time. The state space $(X, \langle .,.\rangle_X)$ as well as the image spaces $(Z_1, \langle .,.\rangle_{Z_1}),\ldots,(Z_n, \langle .,.\rangle_{Z_n})$ of the controls are assumed to be real Hilbert spaces. Let $x_0 \in X$ be any given element. For every $i \in \{1,\ldots,n\}$ let the control u_i be an element of the real linear space

$$L_2([0,t_1], Z_i) := \Big\{ u_i : [0,t_1] \to Z_i \ \Big| \ u_i \text{ is strongly measurable and } \int_0^{t_1} \|u_i(t)\|_{Z_i}^2 \, dt < \infty \Big\}.$$

The map A is assumed to be linear with the domain $D(A) \subset X$ and the range $R(A) \subset X$ and it is assumed to be an infinitesimal generator of a strongly continuous semigroup T_t (for further details see Hille-Phillips [121], Ladas-Lakshmikantham [194], Barbu [19] or Martin

10.3. A Special Cooperative n Player Differential Game

[221]). For every $i \in \{1,\ldots,n\}$ let $B_i : Z_i \to X$ be a continuous linear map. Recall that a map $x : [0, t_1] \to X$ is called a *mild solution* of the system (10.28) with the initial condition (10.29), if

$$x(t) = T_t x_0 + \sum_{i=1}^{n} \int_0^t T_{t-s} B_i u_i(s)\, ds \text{ for all } t \in [0, t_1] \qquad (10.30)$$

(for instance, compare Barbu [19, p. 31]). The integral appearing in (10.30) is a Bochner integral. In order to ensure that the representation (10.30) makes sense we assume $x_0 \in D(A)$. Since, in general, every mild solution is not a solution of (10.28) and (10.29) as well (e.g., see Martin [221, p. 296]), our following investigations are based on the input-output-relation (10.30).

Every player tries to steer the system with minimal effort possibly to the zero state. In other words: The i-th player minimizes the objective map $v_i : S \to Y_i := \mathbb{R}^2$ with

$$v_i(x, u_1, \ldots, u_n, \hat{t}) = \begin{pmatrix} \|x(t_1)\|_X \\ \\ \|u_i\|_{L_2([0,t_1], Z_i)} \end{pmatrix} \text{ for all } (x, u_1, \ldots, u_n, \hat{t}) \in S$$

where the set S of playable $(n+2)$-tuples (compare (10.1)) is defined as

$$S := \{(x, u_1, \ldots, u_n, \hat{t}) \mid \hat{t} = t_1,\ u_i \in L_2([0, t_1], Z_i)$$
$$\text{for every } i \in \{1, \ldots, n\} \text{ and}$$
$$x \text{ satisfies (10.30)}\}.$$

In the case of $n = 1$ the cooperative differential game formulated in Problem 10.12 is known, in a similar form, in optimal control theory as linear-quadratic problem or least squares problem (e.g., compare Brockett [46], Curtain-Pritchard [75], [76] or Jacobson-Martin-Pachter-Geveci [137]).

It is our aim to present optimal controls for the cooperative differential game described in Problem 10.12. But first, we need two technical lemmas.

Lemma 10.13. *Let Problem 10.12 be given, and for every $i \in \{1, \ldots, n\}$ let non-negative real numbers γ_i be fixed. In the class of strongly continuous self-adjoint maps in $B(X, X)$ (real linear space of continuous linear maps from X to X) for which $\langle z, P_t y \rangle_X$ is differentiable by t for all $y, z \in D(A)$, the map P where P_t $(t \in [0, t_1])$ is given as*

$$P_t z = T^*_{t_1-t} T_{t_1-t} z - \sum_{i=1}^{n} \gamma_i \int_t^{t_1} T^*_{s-t} P_s B_i B_i^* P_s T_{s-t} z \, ds \quad \text{for all } z \in X$$

is the unique solution of the Bernoulli differential equation in scalar product form

$$\frac{d}{dt} \langle z, P_t y \rangle_X + \langle P_t z, Ay \rangle_X + \langle Az, P_t y \rangle_X - \sum_{i=1}^{n} \gamma_i \langle P_t B_i B_i^* P_t z, y \rangle_X = 0$$

for all $t \in [0, t_1]$ and all $y, z \in D(A)$ (10.31)

with the terminal condition

$$P_{t_1} = I \text{ (identity)}. \tag{10.32}$$

Proof. The proof of this result can be done in analogy to a proof of Curtain-Pritchard [76, pp. 93]. □

Lemma 10.14. *Let Problem 10.12 be given, and for every $i \in \{1, \ldots, n\}$ let non-negative real numbers γ_i be fixed. If P is the solution of the Bernoulli differential equation (10.31) with the terminal condition (10.32), then we have for all $u_1 \in L_2([0, t_1], Z_1), \ldots, u_n \in L_2([0, t_1], Z_n)$ with the corresponding mild solution x*

$$0 = \langle x_0, P_0 x_0 \rangle_X - \langle x(t_1), x(t_1) \rangle_X$$
$$+ \sum_{i=1}^{n} \int_0^{t_1} [\gamma_i \langle B_i^* P_t x(t), B_i^* P_t x(t) \rangle_{Z_i} + 2 \langle B_i^* P_t x(t), u_i(t) \rangle_{Z_i}] \, dt.$$

10.3. A Special Cooperative n Player Differential Game

Proof. For the proof of this lemma we refer to a proof of a similar result in the book of Curtain-Pritchard [76, pp. 86]. □

The next theorem presents the main result of this section.

Theorem 10.15. *Let Problem 10.12 be given, and assume that for every $i \in \{1,\ldots,n\}$ $C_{Y_i} = \mathbb{R}_+^2$ is the ordering cone in Y_i. Let $(\alpha_i, \beta_i) \in \mathbb{R}_+^2$ ($i \in \{1,\ldots,n\}$) with $\beta_i > 0$ be given vectors. Moreover, let P be the solution of the Bernoulli differential equation (10.31) with the terminal condition (10.32) for $\gamma_i := \frac{\hat{\alpha}}{\beta_i}$ ($i \in \{1,\ldots,n\}$) where $\hat{\alpha} := \sum_{j=1}^n \alpha_j$. Then the feedback control $\bar{u}_i$ given by*

$$\bar{u}_i(t) = -\gamma_i B_i^* P_t x(t) \text{ for all } t \in [0, t_1] \tag{10.33}$$

is an optimal (and also a weakly optimal) control of the i-th player.

Proof. Let $(\alpha_i, \beta_i) \in \mathbb{R}_+^2$ ($i \in \{1,\ldots,n\}$) with $\beta_i > 0$ be arbitrarily given vectors. Furthermore, let $(x, u_1, \ldots, u_n, t_1) \in S$ be any playable $(n+2)$-tuple with $u_i \neq \bar{u}_i$ for at least one $i \in \{1,\ldots,n\}$. Then we conclude with Lemma 10.14 and the positivity of the β_i's:

$$\sum_{i=1}^n (\alpha_i \|x(t_1)\|_X^2 + \beta_i \|u_i\|_{L_2([0,t_1],Z_i)}^2)$$

$$= \hat{\alpha}\langle x(t_1), x(t_1)\rangle_X + \sum_{i=1}^n \beta_i \int_0^{t_1} \langle u_i(t), u_i(t)\rangle_{Z_i} dt$$

$$= \hat{\alpha}\langle x_0, P_0 x_0\rangle_X + \sum_{i=1}^n \int_0^{t_1} [\beta_i \langle u_i(t), u_i(t)\rangle_{Z_i}$$
$$+ \hat{\alpha}\gamma_i \langle B_i^* P_t x(t), B_i^* P_t x(t)\rangle_{Z_i} + 2\hat{\alpha}\langle B_i^* P_t x(t), u_i(t)\rangle_{Z_i}] dt$$

$$= \hat{\alpha}\langle x_0, P_0 x_0\rangle_X + \sum_{i=1}^n \beta_i \int_0^{t_1} [\langle u_i(t), u_i(t)\rangle_{Z_i}$$
$$+ 2\gamma_i \langle B_i^* P_t x(t), u_i(t)\rangle_{Z_i} + \gamma_i^2 \langle B_i^* P_t x(t), B_i^* P_t x(t)\rangle_{Z_i}] dt$$

$$= \hat{\alpha}\langle x_0, P_0 x_0\rangle_X + \sum_{i=1}^{n} \beta_i \int_0^{t_1} \langle u_i(t) + \gamma_i B_i^* P_t x(t), u_i(t) + \gamma_i B_i^* P_t x(t)\rangle_{Z_i} dt$$

$$= \hat{\alpha}\langle x_0, P_0 x_0\rangle_X + \sum_{i=1}^{n} \beta_i \|u_i(\cdot) + \gamma_i B_i^* P.x(\cdot)\|_{L_2([0,t_1],Z_i)}^2$$

$$> \hat{\alpha}\langle x_0, P_0 x_0\rangle_X$$

$$= \sum_{i=1}^{n} \left(\alpha_i \|x(t_1)\|_X^2 + \beta_i \|\bar{u}_i\|_{L_2([0,t_1],Z_i)}^2\right).$$

Finally, an application of Lemma 5.14, (a) and Lemma 5.24 leads to the assertion. □

In the case of only one player ($n = 1$) Theorem 10.15 has a trivial consequence for the two following scalar parametric optimization problems (with $\alpha, \beta > 0$):

$$\left.\begin{array}{l} \inf \|x(t_1)\|_X \\ \text{subject to the constraints} \\ x(t) = T_t x_0 + \displaystyle\int_0^{t_1} T_{t-s} B_1 u_1(s)\, ds \text{ for all } t \in [0, t_1] \\ \|u_1\|_{L_2([0,t_1],Z_1)} \leq \alpha, \end{array}\right\} \quad (10.34)$$

$$\left.\begin{array}{l} \inf \|u_1\|_{L_2([0,t_1],Z_1)} \\ \text{subject to the constraints} \\ x(t) = T_t x_0 + \displaystyle\int_0^{t_1} T_{t-s} B_1 u_1(s)\, ds \text{ for all } t \in [0, t_1] \\ \|x(t_1)\| \leq \beta. \end{array}\right\} \quad (10.35)$$

Corollary 10.16. *Let the assumptions of Theorem 10.15 be satisfied and assume $n = 1$. Let $\bar{u}_1$ be a feedback control given by (10.33) with the associated mild solution $\bar{x}$. Then $(\bar{x}, \bar{u}_1)$ solves the scalar optimization problems (10.34) and (10.35) for $\alpha := \|\bar{u}_1\|_{L_2([0,t_1],Z_1)}$ and $\beta := \|\bar{x}(t_1)\|_X$.*

10.3. A Special Cooperative n Player Differential Game

Proof. This corollary immediately follows from Theorem 10.15 and the definition of optimal controls (one can also use a general result given by Vogel [315, p. 2]). □

The following example shows the applicability of Theorem 10.15.

Example 10.17. Let a thin homogeneous bar with the length 1 and the diffusion coefficient $a > 0$ have an initial temperature distribution x_0. It is the aim to cool down the bar from above (player 1) and from below (player 2) within one time unit with a minimal steering effort (the bar is assumed to be located in a horizontal plane). To be more specific, we consider the heat equation

$$\frac{\partial x}{\partial t}(z,t) = a\frac{\partial x^2}{\partial z \partial z}(z,t) + u_1(z,t) - 2u_2(z,t), \ 0 < z < 1, \ 0 < t < 1$$

with the boundary conditions

$$\frac{\partial x}{\partial z}(0,t) = \frac{\partial x}{\partial z}(1,t) = 0, \ 0 < t < 1,$$

and the initial condition

$$x(z,0) = x_0(z), \ 0 < z < 1.$$

For the determination of optimal controls $\bar{u}_1$ and $\bar{u}_2$ using Theorem 10.15 we have to choose appropriate weights. But we omit that and immediately assume that positive real numbers γ_1 and γ_2 are given. The map A is defined by

$$Ax = a\frac{\partial x^2}{\partial z \partial z}$$

where

$$D(A) = \left\{ x \in L_2([0,1]) \ \Big| \ \frac{\partial x}{\partial z}, \frac{\partial x}{\partial z \partial z} \in L_2([0,1]); \ \frac{\partial x}{\partial z}(0,\cdot) = \frac{\partial x}{\partial z}(1,\cdot) \equiv 0 \right\}.$$

It is known that A is self-adjoint and that it generates an analytical semigroup (compare Curtain-Pritchard [76, p. 45–46]). The eigenvalues of A read

$$\lambda_i = -a\pi^2 i^2 \ \text{for all } i \in \mathbb{N} \cup \{0\}.$$

Every eigenvalue is simple and associated eigenfunctions are, for instance, φ_i with
$$\varphi_0(z) = 1$$
and
$$\varphi_i(z) = \sqrt{2}\cos i\pi z \text{ for all } i \in \mathbb{N}$$
(for eigenvalues and eigenfunctions compare also Triebel [303, p. 301]). The function system $\{\varphi_0, \varphi_1, \ldots\}$ is a complete orthonormal base in $L_2([0,1])$, and every $x \in L_2([0,1])$ can be represented by
$$x = \sum_{i=0}^{\infty} \langle x, \varphi_i \rangle_{L_2([0,1])} \varphi_i$$
(e.g., see Triebel [303, p. 303–304]). Moreover, let us note that $x \in D(A)$ if and only if
$$\sum_{i=0}^{\infty} \lambda_i^2 \langle x, \varphi_i \rangle_{L_2([0,1])}^2 < \infty$$
(compare Triebel [303, p. 273]). For the solution P of the Bernoulli differential equation (10.31) with the terminal condition (10.32) we set (for $t \in [0,1]$)
$$P_t y = \sum_{i=0}^{\infty} p_i(t) \langle y, \varphi_i \rangle_{L_2([0,1])} \varphi_i \text{ for all } y \in L_2([0,1]). \quad (10.36)$$
By coefficient comparison we obtain from (10.31) and (10.32) for all $i \in \mathbb{N} \cup \{0\}$:
$$\left. \begin{array}{l} \dot{p}_i(t) + 2\lambda_i p_i(t) - (\gamma_1 + 4\gamma_2) p_i(t)^2 = 0 \\ p_i(1) = 1. \end{array} \right\} \quad (10.37)$$
For every $i \in \mathbb{N}$ we get the solution of (10.37) as
$$p_0(t) = \frac{1}{1 + (\gamma_1 + 4\gamma_2)(1-t)}$$
and
$$p_i(t) = \frac{-2\lambda_i}{-(\gamma_1 + 4\gamma_2) + (-2\lambda_i + \gamma_1 + 4\gamma_2) e^{-2\lambda_i(1-t)}}.$$

Since for all $i \in \mathbb{N} \cup \{0\}$ and all $t \in [0,1]$

$$0 \leq p_i(t) \leq 1,$$

P_t (by (10.36)) is well-defined and, in fact, it satisfies (10.31) and (10.32). If we set x, $\bar{u}_1$ and $\bar{u}_2$ as

$$x(z,t) = \sum_{i=0}^{\infty} x_i(t)\varphi_i(z),$$

$$\bar{u}_1(z,t) = \sum_{i=0}^{\infty} \bar{u}_{1_i}(t)\varphi_i(z)$$

and

$$\bar{u}_2(z,t) = \sum_{i=0}^{\infty} \bar{u}_{2_i}(t)\varphi_i(z),$$

we obtain because of (10.33) for all $i \in \mathbb{N} \cup \{0\}$

$$\bar{u}_{1_i}(t) = \begin{cases} \frac{-\gamma_1 x_0(t)}{1+(\gamma_1+4\gamma_2)(1-t)} & \text{if } i = 0 \\ \frac{-2a\pi^2 i^2 \gamma_1 x_i(t)}{-(\gamma_1+4\gamma_2)+(2a\pi^2 i^2+\gamma_1+4\gamma_2)e^{2a\pi^2 i^2(1-t)}} & \text{if } i \in \mathbb{N} \end{cases}$$

and

$$\bar{u}_{2_i}(t) = \begin{cases} \frac{2\gamma_2 x_0(t)}{1+(\gamma_1+4\gamma_2)(1-t)} & \text{if } i = 0 \\ \frac{4a\pi^2 i^2 \gamma_2 x_i(t)}{-(\gamma_1+4\gamma_2)+(2a\pi^2 i^2+\gamma_1+4\gamma_2)e^{2a\pi^2 i^2(1-t)}} & \text{if } i \in \mathbb{N}. \end{cases}$$

Notes

Cooperative differential games are described, in a similar way as it is done in the first section, by Vincent-Leitmann [311], Leitmann-Rocklin-Vincent [206], Stalford [289], Blaquière-Juricek-Wiese [28], Leitmann [201], Salz [272] and Vincent-Grantham [310] as well as in the proceedings edited by Blaquière [27], Leitmann-Marzollo [205] and Leitmann [202]. Differential games where one does not have to cooperate exclusively are treated by Juricek [163] and Schmitendorf-Moriarty [277]. Control problems with a vector-valued objective map are also investigated by Stern-Ben-Israel [291], Yu-Leitmann [338], Salukvadze [271] and Leitmann [203].

The maximum principle as a necessary optimality condition is derived along the lines of Girsanov [106]. The approach of Kirsch-Warth-Werner [171] can also be used for the proof of the maximum principle as a necessary optimality condition. Kirsch-Warth-Werner [171] directly use the differential equation as an equality constraint. Furthermore, the initial condition appears in the definition of the set $\hat{S}$. In their book one can also find a detailed proof of the Fréchet differentiability of F and G (compare page 250) in the case of one player. The maximum principle as a sufficient condition (Theorem 10.7) generalizes a similar result of Leitmann [201] and Salz [272]. Example 10.9 is based on a problem formulated by Starr [290] and Leitmann [201]. The sufficient optimality condition given in Theorem 10.10 was already introduced by Leitmann [199] and later modified by Stalford [288] (see also Leitmann [200] and [201]).

The presentation of the cooperative differential game discussed in the last section of this chapter is closely related to investigations of Curtain [72] and Curtain-Pritchard [73], [74] in the case of $n = 1$.

Part IV

Engineering Applications

Nowadays most of the optimization problems arising in technical practice are problems with various objectives which have to be optimized simultaneously. These multiobjective optimization problems are finite dimensional vector optimization problems with the natural partial ordering in the image space of the vector-valued objective function. This part of the book is devoted to the application of the theory of vector optimization to multiobjective optimization problems arising in engineering.

In Chapter 11 we discuss how to specialize the optimality notions defined in Chapter 4 to multiobjective optimization problems. The important scalarization approaches, like the weighted sum and weighted Chebyshev norm approaches, are examined for these special problems because these methods are often used in engineering. Chapter 12 treats numerical methods for the solution of multiobjective optimization problems in engineering. A modified method of Polak, interactive methods and a method for the solution of discrete problems are presented. Finally, special engineering problems are described and solved in Chapter 13. We present the optimal design of antennas in electrical engineering, we investigate the optimization of a FDDI communication network in computer science, and we discuss bicriterial optimization problems in chemical engineering.

Chapter 11

Theoretical Basics of Multiobjective Optimization

This chapter introduces the basic concepts of multiobjective optimization. After the discussion of a simple example from structural engineering in the first section the definitions of several variants of the Edgeworth-Pareto optimality notion are presented: weakly, properly, strongly and essentially Edgeworth-Pareto optimal points. Relationships between these different concepts are investigated and simple examples illustrate these notions. The second section is devoted to the scalarization of multiobjective optimization problems. The weighted sum and the weighted Chebyshev norm approach are investigated in detail.

11.1 Basic Concepts

Optimization problems with several criteria arise in engineering, economics, applied mathematics and physics. As a simple example we discuss a design problem from structural engineering.

Example 11.1. We consider the design of a beam with a rectangular cross-section and a given length l (see Fig. 11.1 and 11.2). The height x_1 and the width x_2 have to be determined.

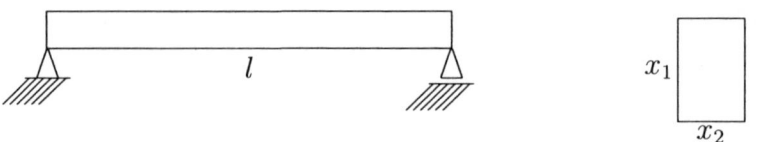

Figure 11.1: Longitudinal section.   Figure 11.2: Cross-section.

The design variables x_1 and x_2 have to be chosen in an area which makes sense in practice. A certain stress condition must be satisfied, i.e. the arising stresses cannot exceed a feasible stress. This leads to the inequality

$$2000 \leq x_1^2 x_2.$$

Moreover, a certain stability of the beam must be guaranteed. In order to avoid a beam which is too slim we require

$$x_1 \leq 4x_2$$

and

$$x_2 \leq x_1.$$

Finally, the design variables should be nonnegative which means

$$x_1 \geq 0, \quad x_2 \geq 0.$$

Among all feasible values for x_1 and x_2 we are interested in those which lead to a light *and* cheap construction. Instead of the weight we can also take the volume of the beam given as lx_1x_2 as a possible criterion (where we assume that the material is homogeneous). As a measure for the costs we take the sectional area of a trunk from which a beam of the height x_1 and the width x_2 can just be cut out. For simplicity this trunk is assumed to be a cylinder. The sectional area is given by $\frac{\pi}{4}(x_1^2 + x_2^2)$ (see Fig. 11.3).

Hence, we obtain a multiobjective optimization problem of the following form:

11.1. Basic Concepts

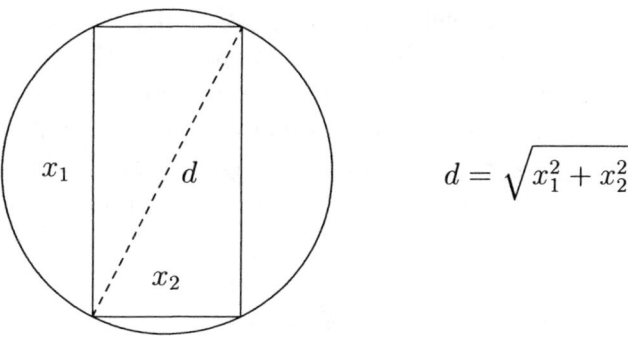

Figure 11.3: Sectional area.

$$\min \begin{pmatrix} lx_1x_2 \\ \frac{\pi}{4}(x_1^2 + x_2^2) \end{pmatrix}$$

subject to the constraints

$$2000 - x_1^2 x_2 \leq 0$$
$$x_1 - 4x_2 \leq 0$$
$$-x_1 + x_2 \leq 0$$
$$-x_1 \leq 0$$
$$-x_2 \leq 0.$$

In this chapter we investigate multiobjective optimization problems in finite dimensional spaces of the general form

$$\min_{x \in S} f(x). \tag{11.1}$$

Here we have the following assumption.

Assumption 11.2. Let S be a nonempty subset of $\mathbb{R}^n$ ($n \in \mathbb{N}$) and let $f : S \to \mathbb{R}^m$ ($m \in \mathbb{N}$) be a given vector function. The image space $\mathbb{R}^m$ is assumed to be partially ordered in a natural way (i.e., $\mathbb{R}^m_+$ is the ordering cone).

In the case of $m = 1$ problem (11.1) reduces to a standard optimization problem with a scalar-valued function f. Since $f_1, \ldots, f_m$

are various objectives to be optimized, one uses the name multiobjective optimization problem for (11.1). Actually it does not matter whether we investigate maximization or minimization problems. In this chapter we consider only minimization problems.

Minimization of a vector-valued function f means that we look for preimages of minimal elements of the image set $f(S)$ with respect to the natural partial ordering (see Fig. 11.4). In practice the minimal

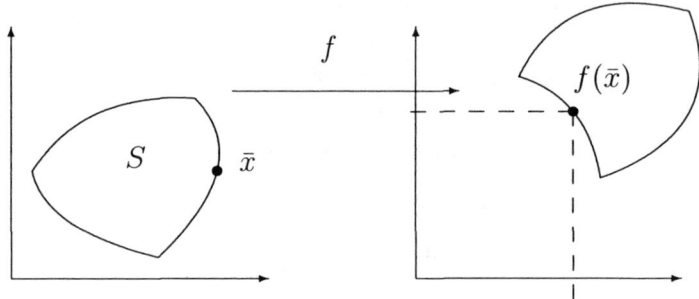

Figure 11.4: Preimage and image set of f.

elements of the image set $f(S)$ do not play the central role but their preimages.

Definition 11.3. Let Assumption 11.2 be satisfied. $\bar{x} \in S$ is called an *Edgeworth-Pareto optimal point* (or an *efficient solution* or a *minimal solution* or a *nondominated point*) of problem (11.1), if $f(\bar{x})$ is a minimal element of the image set $f(S)$ with respect to the natural partial ordering, i.e., there is no $x \in S$ with

$$f_i(x) \leq f_i(\bar{x}) \text{ for all } i \in \{1, \ldots, m\}$$

and

$$f(x) \neq f(\bar{x}).$$

The notion of efficient solutions is often used in economics whereas the notion "Edgeworth-Pareto optimal" can be found in engineering, and in applied mathematics one speaks of minimal solutions.

11.1. Basic Concepts

Example 11.4. Consider the constraint set
$$S := \{(x_1, x_2) \in \mathbb{R}^2 \mid x_1^2 - x_2 \leq 0, \ x_1 + 2x_2 - 3 \leq 0\}$$
and the vector function $f : S \to \mathbb{R}^2$ with
$$f(x_1, x_2) = \begin{pmatrix} -x_1 \\ x_1 + x_2^2 \end{pmatrix} \quad \text{for all} \ (x_1, x_2) \in S.$$
The point $(\frac{3}{2}, \frac{57}{16})$ is the only maximal element of $T := f(S)$, and the set of all minimal elements of T reads
$$\left\{(y_1, y_2) \in \mathbb{R}^2 \mid y_1 \in \left[-1, \frac{1}{2}\sqrt[3]{2}\right] \text{ and } y_2 = -y_1 + y_1^4\right\}.$$
The set of all Edgeworth-Pareto optimal points is given as
$$\left\{(x_1, x_2) \in \mathbb{R}^2 \mid x_1 \in \left[-\frac{1}{2}\sqrt[3]{2}, 1\right] \text{ and } x_2 = x_1^2\right\}$$
(see Fig. 11.5).

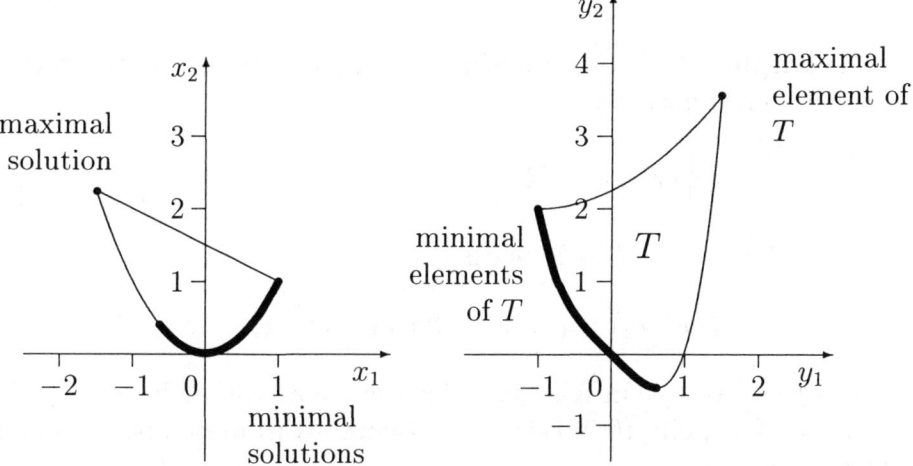

Figure 11.5: Minimal and maximal elements of T.

The Edgeworth-Pareto optimality concept is the main optimality notion used in multiobjective optimization. But there are also other concepts being more weakly or more strongly formulated. First we present a weaker optimality notion.

Definition 11.5. Let Assumption 11.2 be satisfied. $\bar{x} \in S$ is called a *weakly Edgeworth-Pareto optimal point* (or a *weakly efficient solution* or a *weakly minimal solution*) of problem (11.1), if $f(\bar{x})$ is a weakly minimal element of the image set $f(S)$ with respect to the natural partial ordering, i.e., there is no $x \in S$ with

$$f_i(x) < f_i(\bar{x}) \text{ for all } i \in \{1, \ldots, m\}.$$

This weak Edgeworth-Pareto optimality notion is often only used if it is difficult to characterize theoretically Edgeworth-Pareto optimal points or to determine them numerically. In general, in the applications one is not interested in weakly Edgeworth-Pareto optimal solutions; this optimality notion is only of mathematical interest.

Example 11.6. We consider the multiobjective optimization problem (11.1) with the set

$$S := \left\{ (x_1, x_2) \in \mathbb{R}^2 \mid 0 \leq x_1 \leq 1, \ 0 \leq x_2 \leq 1 \right\},$$

and the identity $f : S \to \mathbb{R}^2$ with

$$f(x_1, x_2) = (x_1, x_2) \quad \text{for all} \quad (x_1, x_2) \in S.$$

S describes a square in $\mathbb{R}^2$. Since f is the identity, the image set $f(S)$ equals S. The point $(0,0)$ is the only Edgeworth-Pareto optimal point whereas the set

$$\left\{ (x_1, x_2) \in S \mid x_1 = 0 \ \text{or} \ x_2 = 0 \right\}$$

is the set of all weakly Edgeworth-Pareto optimal points (see Fig. 11.6).

11.1. Basic Concepts

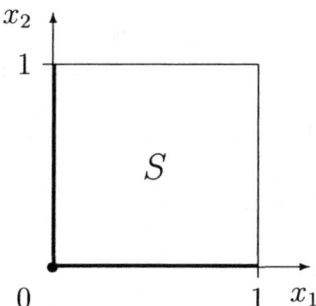

Figure 11.6: Weakly Edgeworth-Pareto optimal points.

With Lemma 4.14 we immediately obtain the following result.

Theorem 11.7. *Let Assumption 11.2 be satisfied. Every Edgeworth-Pareto optimal point of problem (11.1) is a weakly Edgeworth-Pareto optimal point of problem (11.1).*

Notice that the converse statement of Theorem 11.7 is not true in general (compare Example 11.6).

In the following we present a sharper optimality notion.

Definition 11.8. Let Assumption 11.2 be satisfied. $\bar{x} \in S$ is called a *properly Edgeworth-Pareto optimal point* (or a *properly efficient solution* or a *properly minimal solution*) of problem (11.1), if $\bar{x}$ is an Edgeworth-Pareto optimal point and there is a real number $\mu > 0$ so that for every $i \in \{1, \ldots, m\}$ and every $x \in S$ with $f_i(x) < f_i(\bar{x})$ at least one $j \in \{1, \ldots, m\}$ exists with $f_j(x) > f_j(\bar{x})$ and

$$\frac{f_i(\bar{x}) - f_i(x)}{f_j(x) - f_j(\bar{x})} \leq \mu.$$

An Edgeworth-Pareto optimal point which is not properly Edgeworth-Pareto optimal is also called an *improperly Edgeworth-Pareto optimal point*.

In the applications improperly Edgeworth-Pareto optimal points are not desired because a possible improvement of one component leads to a drastic deterioration of another component.

Example 11.9. For simplicity we investigate the multiobjective optimization problem (11.1) with the unit circle

$$S := \left\{ (x_1, x_2) \in \mathbb{R}^2 \mid x_1^2 + x_2^2 \leq 1 \right\},$$

and the identity $f : S \to \mathbb{R}^2$ with

$$f(x_1, x_2) = (x_1, x_2) \quad \text{for all} \quad (x_1, x_2) \in S.$$

The set of Edgeworth-Pareto optimal points reads

$$\left\{ (x_1, x_2) \in \mathbb{R}^2 \mid x_1 \in [-1, 0] \quad \text{and} \quad x_2 = -\sqrt{1 - x_1^2} \right\}$$

(see Fig. 11.7). Except the points $(-1, 0)$ and $(0, -1)$ all other Edge-

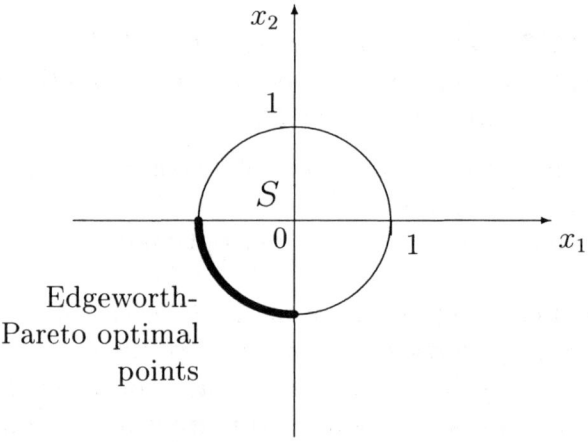

Figure 11.7: Edgeworth-Pareto optimal points in Example 11.9.

worth-Pareto optimal points are also properly Edgeworth-Pareto optimal points. In the following we show that the point $\bar{x} := (-1, 0)$ is an improperly Edgeworth-Pareto optimal point. For an arbitrary $n \in \mathbb{N}$ consider the point $x(n) := \left(-1 + \frac{1}{n}, -\frac{1}{n}\sqrt{2n - 1} \right)$ of the unit circle. For every $n \in \mathbb{N}$ we have $f_1(x(n)) > f_1(\bar{x})$ and $f_2(x(n)) < f_2(\bar{x})$, and we conclude

$$\frac{f_2(\bar{x}) - f_2(x(n))}{f_1(x(n)) - f_1(\bar{x})} = \frac{\bar{x}_2 - x_2(n)}{x_1(n) - \bar{x}_1} = \frac{0 + \frac{1}{n}\sqrt{2n - 1}}{-1 + \frac{1}{n} + 1} = \sqrt{2n - 1}$$

$$\text{for all} \quad n \in \mathbb{N}.$$

11.1. Basic Concepts

It is obvious that an upper bound $\mu > 0$ of this term does not exist. Consequently, $\bar{x} = (-1, 0)$ is an improperly Edgeworth-Pareto optimal point.

Example 11.10. It can be shown that one properly Edgeworth-Pareto optimal point of the design problem discussed in Example 11.1 is, for instance, the point $(10\sqrt[3]{4}, 5\sqrt[3]{4})$. This solution leads to a beam with the height $10\sqrt[3]{4} \approx 15.874$ and the width $5\sqrt[3]{4} \approx 7.937$.

Next we come to a very strong optimality notion.

Definition 11.11. Let Assumption 11.2 be satisfied. $\bar{x} \in S$ is called a *strongly Edgeworth-Pareto optimal point* (or a *strongly efficient solution* or a *strongly minimal solution*) of problem (11.1), if $f(\bar{x})$ is a strongly minimal element of the image set $f(S)$ with respect to the natural partial ordering, i.e.

$$f_i(\bar{x}) \leq f_i(x) \text{ for all } x \in S \text{ and all } i \in \{1, \ldots, m\}.$$

This concept naturally generalizes the standard minimality notion used in scalar optimization. But it is clear that this concept is too strong for multiobjective optimization problems.

Example 11.12. Consider the multiobjective optimization problem in Example 11.6. Here the point (0,0) is a strongly Edgeworth-Pareto optimal point. The problem discussed in Example 11.9 has no strongly Edgeworth-Pareto optimal points.

Theorem 11.13. *Let Assumption 11.2 be satisfied. Every strongly Edgeworth-Pareto optimal point of problem (11.1) is an Edgeworth-Pareto optimal point.*

Proof. Let $\bar{x} \in S$ be a strongly Edgeworth-Pareto optimal point,

i.e.
$$f_i(\bar{x}) \leq f_i(x) \text{ for all } x \in S \text{ and all } i \in \{1,\ldots,m\}.$$
Then there is no $x \in S$ with $f(x) \neq f(\bar{x})$ and
$$f_i(x) \leq f_i(\bar{x}) \text{ for all } i \in \{1,\ldots,m\}.$$
Hence, $\bar{x}$ is an Edgeworth-Pareto optimal point. $\square$

Finally, we come to an optimality concept using the convex hull of the image set $f(S)$.

Definition 11.14. Let Assumption 11.2 be satisfied. $\bar{x} \in S$ is called an *essentially Edgeworth-Pareto optimal point* (or an *essentially efficient solution* or an *essentially minimal solution*) of problem (11.1), if $f(\bar{x})$ is a minimal element of the convex hull of the image set $f(S)$.

Since the image set $f(S)$ is contained in its convex hull it is evident that every essentially Edgeworth-Pareto optimal point $\bar{x} \in S$ is also an Edgeworth-Pareto optimal point. Morover, there is also a relationship to the strong Edgeworth-Pareto optimality concept.

Theorem 11.15. *Let Assumption 11.2 be satisfied. Every strongly Edgeworth-Pareto optimal point is an essentially Edgeworth-Pareto optimal point.*

Proof. Let $\bar{x} \in S$ be a strongly Edgeworth-Pareto optimal point. Then we have
$$f_i(\bar{x}) \leq f_i(x) \text{ for all } x \in S \text{ and all } i \in \{1,\ldots,m\}$$
or
$$f(S) \subset \{f(\bar{x})\} + C$$
with $C := \mathbb{R}_+^m$ ("+" denotes the algebraic sum of sets). Since the set $\{f(\bar{x})\} + C$ is convex, we conclude for the convex hull $\mathrm{co}(f(S))$ of $f(S)$ being the intersection of all convex subsets of $\mathbb{R}^m$ containing $f(S)$
$$\mathrm{co}(f(S)) \subset \{f(\bar{x})\} + C.$$

11.2. Special Scalarization Results

Then there is no $y \in \text{co}(f(S))$ with $y \neq f(\bar{x})$ and
$$y_i \leq f_i(\bar{x}) \text{ for all } i \in \{1, \ldots, m\}.$$
Hence, $f(\bar{x})$ is a minimal element of the set $\text{co}(f(S))$, i.e. $\bar{x} \in S$ is an essentially Edgeworth-Pareto optimal point. □

Example 11.16. Consider the multiobjective optimization problem (11.1) with the discrete constraint set
$$S := \{(0,3), \quad (2,2), \quad (3,0)\},$$
and the identity as objective function f (see Fig. 11.8). Every feasible

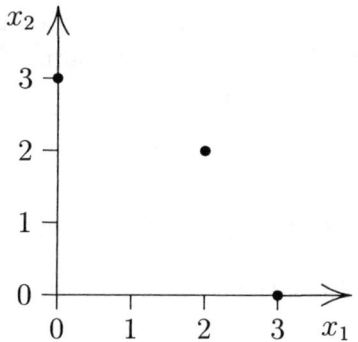

Figure 11.8: Constraint set S.

point is an Edgeworth-Pareto optimal point, but only the points $(3, 0)$ and $(0, 3)$ are essentially Edgeworth-Pareto optimal points.

Summarizing the relationships between the presented optimality concepts we obtain the diagram in Table 11.1. Notice that the converse implications are not true in general.

11.2 Special Scalarization Results

In Chapter 5 scalarization techniques are discussed in detail. In economics and engineering scalarized problems are also called *auxiliary*

> strong EP optimality
> $\Downarrow$
> essential EP optimality
> $\Downarrow$
> proper EP optimality $\Rightarrow$ EP optimality $\Rightarrow$ weak EP optimality

Table 11.1: Relationships between different Edgeworth-Pareto (EP) optimality concepts.

problems, auxiliary programs or *compromise models*. We now present two main approaches for the determination of Edgeworth-Pareto optimal points. We consider the weighted sum of the objectives and a weighted Chebyshev norm approach. Moreover, we investigate special scalar problems.

11.2.1 Weighted Sum Approach

Let Assumption 11.2 be satisfied and consider the multiobjective optimization problem

$$\min_{x \in S} f(x). \tag{11.2}$$

If one formulates a scalar problem using linear functionals (e.g., see Theorem 5.4), then we obtain for this special case the scalarized optimization problem

$$\min_{x \in S} \sum_{i=1}^{m} t_i f_i(x)$$

with appropriate weights $t_1, \ldots, t_m$. This approach uses the weighted sum of the components of the objective vector function. Therefore, one speaks of a *weighted sum approach* (for instance, see [5, 6, 8]). If one specializes the assertions of Theorem 5.18, (a) and Theorem 5.28 for $C := \mathbb{R}^m_+$, then we obtain the scalarization results given in Table 11.2. The result of the following theorem is also considered in this table.

11.2. Special Scalarization Results

Every solution of the scalar optimization problem $$\max_{x \in S} \sum_{i=1}^{m} t_i f_i(x)$$ with			
$t_1, \ldots, t_m > 0$	$t_1, \ldots, t_m \geq 0$, $t_i > 0$ for some $i \in \{1, \ldots, m\}$, where image uniqueness of the solution is given	$t_1, \ldots, t_m \geq 0$, $t_i > 0$ for some $i \in \{1, \ldots, m\}$,
is			
a properly EP optimal point of problem (11.2).	an EP optimal point of problem (11.2).	a weakly EP optimal point of problem (11.2).

Table 11.2: Sufficient conditions for Edgeworth-Pareto (EP) optimal points.

Theorem 11.17. *Let Assumption 11.2 be satisfied, and let* $t_1, \ldots, t_m > 0$ *be given real numbers. If* $\bar{x} \in S$ *is a solution of the scalar optimization problem*

$$\min_{x \in S} \sum_{i=1}^{m} t_i f_i(x), \qquad (11.3)$$

then $\bar{x}$ *is a properly Edgeworth-Pareto optimal point of the multiobjective optimization problem (11.2).*

Proof. By Theorem 5.18, (b), $\bar{x}$ is an Edgeworth-Pareto optimal point of problem (11.2). Assume that $\bar{x}$ is no properly Edgeworth-Pareto optimal point. Then we choose

$$\mu := (m-1) \max_{i,j \in \{1,\ldots,m\}} \left\{ \frac{t_j}{t_i} \right\} \text{ for } m \geq 2,$$

and we obtain for some $i \in \{1, \ldots, m\}$ and some $x \in S$ with $f_i(x) < f_i(\bar{x})$

$$\frac{f_i(\bar{x}) - f_i(x)}{f_j(x) - f_j(\bar{x})} > \mu \text{ for all } j \in \{1, \ldots, m\} \text{ with } f_j(x) > f_j(\bar{x}).$$

This implies

$$f_i(\bar{x}) - f_i(x) > \mu \left(f_j(x) - f_j(\bar{x}) \right) \geq (m-1) \frac{t_j}{t_i} \left(f_j(x) - f_j(\bar{x}) \right)$$

for all $j \in \{1, \ldots, m\} \setminus \{i\}$.

Multiplication with $\frac{t_i}{m-1}$ and summation with respect to $j \neq i$ leads to

$$t_i \left(f_i(\bar{x}) - f_i(x) \right) < \sum_{\substack{j=1 \\ j \neq i}}^{m} t_j (f_j(x) - f_j(\bar{x}))$$

and

$$0 > \sum_{j=1}^{m} t_j (f_j(x) - f_j(\bar{x}))$$

implying

$$\sum_{j=1}^{m} t_j f_j(\bar{x}) > \sum_{j=1}^{m} t_j f_j(x)$$

11.2. Special Scalarization Results

contradicting to the assumption that $\bar{x} \in S$ is a solution of the scalar optimization problem (11.3). □

Example 11.18.

(a) In Example 11.4 we have already investigated the following multiobjective optimization problem (see also Fig. 11.5)

$$\min \begin{pmatrix} -x_1 \\ x_1 + x_2^2 \end{pmatrix}$$
subject to the constraints
$$x_1^2 - x_2 \leq 0 \qquad (11.4)$$
$$x_1 + 2x_2 - 3 \leq 0$$
$$x_1, x_2 \in \mathbb{R}.$$

For the computation of a properly Edgeworth-Pareto optimal point of this problem one can choose, for instance, $t_1 = 1$ and $t_2 = 2$. Then one solves the scalar optimization problem

$$\max x_1 + 2x_2^2$$
subject to the constraints
$$x_1^2 - x_2 \leq 0 \qquad (11.5)$$
$$x_1 + 2x_2 - 3 \leq 0$$
$$x_1, x_2 \in \mathbb{R}.$$

$\bar{x} = (-\frac{1}{2}, \frac{1}{4})$ is the unique solution of the problem (11.5). By Theorem 11.17 $\bar{x}$ is also a properly Edgeworth-Pareto optimal point of the multiobjective optimization problem (11.4).

(b) The application of Theorem 5.18, (c) to discrete problems allows a fast computation of Edgeworth-Pareto optimal points. As a very simple example (see [83, p. 165]) all minimal elements of the discrete set

$$S := \{(-16, -9), (-6, -14), (-11, -13), (-10, -10)\} \subset \mathbb{R}^2$$

are determined. For this purpose we choose the vector function $f : S \to \mathbb{R}^2$ with

$$f(x_1, x_2) = (x_1, x_2) \text{ for all } (x_1, x_2) \in S.$$

The minimal elements of S are exactly the Edgeworth-Pareto optimal points of the problem
$$\min_{x \in S} f(x).$$
For the computation of these Edgeworth-Pareto optimal points one can choose the weight vector $t = (\alpha, 1 - \alpha)$ with $\alpha \in (0, 1)$ and obtains the scalar optimization problem
$$\min_{(x_1, x_2) \in S} \alpha x_1 + (1 - \alpha) x_2$$
for arbitrary $\alpha \in (0, 1)$. The minimal elements of the set S are given in Table 11.3.

α	$\bar{x}$	$\alpha \bar{x}_1 + (1 - \alpha) \bar{x}_2$
$0 < \alpha < \frac{1}{6}$	$(-6, -14)$	$-6\alpha - 14(1 - \alpha)$
$\alpha = \frac{1}{6}$	$(-6, -14)$ or $(-11, -13)$	$-\frac{38}{3}$
$\frac{1}{6} < \alpha < \frac{4}{9}$	$(-11, -13)$	$-11\alpha - 13(1 - \alpha)$
$\alpha = \frac{4}{9}$	$(-11, -13)$ or $(-16, -9)$	$-\frac{109}{9}$
$\frac{4}{9} < \alpha < 1$	$(-16, -9)$	$-16\alpha - 9(1 - \alpha)$

Table 11.3: Minimal elements of the set S for different parameters (Example 11.18, (b)).

(c) The result of Theorem 5.18, (c) can be well applied in linear multiobjective optimization. As an example let us determine all Edgeworth-Pareto optimal points of the following problem (see [80, pp. 155]):
$$\min \begin{pmatrix} -4x_1 - 2x_2 \\ -8x_1 - 10x_2 \end{pmatrix}$$
subject to the constraints
$$x_1 + x_2 \leq 70$$
$$x_1 + 2x_2 \leq 100$$
$$x_1 \leq 60$$
$$x_2 \leq 40$$
$$x_1, x_2 \geq 0.$$

11.2. Special Scalarization Results

The constraint set of this example is illustrated in Fig. 11.9. Again, let the vector t of the weights be given as $t = (\alpha, 1 - \alpha)$

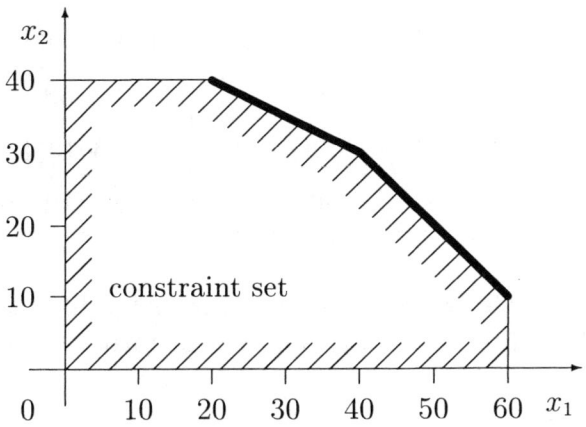

Figure 11.9: Constraint set in Example 11.18, (c).

with $\alpha \in (0, 1)$. Consequently, one obtains for $\alpha \in (0, 1)$ the parametric optimization problem

$$\min \ (-8 + 4\alpha)x_1 + (-10 + 8\alpha)x_2$$
$$\text{subject to the constraints}$$
$$x_1 + x_2 \leq 70$$
$$x_1 + 2x_2 \leq 100$$
$$x_1 \leq 60$$
$$x_2 \leq 40$$
$$x_1, \ x_2 \geq 0.$$

All solutions of this problem are given in Table 11.4. These are also Edgeworth-Pareto optimal points of the considered multiobjective optimization problem.

Notice that for general nonlinear multiobjective optimization problems not *every* Edgeworth-Pareto optimal point can be determined using the weighted sum approach. For instance, Figure 11.10 shows that only two minimal points of the set T can be determined in such

α	$\bar{x}_1$	$\bar{x}_2$
$0 < \alpha < \frac{2}{5}$	20	40
$\frac{2}{5}$	$20\lambda + 40(1-\lambda)$	$40\lambda + 30(1-\lambda)$
$\frac{2}{5} < \alpha < \frac{6}{7}$	40	30
$\frac{6}{7}$	$40\lambda + 60(1-\lambda)$	$30\lambda + 10(1-\lambda)$
$\frac{6}{7} < \alpha < 1$	60	10

Table 11.4: Edgeworth-Pareto optimal points for different parameters (Example 11.18, (c)). $\lambda \in [0,1]$ can be arbitrarily chosen.

a way. Only these two points are supporting points of an appropriate supporting function. The weighted sum approach seems to be

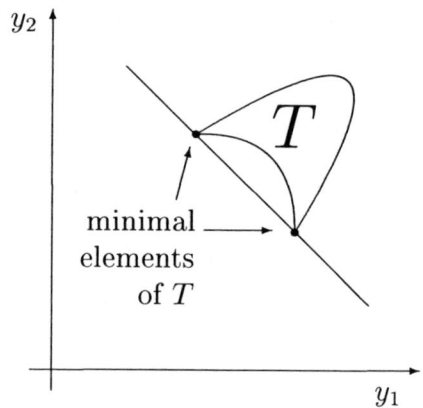

Figure 11.10: Weighted sum approach in the nonconvex case.

only suitable for convex problems, like linear problems. In general, this approach cannot be used for multiobjective optimization problems arising in engineering. For these problems other approaches, for instance, like the weighted Chebyshev norm approach, are more suitable.

We know from Chapter 5 that for multiobjective optimization problems for which the set

11.2. Special Scalarization Results

$$f(S) + \mathbb{R}_+^m$$
$$:= \{y \in \mathbb{R}^m \mid y_i \geq f_i(x) \text{ for some } x \in S \text{ and all } i \in \{1,\ldots,m\}\}$$

is convex, the weighted sum approach is appropriate. The results concerning the weighted sum approach are summarized in Table 11.5. The corresponding mathematical results can be found Theorem 5.18, (b), Theorem 5.4, Corollary 5.29 and the following corollary.

Corollary 11.19. *Let Assumption 11.2 be satisfied, and let the set $f(S) + \mathbb{R}_+^m$ be convex. Then $\bar{x} \in S$ is a properly Edgeworth-Pareto optimal point of the multiobjective optimization problem (11.2) if and only if there are real numbers $t_1, \ldots, t_m > 0$ so that $\bar{x}$ is a solution of the scalar optimization problem (11.3).*

Proof. One part of the assertion is shown in Theorem 11.17. For the converse part assume that $\bar{x}$ is an Edgeworth-Pareto optimal point. Then there is a real number $\mu > 0$ so that for every $i \in \{1,\ldots,m\}$ and every $x \in S$ with $f_i(x) < f_i(\bar{x})$ at least one $j \in \{1,\ldots,m\}$ exists with $f_j(x) > f_j(\bar{x})$ and

$$\frac{f_i(\bar{x}) - f_i(x)}{f_j(x) - f_j(\bar{x})} \leq \mu. \tag{11.6}$$

Consequently, for every $i \in \{1,\ldots,m\}$ the system

$$f_i(x) < f_i(\bar{x}) \tag{11.7}$$
$$f_i(x) + \mu f_j(x) < f_i(\bar{x}) + \mu f_j(\bar{x})$$
$$\text{for all } j \in \{1,\ldots,m\}\setminus\{i\} \tag{11.8}$$

does not have a solution $x \in S$. In order to see this implication assume that for some $i \in \{1,\ldots,m\}$ the system (11.7), (11.8) would have a solution $x \in S$. If there is no $j \in \{1,\ldots,m\}$ with $f_j(x) > f_j(\bar{x})$, $\bar{x}$ cannot be properly Edgeworth-Pareto optimal. On the other hand, if there is some $j \in \{1,\ldots,m\}$ with $f_j(x) > f_j(\bar{x})$, we obtain from (11.8)

$$f_i(\bar{x}) - f_i(x) > \mu(f_j(x) - f_j(\bar{x}))$$

contradicting the inequality (11.6).

Now we procede with the actual proof of the corollary and choose an arbitrary $i \in \{1, \ldots, m\}$. We define the nonempty set

$$M_i := \left\{ \begin{pmatrix} f_i(x) + \alpha_i + \mu(f_1(x) + \alpha_1) \\ \vdots \\ f_i(x) + \alpha_i \\ \vdots \\ f_i(x) + \alpha_i + \mu(f_m(x) + \alpha_m) \end{pmatrix} \in \mathbb{R}^m \;\middle|\; x \in S, \alpha_1, \ldots, \alpha_m \geq 0 \right\}$$

Since $f(S) + \mathbb{R}^m_+$ is assumed to be convex, one can show with simple calculations that the set M is convex as well. If we set

$$\bar{y}_i := \begin{pmatrix} f_i(\bar{x}) + \mu f_1(\bar{x}) \\ \vdots \\ f_i(\bar{x}) \\ \vdots \\ f_i(\bar{x}) + \mu f_m(\bar{x}) \end{pmatrix} \in \mathbb{R}^m$$

and notice that the system (11.7), (11.8) is not solvable, we conclude

$$M_i \cap \text{int}(\{\bar{y}_i\} - \mathbb{R}^m_+) = \emptyset.$$

Then by Eidelheit's separation theorem (Theorem 3.16) there are real numbers $\lambda_1^{(i)}, \ldots, \lambda_m^{(i)}$ with $\lambda^{(i)} \neq 0_{\mathbb{R}^m}$ and

$$\begin{aligned} &\lambda_1^{(i)}(f_i(x) + \alpha_i + \mu(f_1(x) + \alpha_1)) + \cdots + \lambda_i^{(i)}(f_i(x) + \alpha_i) \\ &+ \cdots + \lambda_m^{(i)}(f_i(x) + \alpha_i + \mu(f_m(x) + \alpha_m)) \\ \geq\;& \lambda_1^{(i)}(f_i(\bar{x}) + \mu f_1(\bar{x})) + \cdots + \lambda_i^{(i)} f_i(\bar{x}) \\ &+ \cdots + \lambda_m^{(i)}(f_i(\bar{x}) + \mu f_m(\bar{x})) \end{aligned}$$

for all $x \in S$ and all $\alpha_1, \ldots, \alpha_m \geq 0$. \hfill (11.9)

For $x = \bar{x}$ we immediately obtain $\lambda_1^{(i)}, \ldots, \lambda_m^{(i)} \geq 0$. For $\alpha_1 = \cdots = \alpha_m = 0$ we conclude from (11.9)

$$f_i(x) \sum_{j=1}^{m} \lambda_j^{(i)} + \mu \sum_{\substack{j=1 \\ j \neq i}}^{m} \lambda_j^{(i)} f_j(x) \geq f_i(\bar{x}) \sum_{j=1}^{m} \lambda_j^{(i)} + \mu \sum_{\substack{j=1 \\ j \neq i}}^{m} \lambda_j^{(i)} f_j(\bar{x})$$

11.2. Special Scalarization Results

for all $x \in S$,

and because $\sum_{j=1}^{m} \lambda_j^{(i)} > 0$ we get

$$f_i(x) + \mu \sum_{\substack{j=1 \\ j \neq i}}^{m} \frac{\lambda_j^{(i)}}{\sum_{k=1}^{m} \lambda_k^{(i)}} f_j(x) \geq f_i(\bar{x}) + \mu \sum_{\substack{j=1 \\ j \neq i}}^{m} \frac{\lambda_j^{(i)}}{\sum_{k=1}^{m} \lambda_k^{(i)}} f_j(\bar{x})$$

for all $x \in S$. (11.10)

The inequality (11.10) holds for every $i \in \{1, \ldots, m\}$. Next, we sum up these m inequalities and obtain

$$\sum_{i=1}^{m} \underbrace{\left(1 + \mu \sum_{\substack{j=1 \\ j \neq i}}^{m} \frac{\lambda_j^{(i)}}{\sum_{k=1}^{m} \lambda_k^{(i)}} \right)}_{=:t_i > 0} f_i(x) \geq \sum_{i=1}^{m} \underbrace{\left(1 + \mu \sum_{\substack{j=1 \\ j \neq i}}^{m} \frac{\lambda_j^{(i)}}{\sum_{k=1}^{m} \lambda_k^{(i)}} \right)}_{=t_i} f_i(\bar{x})$$

for all $x \in S$,

Consequently, $\bar{x} \in S$ is a solution of the optimization problem (11.3). □

In economics multiobjective optimization problems are very often linear, i.e. they are of the form

$$\begin{aligned} & \min Cx \\ & \text{subject to the constraints} \\ & Ax \leq b \\ & x \in \mathbb{R}^n \end{aligned} \quad (11.11)$$

where C is a real (m, n) matrix, A is a real (k, n) matrix (with $k \in \mathbb{N}$) and $b \in \mathbb{R}^k$ is a given vector (compare Example 11.18, (c)). For these problems the set $f(S) + \mathbb{R}_+^m$ is always convex and, therefore, the results in Table 11.5 can be applied. Moreover, it can be shown that Edgeworth-Pareto optimal points and properly Edgeworth-Pareto optimal points coincide in this case. This is the result of the following theorem.

If the set $f(S) + \mathbb{R}^m_+$ is convex, then a solution of the scalar optimization problem $$\min_{x \in S} \sum_{i=1}^{m} t_i f_i(x)$$ is		
a properly EP optimal point of problem (11.2)	an EP optimal point of problem (11.2)	a weakly EP optimal point of problem (11.2)
if and only if		
$t_1, \ldots, t_m > 0$.	$\begin{cases} t_1, \ldots, t_m > 0 \\ \text{(suff. cond.)}. \\ t_1, \ldots, t_m \geq 0, \\ t_i > 0 \text{ for some} \\ i \in \{1, \ldots, m\}, \\ \text{(necess. cond.)}. \end{cases}$	$t_1, \ldots, t_m \geq 0,$ $t_i > 0$ for some $i \in \{1, \ldots, m\}$.

Table 11.5: Necessary and sufficient conditions for Edgeworth-Pareto (EP) optimal points.

11.2. Special Scalarization Results

Theorem 11.20. *Let the linear multiobjective optimization problem (11.11) be given where C is a real (m,n) matrix, A is a real (k,n) matrix and $b \in \mathbb{R}^k$ is a given vector. Let the constraint set*

$$S := \{x \in \mathbb{R}^n \mid Ax \leq b\}$$

be nonempty. The image space $\mathbb{R}^m$ is assumed to be partially ordered in a natural way (i.e., $\mathbb{R}^m_+$ is the ordering cone). Then $\bar{x} \in S$ is an Edgeworth-Pareto optimal point of the linear multiobjective optimization problem (11.11) if and only if $\bar{x} \in S$ is a properly Edgeworth-Pareto optimal point of problem (11.11).

Proof. By definition every properly Edgeworth-Pareto optimal point is also an Edgeworth-Pareto optimal point. For the proof of the converse implication fix an arbitrary Edgeworth-Pareto optimal point $\bar{x} \in S$. Then $C\bar{x}$ is a minimal element of the image set

$$T := \{Cx \in \mathbb{R}^m \mid x \in S\},$$

that is

$$(\{C\bar{x}\} - \mathbb{R}^m_+) \cap T = \{C\bar{x}\}$$

or, equivalently,

$$(-\mathbb{R}^m_+) \cap (T - \{C\bar{x}\}) = \{0_{\mathbb{R}^m}\}.$$

Since T is a polytop, the cone generated by $T - \{C\bar{x}\}$ is a polyhedral cone and we conclude

$$(-\mathbb{R}^m_+) \cap \operatorname{cone}(T - \{C\bar{x}\}) = \{0_{\mathbb{R}^m}\}.$$

By a separation theorem for closed convex cones (Theorem 3.22) there are real numbers $t_1, \ldots, t_m$ with $t_i \neq 0$ for at least one $i \in \{1, \ldots, m\}$ so that

$$\sum_{i=1}^m t_i y_i \leq 0 \leq \sum_{i=1}^m t_i z_i \text{ for all } y \in -\mathbb{R}^m_+ \text{ and all } z \in \operatorname{cone}(T - \{C\bar{x}\}) \tag{11.12}$$

and

$$\sum_{i=1}^m t_i y_i < 0 \text{ for all } y \in -\mathbb{R}^m_+ \setminus \{0_{\mathbb{R}^m}\}. \tag{11.13}$$

If we take the unit vectors in $\mathbb{R}^m$, we obtain form the inequality (11.13) $t_1, \ldots, t_m > 0$. The right inequality in (11.12) implies

$$\sum_{i=1}^m t_i(Cx_i - C\bar{x}_i) \geq 0 \text{ for all } x \in S$$

and

$$\sum_{i=1}^m t_i C\bar{x}_i \leq \sum_{i=1}^m t_i Cx_i \text{ for all } x \in S.$$

Consequently, we get with Theorem 11.17 that $\bar{x}$ is a properly Edgeworth-Pareto optimal point. □

11.2.2 Weighted Chebyshev Norm Approach

In this subsection we investigate the scalarization with a weighted Chebyshev norm. For general nonconvex multiobjective optimization problems these norms are much more suitable than linear functionals.

Section 5.3 already presents a discussion of parametric approximation problems used for scalarization. In the case of multiobjective optimization these problems are approximation problems with a weighted Chebyshev norm (see Corollary 5.35). Under Assumption 11.2 we investigate for some $\hat{y} \in \mathbb{R}^m$ the weighted Chebyshev approximation problem

$$\min_{x \in S} \max_{1 \leq i \leq m} \{w_i(f_i(x) - \hat{y}_i)\}. \tag{11.14}$$

For the sake of convenience we reformulate Corollary 5.35 for engineering applications.

Corollary 11.21. *Let Assumption 11.2 be satisfied, and assume that there is a $\hat{y} \in \mathbb{R}^m$ with the property that*

$$\hat{y}_i < f_i(x) \text{ for all } x \in S \text{ and all } i \in \{1, \ldots, m\}.$$

(a) $\bar{x} \in S$ is an Edgeworth-Pareto optimal point of the multiobjective optimization problem (11.2) if and only if there are positive real numbers $w_1, \ldots, w_m$ so that $\bar{x}$ is an image unique solution of the weighted Chebyshev approximation problem (11.14).

11.2. Special Scalarization Results

(b) $\bar{x} \in S$ is a weakly Edgeworth-Pareto optimal point of the multiobjective optimization problem (11.2) if and only if there are positive real numbers $w_1, \ldots, w_m$ so that $\bar{x}$ is a solution of the weighted Chebyshev approximation problem (11.14).

Fig. 11.11 illustrates the result of the previous corollary. In Corol-

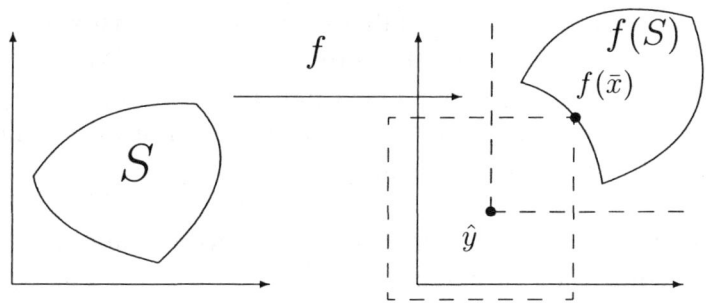

Figure 11.11: Illustration of the weighted Chebyshev norm approach.

lary 11.21 it is assumed that $\hat{y}$ is a strict lower bound of f. But the assertion also remains true if

$$\hat{y}_i \leq f_i(x) \text{ for all } x \in S \text{ and all } i \in \{1, \ldots, m\}.$$

In practice, this is not a critical assumption because the objective funtions $f_1, \ldots, f_m$ are often bounded from below.

For practical purposes we now transform the weighted Chebyshev norm approximation problem (11.14). If we replace the max term by a new variable λ, then problem (11.14) is equivalent to the scalar optimization problem

$$\min \lambda$$
$$\text{subject to the constraints}$$
$$\lambda = \max_{1 \leq i \leq m} \{w_i(f_i(x) - \hat{y}_i)\}$$
$$x \in S, \ \lambda \in \mathbb{R}.$$

This problem is again equivalent to the problem

$$\min \lambda$$
subject to the constraints
$$w_1(f_1(x) - \hat{y}_1) - \lambda \leq 0$$
$$\vdots$$
$$w_m(f_m(x) - \hat{y}_m) - \lambda \leq 0$$
$$x \in S, \ \lambda \in \mathbb{R}.$$

If the set S is described by equality or inequality constraints, then this problem can be solved with standard methods of nonlinear constraint optimization. In the case of a linear multiobjective optimization problem this scalar problem is a linear optimization problem. This fact is illustrated by the following example.

Example 11.22. For simplicity we consider the linear multiobjective optimization problem

$$\min \begin{pmatrix} x_1 - x_2 \\ -x_1 + x_2 \\ -\frac{1}{2}x_1 - \frac{1}{2}x_2 \end{pmatrix}$$
subject to the constraints
$$0 \leq x_1 \leq 1$$
$$0 \leq x_2 \leq 1$$
$$x_1, x_2 \in \mathbb{R}.$$

For the determination of a weakly Edgeworth-Pareto optimal point we choose the weights $w_1 = w_2 = w_3 = 1$ and the point $\hat{y} = (-1, -1, -1)$ being a lower bound of the objective functions. Then we obtain the scalarized optimization problem being equivalent to the weighted Chebyshev approximation problem

$$\min \lambda$$
subject to the constraints
$$x_1 - x_2 - \lambda + 1 \leq 0$$
$$-x_1 + x_2 - \lambda + 1 \leq 0$$
$$-\frac{1}{2}x_1 - \frac{1}{2}x_2 - \lambda + 1 \leq 0$$
$$0 \leq x_1 \leq 1$$
$$0 \leq x_2 \leq 1$$
$$x_1, x_2, \lambda \in \mathbb{R}.$$

11.2. Special Scalarization Results

This is a standard linear optimization problem which can be solved using the simplex method. Solutions are

$$x_1' = 1,\ x_2' = 1,\ \lambda' = 1$$

and

$$x_1'' = \frac{1}{2},\ x_2'' = \frac{1}{2},\ \lambda'' = 1.$$

By Corollary 11.21, (b) (x_1', x_2') and (x_1'', x_2'') are weakly Edgeworth-Pareto optimal points of the considered multiobjective optimization problem. The images of the vector function at these points are $f(x_1', x_2') = (0, 0, -1)$ and $f(x_1'', x_2'') = (0, 0, -\frac{1}{2})$. One can check that (x_1', x_2') is an Edgeworth-Pareto optimal point but (x_1'', x_2'') is not.

Another scalarization approach is similar to the weighted Chebyshev norm approach. Here one uses a scalarizing function $\varphi : \mathbb{R}^m \to \mathbb{R}$ defined by

$$\varphi(y_1, \ldots, y_m) = \max_{i \in \{1, \ldots, m\}} \{y_i\} \text{ for all } (y_1, \ldots, y_m) \in \mathbb{R}^m$$

for the scalarization of the multiobjective optimization problem (11.2). It can be easily seen that φ is strictly monotonically increasing on $\mathbb{R}^m$. For the proof of this assertion choose arbitrary vectors $x,\ y \in \mathbb{R}^m$ with

$$x_i < y_i \text{ for all } i \in \{1, \ldots, m\}.$$

Then it follows

$$\varphi(x) = \max_{i \in \{1, \ldots, m\}} \{x_i\} < \max_{i \in \{1, \ldots, m\}} \{y_i\} = \varphi(y).$$

A scalarization with the function φ has the advantage that one does not need a lower bound of the objective funtions $f_1, \ldots, f_m$.

11.2.3 Special Scalar Problems

In this subsection some special scalar optimization problems are presented which are interesting for applications and which can be derived by known results.

With the result of the next theorem we can examine whether a feasible point is Edgeworth-Pareto optimal.

Theorem 11.23. *Let Assumption 11.2 be satisfied, let $t_1,\ldots,t_m > 0$ be given real numbers, and let $\tilde{x} \in S$ be a given feasible point of the multiobjective optimization problem (11.2). If $\bar{x}$ is a solution of the scalar optimization problem*

$$\min \sum_{i=1}^m t_i f_i(x)$$
subject to the constraints \hfill (11.15)
$$f_i(x) \leq f_i(\tilde{x}) \text{ for all } i \in \{1,\ldots,m\}$$
$$x \in S,$$

then $\bar{x}$ is an Edgeworth-Pareto optimal point of problem (11.2). If $\tilde{x}$ is already an Edgeworth-Pareto optimal point of problem (11.2), then $\tilde{x}$ is also a solution of the scalar optimization problem (11.15).

Proof. For positive real numbers $t_1,\ldots,t_m$ and for a given $\tilde{x} \in S$ let $\bar{x}$ be a solution of the scalar problem (11.15). Suppose that $\bar{x}$ is no Edgeworth-Pareto optimal point. Then there is some $x \in S$ with $f(x) \neq f(\bar{x})$ and

$$f_i(x) \leq f_i(\bar{x}) \text{ for all } i \in \{1,\ldots,m\}.$$

Consequently, we obtain

$$\sum_{i=1}^m t_i f_i(x) < \sum_{i=1}^m t_i f_i(\bar{x})$$

and
$$f_i(x) \leq f_i(\bar{x}) \leq f_i(\tilde{x}) \text{ for all } i \in \{1,\ldots,m\}.$$

But this contradicts the fact that $\bar{x}$ solves the scalar problem (11.15). If $\tilde{x}$ is already an Edgeworth-Pareto optimal point, then there is no $x \in S$ with $f(x) \neq f(\tilde{x})$ and

$$f_i(x) \leq f_i(\tilde{x}) \text{ for all } i \in \{1,\ldots,m\},$$

11.2. Special Scalarization Results

i.e., there is no $x \in S$ with
$$\sum_{i=1}^{m} t_i f_i(x) < \sum_{i=1}^{m} t_i f_i(\tilde{x})$$
and
$$f_i(x) \leq f_i(\tilde{x}) \text{ for all } i \in \{1, \ldots, m\}.$$
Then $\tilde{x}$ is a solution of the scalar optimization problem (11.15). □

Theorem 11.23 can be used, for instance, if one cannot examine the image uniqueness of a solution of a Chebyshev approximation problem as an auxiliary problem. But this theorem should be applied with care. For instance, if $\tilde{x}$ is already an Edgeworth-Pareto optimal point, then the inequality constraints are active, i.e. these are equality constraints. This fact may lead to numerical difficulties (notice that then the known Slater condition is not satisfied).

Example 11.24. Consider again the multiobjective optimization problem (11.4) in Example 11.18, (a). We then investigate the question: Is the point
$$\tilde{x} := (0.65386, 0.46750)$$
an Edgeworth-Pareto optimal point of this multiobjective optimization problem? For an answer of this question we can solve the scalar optimization problem (11.15) for $t = (1,1)$, for instance. The point
$$\bar{x} \approx (0.65386, 0.42753)$$
is the unique solution of the scalar optimization problem (11.15), and by Theorem 11.23 it is also an Edgeworth-Pareto optimal point of the multiobjective optimization problem. Therefore, the point $\tilde{x}$ is no Edgeworth-Pareto optimal point of problem (11.4).

The following scalarization approach is used in economics. It is completely equivalent to an approach using the ℓ_1 norm.

Theorem 11.25. *Let Assumption 11.2 be satisfied, and let a point $\hat{y} \in \mathbb{R}^m$ be given with*
$$\hat{y}_i \leq f_i(x) \text{ for all } x \in S \text{ and all } i \in \{1, \ldots, m\}.$$

Then every solution of the scalar optimization problem

$$\min \sum_{i=1}^{m}(d_i^+ + d_i^-)$$
subject to the constraints
$$f(x) + d^- - d^+ = \hat{y}$$
$$d_i^+, d_i^- \geq 0 \text{ for all } i \in \{1, \ldots, m\}$$
$$x \in S$$
(11.16)

is an Edgeworth-Pareto optimal point of the multiobjective optimization problem (11.2).

Proof. First we show that problem (11.16) is equivalent to the ℓ_1 approximation problem

$$\min_{x \in S} \|f(x) - \hat{y}\|_1. \tag{11.17}$$

Suppose that $(\bar{d}^+, \bar{d}^-, \bar{x})$ is an arbitrary solution of problem (11.16). Let $i \in \{1, \ldots, m\}$ be an arbitrary index. Then we have $\bar{d}_i^+ = 0$ or $\bar{d}_i^- = 0$. For this proof assume that

$$\delta := \min\{\bar{d}_i^+, \bar{d}_i^-\} > 0.$$

In this case the i-th equality constraint is satisfied for $(\bar{d}_i^+ - \delta, \bar{d}_i^- - \delta, \bar{x}) \in \mathbb{R}_+^2 \times S$ but the objective function value decreases by 2δ. This is a contradiction to the assumption that $(\bar{d}^+, \bar{d}^-, \bar{x})$ is a solution of problem (11.16). Then we conclude with $\bar{d}_i := \bar{d}_i^+ - \bar{d}_i^-$

$$\bar{d}_i^+ + \bar{d}_i^- = |\bar{d}_i|.$$

Consequently, $(\bar{d}, \bar{x}) \in \mathbb{R}^m \times S$ is a solution of the optimization problem

$$\min \sum_{i=1}^{m} |d_i|$$
subject to the constraints
$$f(x) - \hat{y} = d$$
$$x \in S$$

and, therefore, it is also a solution of the problem (11.17).
Next, we consider the optimization problem (11.17). For every $i \in \{1,\ldots,m\}$ and every $x \in S$ we define

$$d_i^+ := \max\{0, f_i(x) - \hat{y}_i\} \geq 0$$

and

$$d_i^- := -\min\{0, f_i(x) - \hat{y}_i\} \geq 0.$$

Then we have for every $i \in \{1,\ldots,m\}$ and every $x \in S$

$$|f_i(x) - \hat{y}_i| = d_i^+ + d_i^-$$

and

$$f_i(x) - \hat{y}_i = d_i^+ - d_i^-.$$

Hence, a solution of the optimization problem (11.17) is also a solution of problem (11.16). Because of the equivalence of the problems (11.16) and (11.17) the assertion of this theorem follows from Theorem 5.15, (b) (see also Example 5.2, (b)). □

The proof of the preceding theorem points out that the scalar optimization problem (11.16) is only a reformulated ℓ_1 approximation problem. Problem (11.16) is also called a *goal programming* problem.

Notes

As pointed out at the beginning of Part II of this book the first papers in this research area were published by Edgeworth [87] (1881) and Pareto [244] (1896). Although Pareto gave the definition of the standard optimality notion in multiobjective optimization, his work and his conceptional idea is based on the earlier paper [87] of Edgeworth (cited by Pareto) presenting the basics for Pareto's investigation. Therefore, optimal points are called Edgeworth-Pareto optimal points in the modern special literature.

For remarks on the presented optimality notions we refer to the notes at the end of Chapter 4. The concept of essentially Edgeworth-Pareto optimal points has been proposed by Brucker [47] for discrete problems.

The result in Corollary 11.19 (and Theorem 11.17) has been given by Geoffrion [103]. Theorem 11.20 is based on an early result of Gale [98, Thm. 9.7]. Here we present a proof using a separation theorem for closed convex cones. Theorem 11.23 is based on a result of Charnes-Cooper [55] und Wendell-Lee [321]. The scalarization approach presented in Theorem 11.25 has been proposed by Ijiri [128]. Problems of goal programming are investigated in the book [278]. Example 11.22 is taken from [82, p. 72–73].

Chapter 12

Numerical Methods

During the past 30 years many methods have been developed for the numerical solution of multiobjective optimization problems. Many of these methods are only applicable to special problem classes. In this chapter we present only some few methods which can be applied to general multiobjective optimization problems. These are a method proposed by Polak, a method for discrete problems and in the class of interactive methods we present the STEM method and a method of reference point approximation. In principle, one can also use the scalarization results of Section 11.2 for the determination of an Edgeworth-Pareto optimal point. But then it remains an open question whether the determined Edgeworth-Pareto optimal point is the subjectively best for the decision maker.

12.1 Modified Polak Method

For nonlinear multiobjective optimization problems Polak [252] has proposed a method which can be used for the approximate determination of the whole set of images of Edgeworth-Pareto optimal points. Although one is not interested in this whole image set, this set is very useful for the application of the method of reference point approximation. Moreover, it is then possible to solve the actual decision problem in an effective way. A coupling of the Polak method and the method of reference point approximation makes it possible to solve interactively nonlinear multiobjective optimization problems. In the following we present the Polak method in a simplified form especially for bicriterial optimization problems.

We consider the bicriterial optimization problem

$$\min_{x \in S} \begin{pmatrix} f_1(x) \\ f_2(x) \end{pmatrix} \tag{12.1}$$

where S is a nonempty subset of $\mathbb{R}^n$, and f_1, $f_2 : S \to \mathbb{R}$ are given functions. The image space $\mathbb{R}^2$ is assumed to be partially ordered in a natural way (i.e., $\mathbb{R}^m_+$ is the ordering cone).

Algorithm 12.1. (Modified Polak Method)

Step 1: Determine the numbers

$$a := \min_{x \in S} f_1(x)$$

and

$$b := f_1(\bar{x}) \text{ with } f_2(\bar{x}) := \min_{x \in S} f_2(x).$$

Step 2: For an arbitrary $p \in \mathbb{N}$ determine the discretization points

$$y_1^{(k)} := a + k \frac{b-a}{p} \text{ with } k = 0, 1, 2, \ldots, p.$$

Step 3: For every discretization point $y_1^{(k)}$ ($k = 0, 1, 2, \ldots, p$) compute a solution $x^{(k)}$ of the constrained optimization problem

$$\min f_2(x)$$
$$\text{subject to the constraints}$$
$$x \in S$$
$$f_1(x) = y_1^{(k)},$$

and set

$$y_2^{(k)} := f_2(x^{(k)}) \text{ for } k = 0, 1, 2, \ldots, p.$$

Step 4: Among the numbers $y_2^{(0)}, y_2^{(1)}, \ldots, y_2^{(p)}$ delete those so that the remaining numbers form a strongly monotonically decreasing sequence

$$y_2^{(k_0)} > y_2^{(k_1)} > y_2^{(k_2)} > \ldots .$$

12.1. Modified Polak Method

Step 5: Unite the vectors $x^{(k_0)}$, $x^{(k_1)}$, $x^{(k_2)}$, ... to a set being an approximation of the set of all Edgeworth-Pareto optimal points of the bicriterial optimization problem (12.1).

Figure 12.1 illustrates the approximation of the images of all Edgeworth-Pareto optimal points using the modified Polak method. If all

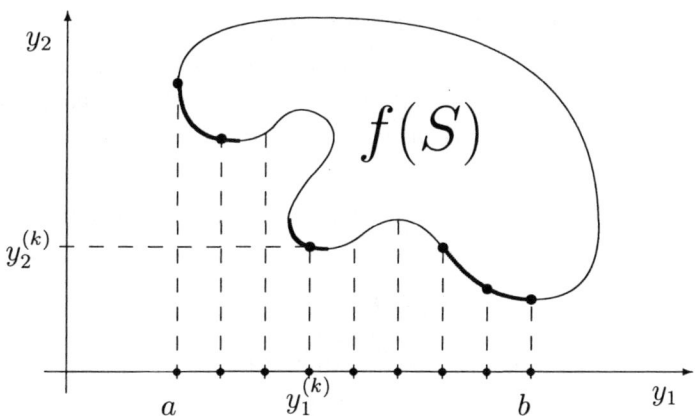

Figure 12.1: Determination of minimal elements of the set $f(S)$ with $f = (f_1, f_2)$.

scalar optimization problems being subproblems in Algorithm 12.1, are solvable, then the discrete set

$$\{x^{(k_0)}, x^{(k_1)}, x^{(k_2)}, \ldots\}$$

is an approximation of the set of all Edgeworth-Pareto optimal points of the bicriterial optimization problem (12.1). The more discretization points are chosen in step 2 the better is this approximation. If the set of minimal elements of the image set $f(S)$ is connected (being not the case in Fig. 12.1), then the points

$$(y_1^{(k_i)}, y_2^{(k_i)}) \text{ for } i = 0, 1, 2, \ldots$$

can be connected by straight lines for a better illustration of the curve being given by the set of minimal elements of $f(S)$. Since this curve is not smooth, in general, it does not make sense to use splines for the approximation.

The modified Polak method is suitable for the approximation of the set of Edgeworth-Pareto optimal points of the bicriterial optimization problem (12.1). We will come back to this method in Subsection 12.2.2. The disadvantage of this method is the high numerical effort because many scalar (nonlinear) constrained optimization problems have to be solved. For the solution of these subproblems known methods of nonlinear optimization can be used.

If one implements the modified Polak method on a computer, the difficulty arises that one needs global solutions of the subproblems. Therefore, we actually have to apply methods of global optimization.

Example 12.2. We investigate the bicriterial optimization problem

$$\min \begin{pmatrix} x_1 \\ x_2 \end{pmatrix}$$

subject to the constraints

$$x_2 - \tfrac{5}{2} \leq 0$$
$$(x_1 - \tfrac{1}{2})^2 - x_2 - \tfrac{9}{2} \leq 0$$
$$-x_1 - x_2^2 \leq 0$$
$$-(x_1 + 1)^2 - (x_2 + 3)^2 + 1 \leq 0$$
$$(x_1, x_2) \in \mathbb{R}^2.$$

Figure 12.2 illustrates the image set of the objective mapping being equal to the constraint set in this special case.

If one solves the scalar optimization problem given in Step 3 of Algorithm 12.1 for the discretization point $y_1^{(k)} := -\tfrac{1}{2}$, then one obtains $x^{(k)} := (-\tfrac{1}{2}, \tfrac{1}{2}\sqrt{2})$ as a local solution of this problem. But the global solution reads $\bar{x}^{(k)} := (-\tfrac{1}{2}, -3 + \tfrac{1}{2}\sqrt{3})$.

The preceding example shows that standard methods of nonlinear optimization must be handled with care for the solution of the subproblems of the modified Polak method. The following simple *tunneling technique* for the solution of a global solution of the scalar subproblems may be useful.

12.1. Modified Polak Method

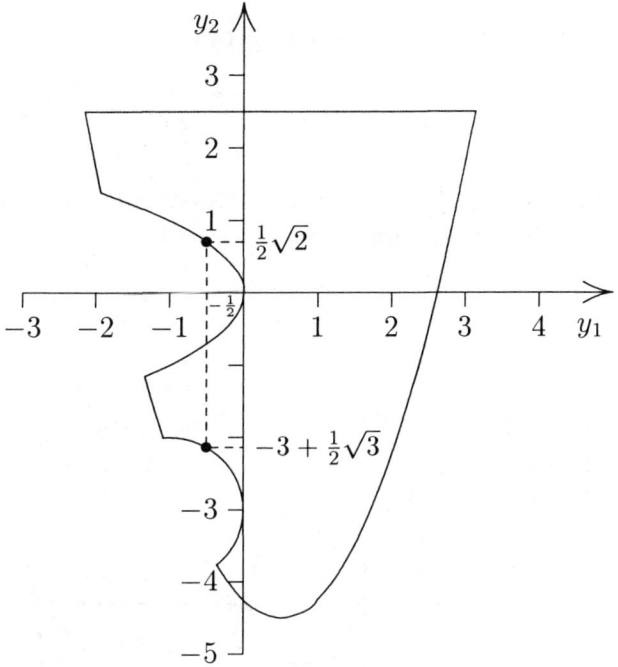

Figure 12.2: Image set in Example 12.2.

Remark 12.3. Let a function $\varphi : \mathbb{R}^n \to \mathbb{R}$ be given. Then we investigate the unconstrained optimization problem

$$\min_{x \in \mathbb{R}^n} \varphi(x). \tag{12.2}$$

We are interested in a global solution of this problem. Such a problem arises, for instance, if one applies a penalty method to the scalar subproblems in Algorithm 12.1.

For the following we assume that we already have an approximation $\hat{x} \in \mathbb{R}^n$ of a global solution of problem (12.2) (for instance, a stationary point or only a local solution). For an arbitrary $\varepsilon > 0$ we then consider the constrained optimization problem

$$\begin{aligned} &\min \; \tfrac{1}{\varphi(\hat{x}) - \varphi(x)} \\ &\text{subject to the constraints} \\ &\quad \varphi(x) \leq \varphi(\hat{x}) - \varepsilon \\ &\quad x \in \mathbb{R}^n. \end{aligned} \tag{12.3}$$

A solution $\bar{x}$ of this problem has the property

$$\varphi(\bar{x}) \leq \varphi(\hat{x}) - \varepsilon < \varphi(\hat{x}).$$

Since we minimize $\frac{1}{\varphi(\hat{x})-\varphi(x)}$, we may expect $\varphi(\bar{x}) \ll \varphi(\hat{x})$. Figure 12.3 illustrates this tunneling effect for $n = 1$.

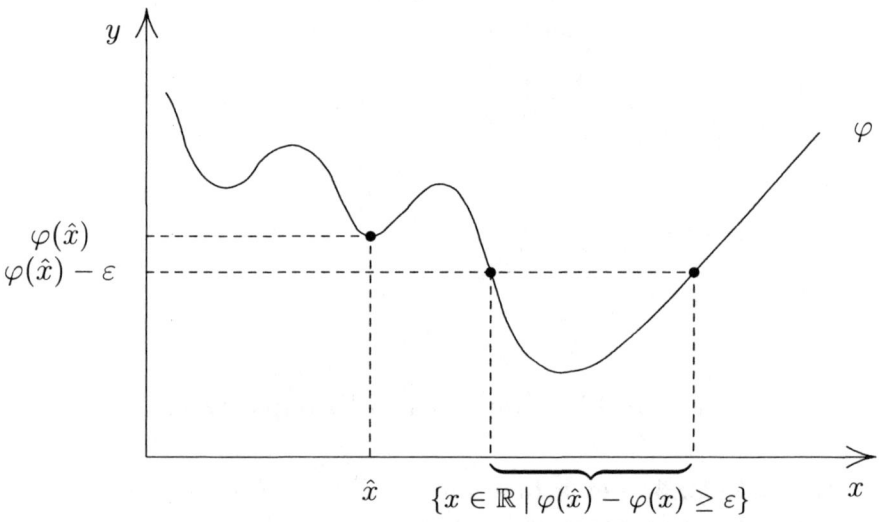

Figure 12.3: Simplified illustration of the tunneling technique.

It is obvious that a solution of problem (12.3) is not a global solution of problem (12.2), in general. But such a solution can be used as a new starting point for a descent method for the solution of a global solution of problem (12.2). This hybrid technique is based on the fact that descent methods (like the BFGS method) compute iteration points in the same "valley" where the starting point is located. If one finds a new starting point by a "tunnel" to another "valley", then this new starting point can lead to another local solution with a smaller value of the objective function.

12.2 Interactive Methods

In principle, the scalarization results presented in Section 11.2 can be used for the numerical solution of multiobjective optimization problems. But these approaches need certain parameters being difficult to choose. Even if one solves these scalar problems for various parameters or even if one approximates the whole set of Edgeworth-Pareto optimal points, a satisfying solution of the actual decision problem is not found. The decision maker has to select an Edgeworth-Pareto optimal point being the subjectively best among all Edgeworth-Pareto optimal points.

During the past 30 years so-called interactive methods have been developed combining the numerical iteration process with subjective thoughts of the decision maker. Therefore, a solution found by an interactive method, is subjectively determined. Such a method is characterized by a permanent change between an objective computation phase and a subjective decision phase.

In this section our investigations are concentrated to a modified STEM method and a method of reference point approximation. Both interactive methods are suitable for the solution of linear as well as nonlinear multiobjective optimization problems.

12.2.1 Modified STEM Method

Already 1971 Benayoun, de Montgolfier, Tergny and Laritchev [22] have proposed a so-called STEM method (step method). This method has been designed for the interactive solution of linear multiobjective optimization problems. In the following we present this method in a modified form so that nonlinear problems can be treated as well.

We consider the multiobjective optimization problem

$$\min_{x \in S} f(x) \qquad (12.4)$$

where S is a nonempty subset of $\mathbb{R}^n$ and $f : S \to \mathbb{R}^m$ is a given vector function with $f = (f_1, \ldots, f_m)$. The image space $\mathbb{R}^m$ is assumed to be partially ordered in a natural way (i.e., $\mathbb{R}^m_+$ is the ordering cone). The following algorithm presents the STEM method in a simplified form.

Algorithm 12.4. (STEM method)

Step 1: For every $i = 1, \ldots, m$ determine the minimal values

$$\hat{y}_i := \min_{x \in S} f_i(x)$$

of the functions $f_1, \ldots, f_m$ on S, and set

$$\hat{y} := (\hat{y}_1, \ldots, \hat{y}_m).$$

Moreover, set $I := \{1, \ldots, m\}$ and $J := \emptyset$.

Step 2: The decision maker chooses the weights $w_1, \ldots, w_m$ of the weighted Chebyshev norm $\|\cdot\|$.

Step 3: Determine a solution $\hat{x}^{(0)} \in S$ of the optimization problem

$$\min_{x \in S} \|\hat{y} - f(x)\|,$$

and set $k := 0$.

Step 4: The decision maker chooses (if possible) an index $i \in I$ indicating that he accepts a deterioration of the value $f_i(\hat{x}^{(k)})$ in order to improve the value $f_j(\hat{x}^{(k)})$ for at least one other objective function f_j. If such a choice is not possible, then the algorithm stops.

Step 5: For the chosen index $i \in I$ the decision maker gives a number Δ_i for which the value $f_i(\hat{x}^{(k)})$ can be maximally increased. Set $\alpha_i := f_i(\hat{x}^{(k)}) + \Delta_i$.

Step 6: Set

$$I := I \setminus \{i\},$$
$$J := J \cup \{i\}$$

and compute a solution $\hat{x}^{(k+1)}$ of the scalar optimization problem

$$\min \lambda$$
subject to the constraints
$$x \in S$$
$$w_i(f_i(x) - \hat{y}_i) \leq \lambda \text{ for all } i \in I$$
$$f_j(x) \leq \alpha_j \text{ for all } j \in J.$$

12.2. Interactive Methods

Step 7: Set $k := k+1$. If $k = m$, then the algorithm stops, otherwise go to step 4.

There are some points which have to be noticed for this method.

Remark 12.5.

- The solutions of the scalar optimization problems in Step 3 and 6 of the preceding algorithm are not always Edgeworth-Pareto optimal points of the multiobjective optimization problem (12.4). Therefore, one should check the Edgeworth-Pareto optimality of these solutions using Theorem 11.23. But this test leads to numerical difficulties (compare the remarks after the proof of Theorem 11.23).

- For a more flexible iteration process one should admit the possibility that after every iteration the value Δ_i can be revised. This can be easily implemented on a computer.

- Instead of using $\hat{y}$ one can also work with another point being smaller than $\hat{y}$.

- If one implements the STEM method on a computer, one should replace the scalar optimization problem in Step 3 by the following problem

$$\min \lambda$$
$$\text{subject to the constraints}$$
$$x \in S$$
$$w_i(f_i(x) - \hat{y}_i) \leq \lambda \text{ for all } i = 1, \ldots, m.$$

If the original multiobjective optimization problem is a linear problem, then the subproblems arising in Algorithm 12.4 can be replaced by linear optimization problems which can be solved by standard methods like the simplex method.

Example 12.6. We investigate the linear multiobjective optimization problem

$$\min \begin{pmatrix} -1 & -1 & -3 & 1 \\ -3 & 1 & 5 & -1 \\ 0 & -1 & -5 & -7 \end{pmatrix} x$$

subject to the constraints

$$\begin{pmatrix} 1 & 1 & 3 & -4 \\ 2 & 0 & 0 & 4 \\ 2 & 0 & 3 & 1 \end{pmatrix} x \leq \begin{pmatrix} 20 \\ 10 \\ 15 \end{pmatrix}$$

$$x \geq 0.$$

In Step 1 of Algorithm 12.4 we obtain

$$\hat{y} := \begin{pmatrix} -27.5 \\ -15.0 \\ -55.833 \end{pmatrix}.$$

In Step 2 the weights of the Chebyshev norm are chosen as 1. A first solution $\hat{x}^{(0)}$ is obtained in Step 3 as

$$\hat{x}^{(0)} := \begin{pmatrix} 0 \\ 0 \\ 2.583 \\ 2.5 \end{pmatrix}.$$

But this point is not an Edgeworth-Pareto optimal point of the original problem. If one solves the scalar problem in Theorem 11.23 (with $t_1 = t_2 = t_3 = 1$), one obtains the Edgeworth-Pareto optimal point

$$\tilde{x}^{(0)} := \begin{pmatrix} 0 \\ 12.917 \\ 0 \\ 2.5 \end{pmatrix}.$$

This point $\tilde{x}^{(0)}$ instead of $\hat{x}^{(0)}$ is used for the following iteration steps. One computes

$$f(\tilde{x}^{(0)}) := \begin{pmatrix} -10.417 \\ 10.417 \\ -30.417 \end{pmatrix}.$$

12.2. Interactive Methods

Next, we assume that in the 4th and 5th step the decision maker decides to deteriorate the third objective function by the value $\Delta_3 := 10$ in order to improve the value of another objective function. A solution of the scalar problem in Step 5 reads

$$\hat{x}^{(1)} := \begin{pmatrix} 1.944 \\ 9.722 \\ 0 \\ 1.528 \end{pmatrix} \text{ with } f(\hat{x}^{(1)}) = \begin{pmatrix} -10.139 \\ 2.361 \\ -20.417 \end{pmatrix}.$$

It turns out that $\hat{x}^{(1)}$ is an Edgeworth-Pareto optimal point of the original problem. Now, we assume that the decision maker accepts a deterioration of the second objective function by the value $\Delta_2 := 5$. Then we get in the 6th step

$$\hat{x}^{(2)} := \begin{pmatrix} 3.662 \\ 19.015 \\ 0 \\ 0.669 \end{pmatrix} \text{ with } f(\hat{x}^{(2)}) = \begin{pmatrix} -22.008 \\ 7.361 \\ -23.699 \end{pmatrix}.$$

$\hat{x}^{(2)}$ is also an Edgeworth-Pareto optimal point of the original problem. Since the first and the second objective function could be improved, we assume that the decision maker terminates the iteration.

12.2.2 Method of Reference Point Approximation

In the following let $S \subset \mathbb{R}^n$ be a given nonempty constraint set, and let $f : S \to \mathbb{R}^m$ be a given vector function. As before, the image space $\mathbb{R}^m$ is assumed to be partially ordered in a natural way. Then we investigate the multiobjective optimization problem

$$\min_{x \in S} f(x). \tag{12.5}$$

Let $M \subset S$ denote the set of all Edgeworth-Pareto optimal points of this problem being assumed to be nonempty.

Assume that the decision maker can give a point $\hat{y} \in \mathbb{R}^m$ (a so-called *reference point*) which should be realized as good as possible by an image $f(x)$ with $x \in M$. Then it makes sense to solve the following approximation problem

$$\min_{x \in M} \|\hat{y} - f(x)\|. \qquad (12.6)$$

In principle, $\|\cdot\|$ may be any norm in $\mathbb{R}^m$. Since the weighted Chebyshev norm can be well interpreted, we use this special norm. Notice that the approximation problem (12.6) is not always solvable. In this case problem (12.6) has to be modified.

In the following we present the resulting *method of reference point approximation* in a simplified form.

Algorithm 12.7. (method of reference point approximation)

Step 1: The decision maker chooses the weights of the weighted Chebyshev norm $\|\cdot\|$.

Step 2: The decision maker chooses an arbitrary reference point $\hat{y}^{(1)} \in \mathbb{R}^m$. Set $i := 1$.

Step 3: Compute a solution of the optimization problem

$$\min_{x \in M} \|\hat{y}^{(i)} - f(x)\|.$$

Step 4: This solution is presented to the decision maker who may stop the algorithm. Otherwise the decision maker chooses another reference point $\hat{y}^{(i+1)}$ and continues the algorithm with $i := i + 1$ in Step 3.

There are different ways in order to extend this algorithm. Therefore, this algorithm describes only a class of methods. For instance, the weights could be varied during the iteration process, and additional information could be provided making the choice of a reference point easier.

Notice from a practical point of view that the set M of all Edgeworth-Pareto optimal solutions of problem (12.5) has to be determined before Algorithm 12.7 can be started. The determination of the set M may be impossible for complicated nonlinear problems.

12.2. Interactive Methods

It can be shown that the approximation problem (12.6) is a complicated semi infinite optimization problem, that is a problem with infinitely many constraints. This approximation problem can be essentially simplified, if the reference point is a strict lower bound of the image set $f(S)$.

Theorem 12.8. *Let the multiobjective optimization problem (12.5) be given with a nonempty set M of Edgeworth-Pareto optimal points, and let a $\hat{y} \in \mathbb{R}^m$ be arbitrarily chosen with*

$$y_i < f_i(x) \text{ for all } x \in S \text{ and all } i \in \{1,\ldots,m\}.$$

If $\bar{x} \in S$ is an image unique solution of the problem

$$\min_{x \in S} \|\hat{y} - f(x)\|,$$

then $\bar{x}$ is a solution of the approximation problem (12.6).

Proof. By Corollary 11.21, (a) $\bar{x} \in S$ is an Edgeworth-Pareto optimal point of the multiobjective optimization problem (12.5), that is $\bar{x} \in M$. Because of $M \subset S$ the point $\bar{x} \in M$ is then a solution of the approximation problem

$$\min_{x \in M} \|\hat{y} - f(x)\|.$$

□

In the following we show that Algorithm 12.7 can be well applied to linear multiobjective and nonlinear bicriterial optimization problems.

The Linear Case

If one applies Algorithm 12.7 to problems of linear multiobjective optimization, then the subproblem in the third step turns out to be very simple. A solution of this subproblem can be determined by solving finitely many linear optimization problems. The following investigations are concentrated only to the third step of Algorithm 12.7.

We consider problem (12.5) in the special form of the linear multiobjective optimization problem

$$\min Cx$$
subject to the constraints
$$Ax \leq b$$
$$x \in \mathbb{R}^n.$$
(12.7)

Let C denote a real (m,n) matrix, let A denote a real (q,n) matrix, and let b be a vector in $\mathbb{R}^q$. The $\leq$ relation has to be understood in a componentwise sense. The constraint set

$$S := \{x \in \mathbb{R}^m \mid Ax \leq b\}$$

is assumed to be nonempty and bounded. Then S describes a bounded convex polytop in $\mathbb{R}^n$. If $\|\cdot\|$ denotes a weighted Chebyshev norm in $\mathbb{R}^m$, i.e.

$$\|y\| := \max_{1 \leq i \leq m} w_i \mid y_i \mid \quad \text{for all } y \in \mathbb{R}^m$$

with appropriate weights $w_1, \ldots, w_m > 0$, then for a given reference point $\hat{y} \in \mathbb{R}^m$ the approximation problem (12.6) can be written as

$$\min_{x \in M} \max_{1 \leq i \leq m} w_i \mid \hat{y}_i - (Cx)_i \mid . \quad (12.8)$$

By Theorem 11.20 the set M of all Edgeworth-Pareto optimal points equals the set of all properly Edgeworth-Pareto optimal points. Therefore, every Edgeworth-Pareto optimal point of the linear multiobjective optimization problem (12.7) is a solution of an appropriate linear optimization problem. These solutions are located on certain facets and edges of the polytop S.

Using a modified simplex method it is possible to determine the vertices of these facets and edges. In other words: A partition of the set M can be determined in such a way that

$$M = M_1 \cup M_2 \cup \ldots \cup M_l,$$

with $l \in \mathbb{N}$, $\emptyset \neq M_j \subset M$ for all $j \in \{1, \ldots, l\}$, and the following holds: For every set M_j ($j = 1, \ldots, l$) there are s_j vertices $x^{(j_1)}, \ldots, x^{(j_{s_j})} \in$

12.2. Interactive Methods

M with

$$M_j = \left\{ x \in S \;\middle|\; x = \sum_{k=1}^{s_j} \lambda_k x^{(j_k)} \text{ with } \lambda_k \geq 0 \text{ and } \sum_{k=1}^{s_j} \lambda_k = 1 \right\}. \tag{12.9}$$

These vertices can be determined by the Isermann method [132], for instance. Figure 12.4 illustrates the partition of the set M.

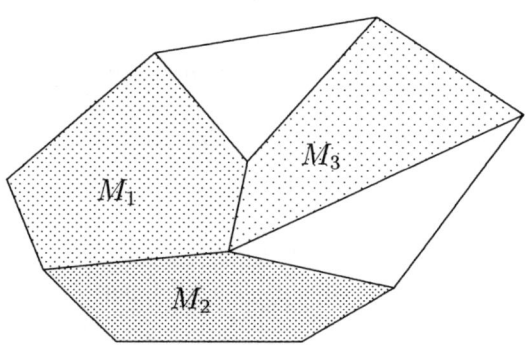

Figure 12.4: Partition of the set $M = M_1 \cup M_2 \cup M_3$.

With the introduced partition of M problem (12.8) can also be written as

$$\min_{x \in M_1 \cup M_2 \cup \ldots \cup M_l} \max_{1 \leq i \leq m} w_i \,|\, \hat{y}_i - (Cx)_i \,|.$$

One obtains a solution of this problem, if one solves for every $j = 1, \ldots, l$ an approximation problem of the form

$$\min_{x \in M_j} \max_{1 \leq i \leq m} w_i \,|\, \hat{y}_i - (Cx)_i \,|. \tag{12.10}$$

Among all solutions of these l problems one chooses the solution with the smallest minimal value. Using the equation (12.9) problem (12.10) can also be written as

$$\min \max_{1\leq i\leq m} w_i \left| \hat{y}_i - \sum_{k=1}^{s_j} \lambda_k (Cx^{(j_k)})_i \right|$$

subject to the constraints

$$\sum_{k=1}^{s_j} \lambda_k = 1$$

$$\lambda_1, \ldots, \lambda_{s_j} \geq 0.$$

This problem is equivalent to the problem

$$\min \lambda_0$$

subject to the constraints

$$\lambda_0 = \max_{1\leq i\leq m} w_i \left| \hat{y}_i - \sum_{k=1}^{s_j} \lambda_k (Cx^{(j_k)})_i \right|$$

$$\sum_{k=1}^{s_j} \lambda_k = 1$$

$$\lambda_0 \in \mathbb{R}, \quad \lambda_1, \ldots, \lambda_{s_j} \geq 0$$

being equivalent to

$$\min \lambda_0$$

subject to the constraints

$$\lambda_0 \geq w_i \left| \hat{y}_i - \sum_{k=1}^{s_j} \lambda_k (Cx^{(j_k)})_i \right| \quad \text{for all } i = 1, \ldots, m$$

$$\sum_{k=1}^{s_j} \lambda_k = 1$$

$$\lambda_0 \in \mathbb{R}, \quad \lambda_1, \ldots, \lambda_{s_j} \geq 0.$$

Using the definition of the absolute value this problem can be written as

$$\min \lambda_0$$

subject to the constraints

$$\left. \begin{array}{l} -\dfrac{1}{w_i}\lambda_0 - \sum_{k=1}^{s_j} \lambda_k (Cx^{(j_k)})_i \leq -\hat{y}_i \\[2ex] -\dfrac{1}{w_i}\lambda_0 + \sum_{k=1}^{s_j} \lambda_k (Cx^{(j_k)})_i \leq \hat{y}_i \end{array} \right\} \quad \text{for all } i \in \{1, \ldots, m\}$$

12.2. Interactive Methods

$$\sum_{k=1}^{s_j} \lambda_k = 1$$
$$\lambda_0 \in \mathbb{R}, \ \lambda_1, \ldots, \lambda_{s_j} \geq 0.$$

This is a linear optimization problem which can be easily solved with the simplex method

Summarising our investigations we obtain the following method: If one applies the method of reference point approximation to a linear multiobjective optimization problem with bounded constraint set, first of all one determines all facets, edges and the corresponding vertices describing the set of all Edgeworth-Pareto optimal points, and then one solves l linear optimization problems for every iteration and one chooses the solution with the smallest minimal value in order to get a solution of the subproblem in the third step of Algorithm 12.7.

In the following we discuss various examples being solved with Algorithm 12.7. Here the weights of the weighted Chebyshev norm are chosen as 1. The information concerning the vertices of the set M are taken from [135].

Example 12.9. We investigate the linear multiobjective optimization problem

$$\min \begin{pmatrix} -4 & -1 & -2 \\ -1 & -3 & 1 \\ 1 & -1 & -4 \end{pmatrix} x$$

subject to the constraints

$$\begin{pmatrix} 1 & 1 & 1 \\ 2 & 2 & 1 \\ 1 & -1 & 0 \end{pmatrix} x \leq \begin{pmatrix} 3 \\ 4 \\ 0 \end{pmatrix}$$

$$x \geq 0_{\mathbb{R}^3}.$$

The set M of all Edgeworth-Pareto optimal points of this problem contains 5 vertices and is generated by 2 facets or edges. For different reference points we obtain solutions of the approximation problem given in Table 12.1.

reference point	minimal solution	minimal value
$\hat{y}^{(1)}$=(-10, -8, -15)	(0, 1.0833, 1.8333)	6.5833
$\hat{y}^{(2)}$=(-8, -6, -13)	(0, 1.0833, 1.8333)	4.5833
$\hat{y}^{(3)}$=(-7, -5, -12)	(0, 1.0833, 1.8333)	3.5833
$\hat{y}^{(4)}$=(-7, -5, -11)	(0, 1.1667, 1.6667)	3.1667
$\hat{y}^{(5)}$=(-7, -5, -10)	(0, 1.25, 1.5)	2.75

Table 12.1: Compromise solutions (Example 12.9).

Example 12.10. We now consider the linear multiobjective optimization problem

$$\min \begin{pmatrix} -1 & -3 & 2 & 0 & -1 \\ -3 & 1 & 0 & -3 & -1 \\ -1 & 0 & -2 & 0 & -3 \end{pmatrix} x$$

subject to the constraints

$$\begin{pmatrix} 2 & 4 & 0 & 0 & 3 \\ 0 & 0 & 2 & 5 & 4 \\ 5 & 0 & 0 & 0 & 0 \\ 0 & 0 & 0 & 2 & 0 \\ 5 & 5 & 2 & 0 & 0 \end{pmatrix} x \leq \begin{pmatrix} 27 \\ 35 \\ 26 \\ 24 \\ 36 \end{pmatrix}$$

$$x \geq 0_{\mathbb{R}^5}.$$

The set M of all Edgeworth-Pareto optimal points has 11 vertices and 4 facets or edges. Table 12.2 gives some minimal solutions of the approximation problem.

Example 12.11. Finally we discuss the linear multiobjective optimization problem

$$\min \begin{pmatrix} 0 & 0 & 1 & 0 & 0 & 0 \\ 0 & 0 & 0 & 1 & 0 & 0 \\ 0 & 0 & 0 & 0 & 1 & 0 \\ 0 & 0 & 0 & 0 & 0 & 1 \end{pmatrix} x$$

subject to the constraints

reference point	minimal solution	minimal value
$\hat{y}^{(1)}$=(-15, -40, -20)	(5.2, 0.0526, 0, 4.9368, 2.5789)	7.0632
$\hat{y}^{(2)}$=(-13, -39, -19)	(5.2, 0, 0, 4.9273, 2.5909)	6.0273
$\hat{y}^{(3)}$=(-12, -38, -19)	(5.2, 0, 0, 4.7455, 2.8182)	5.3455
$\hat{y}^{(4)}$=(-12, -37, -19)	(5.2, 0, 0, 4.5636, 3.0455)	4.6636
$\hat{y}^{(5)}$=(-12, -37, -18)	(5.2, 0, 0, 4.7455, 2.8182)	4.3455
$\hat{y}^{(6)}$=(-12, -37, -17)	(5.2, 0.0526, 0, 4.9368, 2.5789)	4.0632
$\hat{y}^{(7)}$=(-11, -35, -16.5)	(5.2, 0.1053, 0, 4.6737, 2.9079)	2.5763

Table 12.2: Compromise solutions (Example 12.10).

$$\begin{pmatrix} 2 & 4 & 0 & 0 & 0 & 0 \\ 4 & 3 & 0 & 0 & 0 & 0 \\ 1 & 1 & 0 & 0 & 0 & 0 \\ 0 & -1 & 0 & 0 & -1 & -1 \\ -1 & -1 & 0 & 0 & 0 & 0 \end{pmatrix} x \leq \begin{pmatrix} 24 \\ 28 \\ 8 \\ -5.75 \\ -7 \end{pmatrix}$$

$$(1, 0, 1, -1, 0, 0)x = 6$$

$$x \geq 0_{\mathbb{R}^6}.$$

For this problem the set M of all Edgeworth-Pareto optimal points consists of 3 vertices and 2 facets or edges. Numerical results are given in Table 12.3.

reference point	minimal solution	min. value
$\hat{y}^{(1)}$=(-5, -5, -2, -2)	(5.3929, 2.1429, 0.6071, 0, 3.6071, 0)	5.6071
$\hat{y}^{(2)}$=(-2, -3, 0, -2)	(4.9643, 2.7143, 1.0357, 0, 3.0357, 0)	3.0357
$\hat{y}^{(3)}$=(1, -1, 0.5, 0)	(3.8333, 4.0833, 2.1667, 0, 1.6667, 0)	1.1667

Table 12.3: Compromise solutions (Example 12.11).

The Bicriterial Nonlinear Case

For a nonlinear multiobjective optimization problem it is difficult to solve the subproblem in Step 3 of Algorithm 12.7. But in the bicriterial case it is possible to approximate the set M of all Edgeworth-Pareto optimal points by a discrete set (for instance, using the modified Polak method). Then the approximation problem in the third step can be replaced by a related problem with discrete constraint set. This modified problem is easy to solve.

In the following we consider the bicriterial optimization problem

$$q \min_{x \in S} \begin{pmatrix} f_1(x) \\ f_2(x) \end{pmatrix} \qquad (12.11)$$

where S is a nonempty subset in $\mathbb{R}^n$ and $(f_1, f_2) : S \to \mathbb{R}^2$ is a given vector function. Again, we use the componentwise ordering in $\mathbb{R}^2$. If we combine the method of reference point approximation (Algorithm 12.7) with the modified Polak method (Algorithm 12.1), then we obtain the following interactive method for the solution of problem (12.11).

Algorithm 12.12 (method of reference point approximation in the bicriterial case).

Part I. Computation phase

Step 1: Determine the numbers

$$a := \min_{x \in S} f_1(x)$$

and

$$b := f_1(\tilde{x}) \text{ with } f_2(\tilde{x}) = \min_{x \in S} f_2(x).$$

Step 2: For an arbitrary $p \in \mathbb{N}$ determine the discretization points

$$y_1^{(k)} := a + k \frac{b-a}{p} \text{ with } k = 0, 1, 2, \ldots, p.$$

Step 3: For every discretization point $y_1^{(k)}$ ($k = 0, 1, \ldots, p$) compute a (global) solution $x^{(k)}$ of the constrained optimization problem

$$\begin{aligned} f_2(x^{(k)}) \;=\; & \min \; f_2(x) \\ & \text{subject to the constraints} \\ & x \in S \\ & f_1(x) = y_1^{(k)}, \end{aligned}$$

and set
$$y_2^{(k)} := f_2(x^{(k)}) \text{ for } k = 0, 1, 2, \ldots, p$$
(remark: It is important to work with a numerical method of global optimization).

Step 4: Among the numbers $y_2^{(0)}, y_2^{(1)}, \ldots, y_2^{(p)}$ delete those so that the remaining numbers form a strongly monotonically decreasing sequence
$$y_2^{(k_0)} > y_2^{(k_1)} > y_2^{(k_2)} > \cdots,$$
and set
$$\tilde{M} := \{x^{(k_0)}, x^{(k_1)}, x^{(k_2)}, \ldots\}.$$

Part II. Decision phase

Step 5: The decision maker chooses the weights $t_1, t_2 > 0$ of the weighted Chebyshev norm in $\mathbb{R}^2$.

Step 6: The decision maker chooses an arbitrary reference point $\hat{y}^{(1)} \in \mathbb{R}^2$.

Part III. Computation phase

Step 7: Set $i := 1$.

Step 8: Compute a point $\bar{x}^{(i)} \in \tilde{M}$ with the property
$$\max_{j=1,2} \{t_j |\hat{y}_j^{(i)} - f_j(\bar{x}^{(i)})|\} \leq \max_{j=1,2} \{t_j |\hat{y}_j^{(i)} - f_j(x)|\}$$
for all $x \in \tilde{M}$.

Part IV. Decision phase

Step 9: The point $\bar{x}^{(i)} \in \tilde{M}$ is presented to the decision maker. If the decision maker accepts this point as the subjectively best, then the algorithm stops; otherwise continue with the next step.

Step 10: Using additional information about the original problem and numerical results obtained in the third step, the decision maker proposes a new reference point $\hat{y}^{(i+1)} \in \mathbb{R}^2$.

Part V. Computation phase

Step 11: Set $i := i + 1$, and go to Step 8.

Part I of Algorithm 12.12 is the computationally intensive part whereas the parts II–V may run very fast online. On can apply this algorithm in such a way that the first part is done offline, independently from the other parts. The actually interactive part begins with the set $\tilde{M}$.

It is not necessary to choose equidistant discretization points $y_1^{(k)}$ in the second step of this algorithm. In some cases another choice of descretization may be better.

In the fifth step the decision maker can choose the weights of the weigthed Chebyshev norm. This is of importance in order to be able to compare the two objectives f_1 and f_2 without scaling the function values. In the tenth step it should be possible to provide the decision maker with all information being available during the computation phases. An essential aid is the graphical illustration of the image set of the Edgeworth-Pareto optimal points. Then it is simpler to choose an appropriate reference point.

Example 12.13. Again, we consider the bicriterial optimization problem given in Example 12.2

$$\min \begin{pmatrix} x_1 \\ x_2 \end{pmatrix}$$

subject to the constraints

$$x_2 - \tfrac{5}{2} \leq 0$$
$$(x_1 - \tfrac{1}{2})^2 - x_2 - \tfrac{9}{2} \leq 0$$
$$-x_1 - x_2^2 \leq 0$$
$$-(x_1 + 1)^2 - (x_2 + 3)^2 + 1 \leq 0$$
$$(x_1, x_2) \in \mathbb{R}^2.$$

Since the objective mapping is the identity, the constraint set and the image set are equal. This set is illustrated in Figure 12.2. If one applies Algorithm 12.12 to this problem, one obtains in Part I the elements of the set $\tilde{M}$ given in Table 12.4 (notice that we have not chosen equidistant discretization points).

12.2. Interactive Methods

k	$f(x^{(k)})$
0	(-2.131087, 2.422933)
1	(-1.930941, 1.409813)
2	(-1.731149, 1.315711)
3	(-1.531138, 1.238715)
4	(-1.331107, 1.156983)
5	(-1.131001, -1.207556)
6	(-0.931086, -2.002503)
7	(-0.731135, -2.036241)
8	(-0.526106, -2.117397)
9	(-0.315858, -3.729672)
10	(-0.115894, -4.013447)
11	(0.084136, -4.327440)
12	(0.500000, -4.500500)

Table 12.4: Elements of the set $\tilde{M}$.

If one connects the points given in Table 12.4 by straight lines and if one notices that the set of all minimal points consists of three non-connected parts, then one obtains a set illustrated in Figure 12.5.

If we choose the weights $t_1 = t_2 = 1$ in the fifth step of Algorithm 12.12, we get for various reference points the Edgeworth-Pareto optimal points given in Table 12.5.

reference point	minimal solution	minimal value
$\hat{y}^{(1)}$=(-2, 0)	(-1.331107, 1.156983)	1.156983
$\hat{y}^{(2)}$=(0, 0)	(-1.131001, -1.207556)	1.207556
$\hat{y}^{(3)}$=(-1, 3)	(-2.131087, 2.422933)	1.131087
$\hat{y}^{(4)}$=(-2, 2)	(-2.131087, 2.422933)	0.422933

Table 12.5: Compromise solutions.

The method described in Algorithm 12.12 is an interactive method being useful in practice because the decision maker has to provide

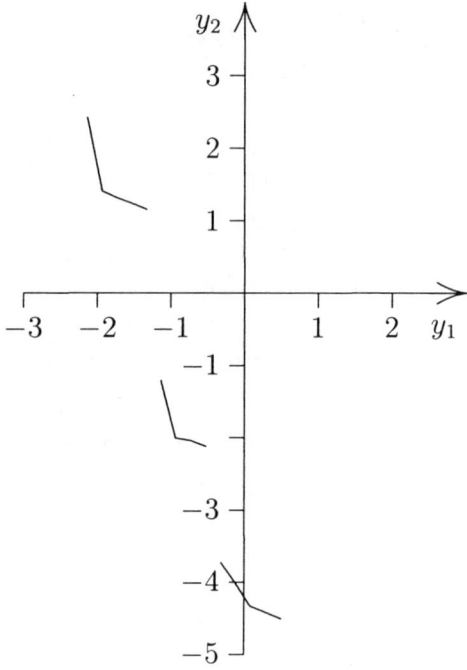

Figure 12.5: Approximated set of all minimal elements.

only simple information. In Sections 13.3 and 13.4 this algorithm is applied to concrete problems from chemical engineering.

12.3 Method for Discrete Problems

In this section we investigate the special case that we want to determine the minimal elements of a set of finitely many points. In practice, such a set consists of many points so that it is not possible to use only the definition of minimality. Here we present a reduction approach which can be used for the elemination of non-minimal elements in such a set.

In the following let S be a nonempty discrete subset of $\mathbb{R}^n$ being partially ordered in a natural way. Let S consist of many vectors. We are interested in the determination of all minimal elements of S.

12.3. Method for Discrete Problems

Example 12.14. For instance, one obtains such a discrete problem by discretization of the image set of a continuous multiobjective optimization problem. The discrete set generated in this way typically contains many elements.

For complexity reasons it does not make sense to determine all minimal elements using the definition. Therefore, one tries to reduce the set S, that is to eliminate those elements in S which cannot be minimal. Such a reduction of S can be carried out with the Graef-Younes method.

Algorithm 12.15. (Graef-Younes method)

Input: $S := \{x^1, x^2, \ldots, x^k\} \subset \mathbb{R}^n$
Step 1: Set $T := \{x^1\}$ and $j := 2$.
Step 2: If for $x^j \in S \setminus T$
$$x^j \not\geq x \text{ for all } x \in T,$$
then set
$$T := T \cup \{x^j\}.$$
Step 3: If $j < k$, then set $j := j+1$ and go to Step 2; otherwise the algorithm stops.
Output: T

It is important to note that the if-condition in the second step of the preceding algorithm is not so hard because one compares x^j with all points in T and not in S. In practice, the set T has much less elements than S.

The following theorem shows that Algorithm 12.15 is really a reduction method.

Theorem 12.16. *Under the assumptions of this subsection we assert:*

(a) Algorithm 12.15 is well-defined.

(b) Algorithm 12.15 generates a nonempty set $T \subset S$.

(c) Every minimal element of the set S is also contained in the set T generated by Algorithm 12.15.

Proof. The assertions under (a) and (b) are obvious. For the proof of part (c) let x^j be an arbitrary minimal element of S, and assume that $x^j \notin T$. Then there does not exist any $x \in S\setminus\{x^j\}$ with $x \leq x^j$. Consequently, we have

$$x \not\leq x^j \text{ for all } x \in S\setminus\{x^j\},$$

and since $T \subset S\setminus\{x^j\}$ we conclude

$$x \not\leq x^j \text{ for all } x \in T.$$

Hence, x^j satisfies the condition in the second step of Algorithm 12.15, and x^j is added to the set T. This is a contradiction to our assumption. □

Remark 12.17. The set of all minimal elements of a discrete set S is computed as follows: First, one applies Algorithm 12.15 in order to reduce the original set S. By Theorem 12.16, (c) the minimal elements are not lost by this reduction. Then all minimal elements of the reduced set are determined using the definition of minimality.

The following example points out that the reduction gains of the Graef-Younes method may be very large.

Example 12.18. Again, we consider the multiobjective optimization problem discussed in Example 12.2. Using a random generator points in the constraint set being equal to the image set of the objective mapping, are produced. As described in Remark 12.17 all minimal elements of the generated discrete set are determined. It is documented in [333] that the Graef-Younes method reduces a set containing 500,000 points to a set T containing only 1,001 points. A total number of 471 points are minimal elements. If one generates 5,000,000 points (see [333]), then the Graef-Younes method reduces these points to only 3,067 points. Among these points, only 1,497 are minimal elements. Hence, in both cases about every second element of the set T is minimal.

Notes

This chapter makes only a selection of the numerical methods currently used in multiobjective optimization. For a survey of standard methods for the solution of nonlinear problems we refer to the recent book of Hillermeier [122]. This book also describes a new generalized homotopy method for the solution of nonlinear multiobjective optimization problems.

The modified Polak method is based on a method proposed by Polak [252] for nonlinear multiobjective optimization problems with several and not only two objectives. The presentation of Section 12.1 follows the lines in [154]. Example 12.2 is taken from [224]. The tunneling technique discussed in Remark 12.3 has been proposed in [154]. Although Monte-Carlo methods may be used for global optimization, they have the disadvantage that they can only be applied to problems with some few variables. Nowadays there are modern methods in global optimization for the solution of the scalar subproblems in the modified Polak method. For these methods of global optimization we refer to Schäffler [274]. In [275] a new stochastic method for global unconstrained multiobjective optimization is given.

The investigations of Subsection 12.2.1 are based on the paper [85]. Example 12.6 is taken from [85]. The discussion of the method of reference point approximation follows the article [154]. The linear case is treated in [147]. Example 12.9 is taken from [339], [132] and [135]. Example 12.10 can be found in [135] and [293]. It is cited in [135] that Example 12.11 has been proposed in [292]. Algorithm 12.12 for the bicriterial nonlinear case has been published in [154]. Example 12.13 is taken from [154].

The discussion of a method for the solution of discrete multiobjective optimization problems in Section 12.3 follows the dissertation [333] of Younes. The algorithmic conception of the Graef-Younes method has been originally proposed by Graef [111]. Algorithm 12.15 is taken from the dissertation of Younes [333].

Chapter 13

Multiobjective Design Problems

Multiobjective optimization problems turn up in almost all fields of the science of engineering. The application areas range from designs of electrical switching circuits, machine parts, airplanes and weight-bearing structures (bridges, pylons etc.) to planning and controlling of water-supply systems.

The configuration of industrial systems is an optimization task whose multiobjective character is particularly obvious. Consider, e.g., the design of a vacuum pump. Such a pump should simultaneously have maximum suction capacity, minimal power demand and minimal demand for operating liquid. Optimizing the variables which characterize the geometry of a vacuum pump is, therefore, an multiobjective optimization problem. Typical conflicting objectives within industrial system design are the maximization of efficiency (or plant productivity), the minimization of failure and the minimization of the investment funds to be raised for the acquisition of the plant.

Another illustrative example is the search for optimal operating points of internal combustion engines (see [280]). Here, one strives for the simultaneous minimization of the specific fuel consumption, the emission of NO_x and the opacity of the exhaust gas as a measure for the production of polluting particles.

In modern applications the mathematical modeling of multiobjective optimization problems plays an important role. Often, comprehensive simulations have to be carried out for the evaluation of the complicated objective functions and constraint functions.

In this chapter we present a detailed discussion of nonlinear multiobjective optimization problems arising in engineering. As an application from electrical engineering we describe the optimal design of rod antennas. The optimization of a FDDI communication network in computer science is also discussed. From chemical engineering we analyse a fluidized reactor-heater system and a cross-current multistage extraction process. After a description of the used mathematical model we present the constraints and objectives of these design problems. Solutions are computed using the modified Polak method, the weighted Chebyshev norm approach or the method of reference point approximation.

13.1 Design of Antennas

Antennas, i.e. devices for transmitting or receiving electromagnetic energy, take on a variety of different forms. They can be as simple as single dipols or arrays of dipols, or far more complicated structures consisting of solid surfaces. It is a basic problem in antenna design to construct the shape or choose the "feeding" of the antenna to optimize the performance of the antenna. Many of the performance criteria used in the literature are "conflicting": Improving one criterion is only possible to the cost of others. Therefore, in classical antenna theory one tries to optimize one criterion and keeps the other restricted. This leads to constrained scalar optimization problems.

In many cases it doesn't seem to be clear a priori which performance criterion has to be optimized and which to be restricted. Therefore, we are in a classical case of a multiobjective optimization problem. In this section we assume that the geometry of the antenna is fixed and that we are able to vary the feeding of the antenna. It is our aim to show how the modified Polak method (Algorithm 12.1) can successfully be used to compute the set of Edgeworth-Pareto optimal points. We demonstrate this for a simple problem which arises naturally in directing the power of the antenna in a specific direction.

13.1. Design of Antennas

Now we describe the geometry of the antenna. Let the antenna be a hollow infinite cylinder in x_3-direction with constant cross-section $\Omega \subset \mathbb{R}^2$. We assume that Ω is open, bounded and simply connected with C^∞-boundary Γ. Let $j = j(x_1, x_2)$ be the x_3-component of a current distribution which is assumed to be constant along the infinite axis of the antenna. The physical current distribution is thus given by the real part of $j(x) e^{-i\omega t} \hat{z}$ where $\hat{z}$ denotes the unit vector in x_3-direction and ω describes the used frequency.

Now we define the performance criteria. The *radiation efficiency* $G(\hat{x})$ is defined as the ratio of the power radiated in a particular direction $\hat{x}$ to the total power fed to the antenna (ignoring normalizing constants):

$$G(\hat{x}) = \frac{|u_\infty(\hat{x})|^2}{\int_\Gamma |j|^2 ds}, \quad \hat{x} \in S_1.$$

Here S_1 denotes the unit sphere, j is the chosen surface current and

$$u_\infty(\hat{x}) := \int_\Gamma j(y) e^{-iky^T \hat{x}} ds(y)$$

is the so-called *far field pattern* or *radiation pattern* of the single layer potential u. The wave number is denoted by $k = \omega \sqrt{\varepsilon \mu}$ where ε and μ are the permittivity and permeability respectively in free space. Certainly, we wish to maximize this efficiency in a particular direction $\hat{\vartheta}$. We take a slightly different point of view and maximize the power in the direction $\hat{\vartheta}$ under the constraint $\int_\Gamma |j|^2 ds \leq 1$.

On the other hand we would like to minimize the power radiated into other directions, given by some subset T of S_1 (see Fig. 13.1). Therefore we like to minimize the function

$$\max_{\hat{x} \in T} |u_\infty(\hat{x})|.$$

Inserting the forms of u_∞ this leads to the following bicriterial optimization problem

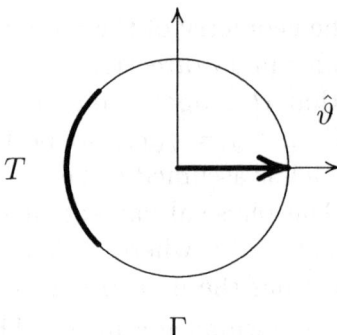

Figure 13.1: Radiation areas of the antenna.

$$\min \begin{pmatrix} -\left| \int_\Gamma j(y) e^{-ik\hat{\vartheta}^T y} ds(y) \right|^2 \\ \max_{\hat{x} \in T} \left| \int_\Gamma j(y) e^{-ik\hat{x}^T y} ds(y) \right|^2 \end{pmatrix}$$

subject to the constraints (13.1)

$$j \in L^2(\Gamma, \mathbb{C}),$$
$$\|j\|_{L^2(\Gamma,\mathbb{C})} \leq 1$$

where we assume that $\hat{\vartheta} \in S_1$ is a given direction, T is a nonempty closed subset of S_1 with positive measure (with respect to S_1) and $k > 0$ denotes the wave number. For a detailed description of the mathematical model we refer to [162]. It can be shown that this general multiobjective optimization problem (13.1) is solvable.

As a next step we take for Γ the unit circle in $\mathbb{R}^2$. Using polar coordinates we can replace $L^2(\Gamma, \mathbb{C})$ by $L^2([0, 2\pi], \mathbb{C})$. If we assume, in addition, that the subset T of S_1 appearing in the original problem (13.1) is connected, then T can now be identified with a closed interval $[t_1, t_2] \subset [0, 2\pi]$ ($t_1 < t_2$), and the direction $\hat{\vartheta} \in S_1$ corresponds to a point $\hat{t} \in [0, 2\pi]$. Then the original problem (13.1) is equivalent to the continuous problem

13.1. Design of Antennas

$$\min \begin{pmatrix} -\left| \int_0^{2\pi} \varphi(s) e^{-ik\cos(\hat{t}-s)} ds \right|^2 \\ \max_{t_1 \leq t \leq t_2} \left| \int_0^{2\pi} \varphi(s) e^{-ik\cos(t-s)} ds \right|^2 \end{pmatrix}$$

subject to the constraints (13.2)

$\varphi \in L^2([0, 2\pi], \mathbb{C})$,

$\|\varphi\|_{L^2([0,2\pi],\mathbb{C})} \leq 1$

where $\hat{t} \in [0, 2\pi]$ is given and $k > 0$ denotes the wave number.

It is well known that every function of $L^2([0, 2\pi], \mathbb{C})$ can be represented by its Fourier series of the form $\sum_{\nu=-\infty}^{\infty} z_\nu e^{i\nu t}$ ($t \in [0, 2\pi]$) with appropriate Fourier coefficients $z_\nu \in \mathbb{C}$. The truncation of these Fourier series leads to finitely many Fourier coefficients and, therefore, to a finite dimensional bicriterial optimization problem. Then the discretized version of the continuous problem (13.2) for an arbitrary $n \in \mathbb{N}_0$ reads as follows:

$$\min \begin{pmatrix} -\left| \int_0^{2\pi} \varphi(s) e^{-ik\cos(\hat{t}-s)} ds \right|^2 \\ \max_{t_1 \leq t \leq t_2} \left| \int_0^{2\pi} \varphi(s) e^{-ik\cos(t-s)} ds \right|^2 \end{pmatrix}$$

subject to the constraints (13.3)

$\varphi \in X_n$,

$\|\varphi\|_{L^2([0,2\pi],\mathbb{C})} \leq 1$

where $\hat{t} \in [0, 2\pi]$ is given and $k > 0$ denotes the wave number. Here

$$X_n := \text{span}\{e^{i\nu t} \mid t \in [0, 2\pi], \nu \in \mathbb{Z}, |\nu| \leq n\}$$

denotes a finite dimensional subspace of $L^2([0, 2\pi]), \mathbb{C})$. It is shown in [162] that the multiobjective optimization problem (13.3) is solvable for arbitrary $n \in \mathbb{N}_0$.

Using the Jacobi-Anger expansion (see [13]) and Parseval's equation the finite dimensional bicriterial problem (13.3) can be written as

$$\min \begin{pmatrix} -\left|2\pi \sum_{\nu=-n}^{n}(-i)^{\nu}\mathcal{J}_{\nu}(k)z_{\nu}e^{i\nu\hat{t}}\right|^{2} \\ \max_{t_1 \le t \le t_2}\left|2\pi \sum_{\nu=-n}^{n}(-i)^{\nu}\mathcal{J}_{\nu}(k)z_{\nu}e^{i\nu t}\right|^{2} \end{pmatrix}$$

subject to the constraints (13.4)

$$z_{\nu} \in \mathbb{C} \quad (\nu \in \mathbb{Z},\ |\nu| \le n),$$
$$2\pi \sum_{\nu=-n}^{n} |z_{\nu}|^2 \le 1$$

(here $\mathcal{J}_{\nu}$ denotes the Bessel function of ν-th order). Because the max term in the second objective is numerically complicated in this form, we replace it by a slack variable (as it is done in Chebyshev approximation). The disadvantage is that we obtain infinitely many constraints and, therefore, the following semi-infinite bicriterial optimization problem:

$$\min \begin{pmatrix} -4\pi^2\left|\sum_{\nu=-n}^{n}(-i)^{\nu}\mathcal{J}_{\nu}(k)z_{\nu}e^{i\nu\hat{t}}\right|^{2} \\ \delta \end{pmatrix}$$

subject to the constraints (13.5)

$$z_{\nu} \in \mathbb{C} \quad (\nu \in \mathbb{Z},\ |\nu| \le n),$$
$$2\pi \sum_{\nu=-n}^{n} |z_{\nu}|^2 \le 1,$$
$$4\pi^2 \left|\sum_{\nu=-n}^{n}(-i)^{\nu}\mathcal{J}_{\nu}(k)z_{\nu}e^{i\nu t}\right|^{2} \le \delta \quad \text{for all } t \in [t_1, t_2].$$

It is evident that the problems (13.4) and (13.5) are equivalent. In order to get a problem with finitely many constraints we select finitely

13.1. Design of Antennas

many points $s_\eta \in [t_1, t_2]$ ($\eta = 0, 1, \ldots, N$) with $N \in \mathbb{N}$. If we also write $z_\nu = x_\nu + iy_\nu$ with $x_\nu, y_\nu \in \mathbb{R}$, the problem (13.5) can be replaced by the simpler problem

$$\min \begin{pmatrix} -4\pi^2 \left| \sum_{\nu=-n}^{n} (-i)^\nu J_\nu(k)(x_\nu + iy_\nu) e^{i\nu\hat{t}} \right|^2 \\ \delta \end{pmatrix}$$

subject to the constraints (13.6)

$$x_\nu, y_\nu \in \mathbb{R} \quad (\nu \in \mathbb{Z}, |\nu| \leq n),$$

$$2\pi \sum_{\nu=-n}^{n} (x_\nu^2 + y_\nu^2) \leq 1,$$

$$4\pi^2 \left| \sum_{\nu=-n}^{n} (-i)^\nu J_\nu(k)(x_\nu + iy_\nu) e^{i\nu s_\eta} \right|^2 \leq \delta \quad \text{for } \eta = 0, \ldots, N.$$

This bicriterial optimization problem has $4n + 3$ real variables and $N + 2$ inequality constraints. The arising functions are quadratic or even linear. The first objective is quadratic and concave and the second one is linear.

The bicriterial optimization problem (13.6) can be solved for different parameters. The following numerical results are obtained for the special values $\hat{t} = 0$, $t_1 = \frac{3}{4}\pi$, $t_2 = \frac{5}{4}\pi$, $N = 5$, $k = 10$ and $n = 10$ with the special discretization points

$$s_\eta := \frac{3}{4}\pi + \eta \frac{\pi}{10} \quad (\eta = 0, \ldots, 5).$$

The figures are organized in such a way that the image set of all Edgeworth-Pareto optimal points is approximated by 100 discretization points obtained by the modified method of Polak (Fig. 13.2) and then the radiation characteristics represented by some of these points is illustrated (Fig. 13.3). Here the radiation intensity

$$I(\vartheta) = 4\pi^2 \left| \sum_{\nu=-n}^{n} (-i)^\nu J_\nu(k)(x_\nu + iy_\nu) e^{i\nu\vartheta} \right|^2, \quad \vartheta \in [0, 2\pi],$$

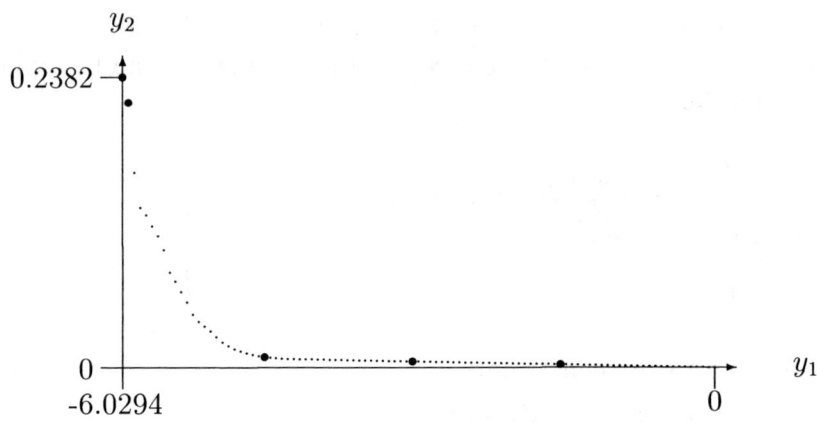

Figure 13.2: Approximation of the images of Edgeworth-Pareto optimal points.

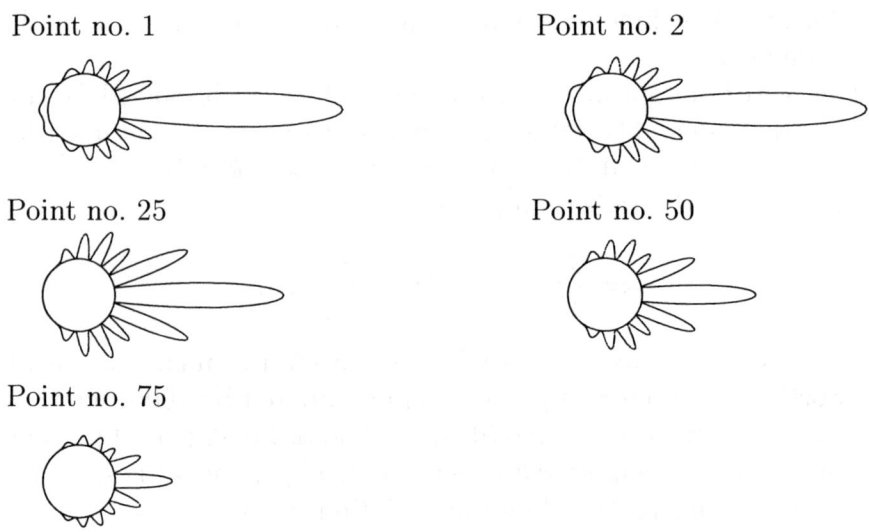

Figure 13.3: Radiation characteristics of Edgeworth-Pareto optimal points given in Fig. 13.2.

is illustrated with respect to the unit circle. This circle is then the zero curve of the radiation characteristics. The graph of I is radially drawn. Moreover, the Fourier coefficients in the form of $z = \left(z_{-n}, z_{-(n-1)}, \ldots, z_0, \ldots, z_{n-1}, z_n\right) \in \mathbb{C}^{2n+1}$ are computed for the discretization point no. 25 as follows

$$\begin{aligned}
z = (\quad &-0.009276 + 0.025612i, & &-0.051782 + 0.142982i, \\
&+0.016229 - 0.044812i, & &+0.024146 - 0.066672i, \\
&+0.002190 - 0.006046i, & &+0.002254 - 0.006224i, \\
&-0.010658 + 0.029429i, & &+0.007835 - 0.021635i, \\
&-0.004423 + 0.012214i, & &-0.011581 + 0.031978i, \\
&+0.001747 + 0.004823i, & &-0.011581 + 0.031978i, \\
&-0.004423 + 0.012214i, & &+0.007835 - 0.021635i, \\
&-0.010658 + 0.029429i, & &+0.002254 - 0.006224i, \\
&+0.002190 - 0.006046i, & &+0.024146 - 0.066672i, \\
&+0.016229 - 0.044812i, & &-0.051782 + 0.142982i, \\
&-0.009276 + 0.025612i & &\quad).
\end{aligned}$$

13.2 Design of FDDI Computer Networks

Communication networks are the essential base for distributed computation and the exchange of information among computers. Besides the ATM standard the FDDI (fiber distributed data interface) communication protocol is often used for the network of high speed computers. It is physically realized as a fiber optics backbone network configured as a ring, and it is commonly part of a complex hierarchy of different bus systems (see Fig. 13.4).

Because of the increasing demands made on such a network (multimedia, internet, transfer speed, change of loads) there is a need for the improvement of the performance of the FDDI network. This performance can only be improved by optimization of the protocol parameters. In the future an optimal network management is required, since the performance of the FDDI fiber optics ring is physically limited. Computer experiments with the FDDI ring have already shown that very remarkable performance gains are possible ([257]).

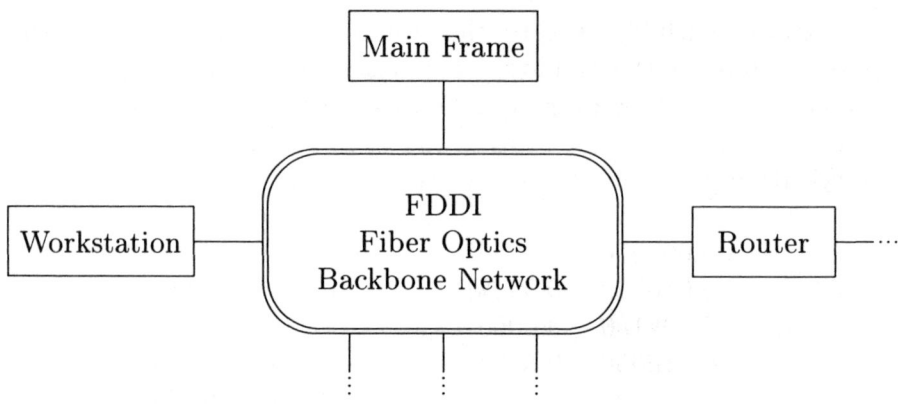

Figure 13.4: Typical FDDI Network.

Based on stochastic models Tangemann [296] and Klehmet [174] developed formulas for the evaluation of the mean waiting time in a station belonging to a FDDI network. Using these formulas it is possible to minimize the mean waiting times in FDDI rings. For realistic applications the throughput of a station or the total throughput in the FDDI ring is also of great importance as an optimization criterion. An improvement of waiting times generally leads to a deterioration of throughputs which is not desirable. Therefore, the consideration of multiobjective functions has to be included in the developed mathematical models. Since we can formulate an objective function for every computer belonging to such a FDDI ring, it makes sense to investigate the whole system as a game where the stations "play cooperatively".

13.2.1 A Cooperative Game

A FDDI fiber optics ring connects several different computers (see Fig. 13.4) called stations in this context. We assume that the network consists of $n \in \mathbb{N}$ stations. The FDDI medium access protocol allows to set certain variables for the management of this ring. For instance, the target token rotation time (TTRT) parameter controlling the maximal allowable time delay of message sending is an important design variable. Another variable used in a synchronous mode is the

13.2. Design of FDDI Computer Networks

so-called token holding time (THT). And the applied loads in the stations are possible parameters as well. Here we assume for simplicity that we have a vector $x \in \mathbb{R}^m$ (with $m \in \mathbb{N}$) representing all possible variables. But this vector should also satisfy technological and model theoretical constraints defining a feasible set $S \subset \mathbb{R}^m$.

On the set of feasible points we want to minimize, for every station i, a vector-valued function $\varphi_i : S \to \mathbb{R}^{p_i}$ with $p_i \in \mathbb{N}$ (where we assume that the space $\mathbb{R}^{p_i}$ is partially ordered in a componentwise sense). For instance, one could minimize the mean waiting time of a message to be sent to a specific station in a network and one could simultaneously maximize the throughput in the whole ring.

Since we obtain an optimization problem for every station, we have the typical situation of a game where a station is to be understood as a "player". An improvement of the performance of the whole FDDI ring can be reached, if the n players cooperate and are not only interested to maximize their own profit. In this case we have a cooperative n player game which reads as follows: Determine a feasible point $x \in S$ which is "preferred" by all players because of their cooperation.

Noncooperative games are also possible for such a network. For instance, if we consider only the minimization of the mean waiting time for one specific station, we obtain an improvement of the performance of this station at the cost of the other stations. But if we think of the total performance of the whole net, we have a cooperative game.

Such cooperative games (in control theory) are investigated in Chapter 10. For the description of the solution concept we introduce an objective map $f : S \to \mathbb{R}^p$ with $p := p_1 + \cdots + p_n$ and

$$f(x) = \begin{pmatrix} \varphi_1(x) \\ \vdots \\ \varphi_n(x) \end{pmatrix} \quad \text{for all } x \in S.$$

Following Section 10.1 this cooperative n player game can be formulated as a vector optimization problem of the following type:

$$\min_{x \in S} f(x). \tag{13.7}$$

Using the componentwise ordering in $\mathbb{R}^p$ we get an adequate description of the cooperation, since feasible vectors are "preferred" if and only if they are "preferred" by each player. Optimal solutions of this problem are defined as in Definition 10.1.

13.2.2 Minimization of Mean Waiting Times

Although the derivation of the general n players game developed in the previous section does not use special properties of the FDDI medium access protocol (IEEE 802.8), we now need an exact protocol definition for the description of the mean waiting times in such a ring. We restrict ourselves to a short description of the functionality of this network.

A token being a special sequence of bits rotates within the FDDI ring. If a station obtains the token, it has the right to send data. The FDDI protocol allows the transfer of synchronous and asynchronous messages. For the asynchronous messages eight priority classes are possible. These classes are served hierarchically: first the messages of a high priority are sent, then the data of lower priority. This procedure is stopped, if there are no further data available or the submission time is terminated.

For the mathematical modeling one distinguishes two processes. In every station one has arrival processes (for each priority class) assumed to be Poisson distributed. In the ring one considers a server model where the token plays the role of the server. Since the service times are arbitrarily distributed, we have the problem of a M/G/1 queue. The resulting stochastical model allows to present formulas for the mean waiting times in the stochastical mean. Estimates for these mean waiting times have been developed by Tangemann [296] and Klehmet [174].

The mean waiting time for synchronous messages at a station i (with $1 \leq i \leq n$) developed by Tangemann is given as

$$W_i^{T,\text{syn}} := \frac{1 - \rho_i \left(1 - \frac{\beta_i(2-\rho)}{\text{THT}_i(1-\rho)}\right)}{1 - \frac{\rho_i C}{\text{THT}_i}} \cdot \frac{A + \sum_{j=1}^{n} \frac{\rho_j C}{\text{THT}_j} \left[\rho_j - (1-\rho_j)\frac{\lambda_j C}{2}\left(1 - \frac{\beta_j}{\text{THT}_j}\right)\right]\beta_j}{\sum_{j=1}^{n} \rho_j \left[1 - \rho_j \left(1 - \frac{\beta_j(2-\rho)}{\text{THT}_j(1-\rho)}\right)\right]}$$

13.2. Design of FDDI Computer Networks

and the formula for asynchronous messages reads

$$W_i^{T,\text{asyn}} := \frac{1 - \rho_i \left(1 - \frac{\beta_i(2-\rho)}{(\text{THT}_i - s)(1-\rho)}\right)}{1 - \frac{(\rho_i + \rho)C}{\text{THT}_i - s}} \cdot \frac{A + \sum_{j=1}^{n} \frac{(\rho_j + \rho)C}{\text{THT}_j - s} \left[\rho_j - (1 - \rho_j)\frac{\lambda_j C}{2}\left(1 - \frac{\beta_j}{\text{THT}_j - s}\right)\right] \beta_j}{\sum_{j=1}^{n} \rho_j \left[1 - \rho_j \left(1 - \frac{\beta_j(2-\rho)}{(\text{THT}_j - s)(1-\rho)}\right)\right]}.$$

Here we set

$$C = \frac{s}{1-\rho}, \tag{13.8}$$

$$A = \frac{\rho \sum_{j=1}^{n} \lambda_j \beta_j^{(2)}}{2(1-\rho)} + \frac{\rho s^{(2)}}{2s} + \frac{s}{2(1-\rho)}\left(\rho^2 - \sum_{j=1}^{n} \rho_j^2\right) \tag{13.9}$$

and the terms have the following meaning:

ρ_i ≡ applied load at station i,
$\rho = \sum_{i=1}^{n} \rho_i$ ≡ throughput,
λ_i ≡ Poisson arrival rate at station i,
β_i ≡ average service time at station i (notice $\rho_i = \lambda_i \beta_i$),
$\beta_i^{(2)}$ ≡ second moment of β_i,
s ≡ total switch over time,
$s^{(2)}$ ≡ second moment of s.

The formula for the mean waiting time at a station i (with $1 \leq i \leq n$) given by Klehmet for the case of synchronous and asynchronous messages reads

$$W_i^K := \frac{1 - \rho_i + \frac{\rho_i}{k_i}\left(1 + \frac{1}{1-\rho}\right)}{1 - \frac{\lambda_i C}{k_i}} \cdot \frac{A + \frac{s}{1-\rho}\sum_{j=1}^{n}\frac{\rho_j^2}{k_j} - \frac{\rho_i C}{\text{TTRT} - C}\sum_{j=1}^{n}\frac{\beta_j(1-\rho_j)(\lambda_j C)^2}{2k_j}}{\sum_{j=1}^{n} B_j}$$

with C and A as in (13.8) and (13.9), respectively, and

$$B_j = \rho_j \left(1 - \frac{\lambda_j s}{k_j(1-\rho)}\right) \frac{1 - \rho_j + \frac{\rho_j}{k_j}\left(1 + \frac{1}{1-\rho}\right)}{1 - \frac{\lambda_j C}{k_j}},$$

$$k_i = \begin{cases} \left\lfloor \frac{\text{THT}_i}{\beta_i} \right\rfloor & \text{for synchronous messages} \\ \left\lfloor \frac{\text{TTRT}_i - C}{\beta_i} + 0.5 \right\rfloor & \text{for asynchronous messages} \end{cases}.$$

Here TTRT_i represents the target token rotation time and means the maximal allowable time delay of message sending with respect to the i-th station.

In these formulas for the mean waiting times the loads $\rho_1, \ldots, \rho_n$, the times $\text{TTRT}_1, \ldots, \text{TTRT}_n$ and $\text{THT}_1, \ldots, \text{THT}_n$ are possible variables. And these variables have to satisfy certain technological and model theoretical constraints. For the mean waiting times given by Tangemann we have the constraints

$$\rho_i \geq \alpha_i \quad \text{for all } i \in \{1, \ldots, n\},$$

$$\sum_{i=1}^{n} \rho_i \geq \rho_{\min},$$

$$\text{THT}_i \leq \text{THT}_{\max} \quad \text{for all } i \in \{1, \ldots, n\}$$

and, in addition, for synchronous messages

$$\left(1 + \frac{s}{\text{THT}_i}\right)\rho_i + \sum_{\substack{j=1 \\ j \neq i}}^{n} \rho_j \leq 1 \quad \text{for all } i \in \{1, \ldots, n\}$$

and for asynchronous messages

$$\left(1 + \frac{s}{\text{THT}_i}\right)\rho_i + \sum_{\substack{j=1 \\ j \neq i}}^{n} \rho_j \leq \frac{\text{THT}_i - s}{\text{THT}_i} \quad \text{for all } i \in \{1, \ldots, n\}.$$

$\alpha_i \geq 0$ denotes the lower bound for the load ρ_i, $\rho_{\min}$ means the minimal throughput and $\text{THT}_{\max}$ describes the maximal token holding time.

13.2. Design of FDDI Computer Networks

The constraints given by Klehmet are slightly different:

$$\rho_i \geq \alpha_i \text{ for all } i \in \{1,\ldots,n\},$$

$$\sum_{i=1}^{n} \rho_i \geq \rho_{\min},$$

$$\text{THT}_i \leq \text{THT}_{\max} \text{ for all } i \in \{1,\ldots,n\}$$

and, in addition, for synchronous messages

$$\left(1 + \frac{s}{\text{THT}_i}\right)\rho_i + \sum_{\substack{j=1 \\ j \neq i}}^{n} \rho_j \leq 1 \text{ for all } i \in \{1,\ldots,n\}$$

and for asynchronous messages

$$\left(1 + \frac{s}{\text{TTRT}_i + 0.5\beta_i}\right)\rho_i + \sum_{\substack{j=1 \\ j \neq i}}^{n} \rho_j \leq \frac{\text{TTRT}_i + 0.5\beta_i - s}{\text{TTRT}_i + 0.5\beta_i}$$

for all $i \in \{1,\ldots,n\}$.

We use the same constants as before.

One can formulate games for which the objective functions of the players are the mean waiting times or also games with vector-valued objectives with the throughput $\rho = \sum_{i=1}^{n} \rho_i$ as an additional objective.

13.2.3 Numerical Results

We consider a special FDDI ring investigated by Tangemann [296] and Klehmet [174] and assume that it is a symmetric system, i.e. for every station we have the same parameters. This ring consists of $n = 10$ stations with the total switch over time $s = 0.1$ ms. The second moment of s is assumed to be $s^{(2)} = 0.01$ ms. The average service times are $\beta_1 = \ldots = \beta_{10} = 0.01$ ms with the second moments $\beta_1^{(2)} = \ldots = \beta_{10}^{(2)} = 0.001$ ms. The upper bounds for the token holding times and target token rotation times are $\text{THT}_{\max} = 0.8$ ms and $\text{TTRT}_{\max} = 4$ ms. The minimal throughput is given as $\rho_{\min} = 0.1$. Moreover, we set $\alpha_1 = \ldots = \alpha_{10} = 0.01$.

With these constants we investigate the cooperative game of the simultaneous minimization of the mean waiting times of every station, i.e. we have the multiobjective optimization problem (13.7) where the i-th component of f is the mean waiting time at the i-th station. For the determination of Edgeworth-Pareto optimal points of problem (13.7) we use the weighted Chebyshev norm approach presented in Subsection 11.2.2. In Cor. 11.21 we assume that the mean waiting times have a lower bound. Since times are nonnegative, such a lower bound can be assumed but, in general, it cannot be shown that the used mean waiting times formulas are nonnegative. But nevertheless we use the weighted Chebyshev norm approach which means that we minimize the worst case. For simplicity we choose the weights $w_1 = \ldots = w_{10} = 1$ and set $\hat{y}_1 = \ldots = \hat{y}_{10} = 0$. Numerical results are given in the tables 13.1 and 13.2.

$\rho_{1_{start}}$ $\rho_{1_{opt}}$	$\rho_{2_{start}}$ $\rho_{2_{opt}}$	$\rho_{3_{start}}$ $\rho_{3_{opt}}$	$\cdots$ $\cdots$	$\rho_{10_{start}}$ $\rho_{10_{opt}}$	THT_{start} THT_{opt}	min max$_{start}$ min max$_{opt}$ $w_i^{T,\ syn}$	min max$_{start}$ min max$_{opt}$ w_i^{K}
0.010	0.010	0.010	$\ldots$	0.010	0.20	0.06066	0.06095
0.010	0.010	0.010	$\ldots$	0.010	0.80	0.06058	0.06066
0.030	0.030	0.030	$\ldots$	0.030	0.20	0.09148	0.09291
0.010	0.010	0.010	$\ldots$	0.010	0.80	0.06058	0.06066
0.050	0.050	0.050	$\ldots$	0.050	0.20	0.14841	0.15311
0.010	0.010	0.010	$\ldots$	0.010	0.80	0.06058	0.06066
0.100	0.044	0.044	$\ldots$	0.044	0.20	0.14777	0.15282
0.010	0.010	0.010	$\ldots$	0.010	0.80	0.06058	0.06066
0.044	0.100	0.044	$\ldots$	0.044	0.20	0.14777	0.15282
0.010	0.010	0.010	$\ldots$	0.010	0.80	0.06058	0.06066
0.040	0.060	0.040	$\ldots$	0.060	0.20	0.14856	0.15344
0.010	0.010	0.010	$\ldots$	0.010	0.80	0.06058	0.06066
0.060	0.040	0.060	$\ldots$	0.040	0.20	0.14856	0.15344
0.010	0.010	0.010	$\ldots$	0.010	0.80	0.06058	0.06066
0.041	0.051	0.051	$\ldots$	0.051	0.20	0.14842	0.15314
0.010	0.010	0.010	$\ldots$	0.010	0.80	0.06058	0.06066

Table 13.1: Minmax approach in the case of synchronous messages.

The numerical results in these two tables show for several configurations that we get the same optimal solution of the minmax problem. An optimal network configuration can be reached, if we decrease the loads at every station to the smallest possible load and increase the target token rotation time or the token holding time to the largest possible time. This seems to be a general optimization rule for these cooperative games with respect to FDDI computer networks.

It is also interesting to see that in the case of small applied loads $\rho_{i_{start}}$ we only get a small improvement of the waiting times whereas

ρ_{1start} ρ_{1opt}	ρ_{2start} ρ_{2opt}	ρ_{3start} ρ_{3opt}	...	$\rho_{10start}$ ρ_{10opt}	$THT_{start}/TTRT_{start}$ $THT_{opt}/TTRT_{opt}$	min max$_{start}$ min max$_{opt}$ $W_i^{T, asyn}$	min max$_{start}$ min max$_{opt}$ W_i^K
0.010 0.010	0.010 0.010	0.010 0.010	...	0.010 0.010	0.20 / 1.30 0.80 / 4.00	0.06349 0.06085	0.06062 0.06058
0.030 0.010	0.030 0.010	0.030 0.010	...	0.030 0.010	0.20 / 1.30 0.80 / 4.00	0.12493 0.06085	0.09109 0.06058
0.050 0.010	0.050 0.010	0.050 0.010	...	0.050 0.010	0.20 / 1.30 0.80 / 4.00	0.22414 0.06085	0.14641 0.06058
0.100 0.010	0.044 0.010	0.044 0.010	...	0.044 0.010	0.20 / 1.30 0.80 / 4.00	0.23431 0.06085	0.14584 0.06058
0.044 0.010	0.100 0.010	0.044 0.010	...	0.044 0.010	0.20 / 1.30 0.80 / 4.00	0.23431 0.06085	0.14584 0.06058
0.040 0.010	0.060 0.010	0.040 0.010	...	0.060 0.010	0.20 / 1.30 0.80 / 4.00	0.22705 0.06085	0.14774 0.06058
0.060 0.010	0.040 0.010	0.060 0.010	...	0.040 0.010	0.20 / 1.30 0.80 / 4.00	0.22705 0.06085	0.14774 0.06058
0.041 0.010	0.051 0.010	0.051 0.010	...	0.051 0.010	0.20 / 1.30 0.80 / 4.00	0.22443 0.06085	0.14630 0.06058

Table 13.2: Minmax approach in the case of asynchronous messages.

the optimization of configurations with large applied loads $\rho_{i_{start}}$ leads to a significant decrease of the mean waiting times. But this decrease is obtained at the cost of a decrease of applied loads $\rho_{i_{opt}}$, i.e., a decrease of the throughput of the network. Since the throughput is also an important performance criterion, it should be used as an additional objective.

13.3 Fluidized Reactor-Heater System

Kitagawa et al. consider in [173] a bicriterial optimization problem which arises from minimizing simultaneously the total investment and net operating costs of a fluidized reactor-heater system for an exothermic chemical reaction. This system consists of a reactor, a heat exchanger and a cooler.

The design variables are the extent of reaction (x_1) and the temperature (x_2); in the next section it is shown that the third variable x_3 is not needed as a design variable. The aim is to minimize the function $f : \mathbb{R}^3 \to \mathbb{R}^2$ with

$$f(x_1, x_2, x_3) = \begin{pmatrix} p_{01} V(x_1, x_2)^\alpha x_2^\beta + p_{02} x_1^\gamma + p_{04} 1_{\{x_1 > \xi(x_2)\}} + p_{06,g}(x_1 - \xi(x_2))^\zeta \\ p_{03}/x_1 - p_{05}(x_1 - \xi(x_2)) \end{pmatrix}$$

under the constraints

$$p_{11}x_3^2 x_1 \leq x_2^2 \qquad (13.10)$$
$$p_{21}(1+x_1^{-1})x_2 \leq x_3^2 \qquad (13.11)$$
$$p_{31}V(x_1,x_2) \leq x_3^3 \qquad (13.12)$$
$$x_2 \in [0, \frac{1}{p_{41}}] \qquad (13.13)$$
$$(p_{51}x_1 + p_{52})x_2 \leq 1 + p_{53}x_1 \qquad (13.14)$$
$$\phi_e(x_1,x_2) > 0 \qquad (13.15)$$
$$x_1 \in [0,1]. \qquad (13.16)$$

Here f_1 measures the total investment cost; due to the term

$$1_{\{x_1 > \xi(x_2)\}} := \begin{cases} 1, & x_1 > \xi(x_2) \\ 0, & x_1 \leq \xi(x_2) \end{cases}$$

(which specifies the investment cost for an auxiliary cooler) it is discontinuous. f_2 measures the net operating costs. The constants and functions mentioned above are

$$\begin{aligned}
&p_{01} = 1750, &&p_{02} = 15000, &&p_{03} = 6550, &&p_{04} = 3000 \\
&p_{06,g} = 33000, &&p_{11} = 63.8, &&p_{21} = 0.00036, &&p_{31} = 0.85 \\
&p_{41} = 0.00121, &&p_{51} = 0, &&p_{52} = 0.00206, &&p_{53} = 2.76 \\
&p_{61} = 0.0362,
\end{aligned}$$

$$\alpha = \beta = \gamma = \zeta = 0.6$$

and

$$\phi_e(x_1,x_2) = 55\left(\frac{1-x_1}{2-x_1}\right)^2 e^{-4770/x_2} - 0.000014\frac{x_1}{2-x_1}e^{-19270/x_2},$$

$$\xi(x_2) = \frac{p_{52}x_2 - 1}{p_{53} - p_{51}x_2},$$

$$V(x_1,x_2) = \frac{p_{61}}{x_1}\int_0^{x_1}\frac{1}{\phi_e(u,x_2)}du.$$

ϕ_e is the net reaction rate, V denotes the volume, and $\xi(x_2)$ is the value for x_1 which guarantees that no auxiliary cooler is necessary.

13.3.1 Simplification of the Constraints

Because of the special character of the constraints and the independence of f from x_3 we are able to eliminate the variable x_3: The condition "$\exists x_3 \in \mathbb{R} : (13.10) - (13.12)$ is valid" is equivalent to

$$(p_{31}V(x_1, x_2))^{2/3} \leq \frac{x_2^2}{p_{11}x_1} \tag{13.17}$$

$$p_{11}p_{21}(1 + x_1) \leq x_2. \tag{13.18}$$

The inequality (13.18) ensures $x_2 \geq 0$ because $x_1 \in [0, 1]$. Thus we can drop the condition $x_2 \geq 0$ in (13.13).

For values of x_1 near to 1 $V(x_1, x_2)$ rapidly increases. This is the reason why Kitagawa et al. sharpen the constraint $x_1 \leq 1$ to $x_1 \leq 0.99999$.

Now we can show that under this constraint $x_1 \leq 0.99999$ we have $\phi_e(x_1, x_2) > 0$:

Let $a := 55e^{-4770/x_2}$ and $b := 0.000014e^{-19270/x_2}$ for an arbitrarily fixed x_2. Then we have

$$
\begin{aligned}
&\phi_e(x_1, x_2) \geq 0 \\
\Longleftrightarrow\quad & a\left(\frac{1-x_1}{2-x_1}\right)^2 - b\frac{x_1}{2-x_1} \geq 0 \\
\Longleftrightarrow\quad & \frac{a(1-x_1)^2 - bx_1(2-x_1)}{(2-x_1)^2} \geq 0 \\
\Longleftrightarrow\quad & a(1-x_1)^2 - bx_1(2-x_1) \geq 0 \\
\Longleftrightarrow\quad & (a+b)x_1^2 - 2(a+b)x_1 + a \geq 0.
\end{aligned}
$$

The quadratic equation $(a+b)x_1^2 - 2(a+b)x_1 + a = 0$ has the zeros $x_1 = 1 \pm \sqrt{\frac{b}{a+b}}$. Now

$$
\begin{aligned}
\sqrt{\frac{b}{a+b}} &= \sqrt{\frac{0.000014e^{-19270/x_2}}{0.000014e^{-19270/x_2} + 55e^{-4770/x_2}}} \\
&= \sqrt{0.000014}\sqrt{\frac{1}{0.000014 + 55e^{14500/x_2}}} \\
&\leq \sqrt{0.000014}\sqrt{\frac{1}{55e^{14500}p_{41}}}
\end{aligned}
$$

$$= \sqrt{\frac{0.000014}{55e^{77.545}}}$$
$$< 10^{-5}.$$

For values $x_1 \leq 0.99999$ we have shown $\phi_e(x_1, x_2) > 0$, consequently we can drop the inequality (13.15).

In view of these calculations we can substitute the constraints (13.10) to (13.16) by the system of inequalities

$$\begin{aligned}
(p_{31}V(x_1, x_2))^{2/3} &\leq \frac{x_2^2}{p_{11}x_1} \\
p_{11}p_{21}(1 + x_1) &\leq x_2 \\
x_2 &\leq \frac{1}{p_{41}} \\
(p_{51}x_1 + p_{52})x_2 &\leq 1 + p_{53}x_1 \\
x_1 &\geq 0 \\
x_1 &\leq 0.99999.
\end{aligned}$$

The integral in the definition of V can be solved symbolically rather than using a numerical integration formula:

Defining $\psi(x) := 1/\phi_e(x)$ we have

$$\begin{aligned}
\psi(x) &= \frac{1}{a(\frac{1-x_1}{2-x_1})^2 - \frac{x_1}{2-x_1}b} \\
&= \frac{(x_1 - 2)^2}{a(1 - x_1)^2 - bx_1(2 - x_1)} \\
&= \frac{(x_1 - 2)^2}{a(x_1 - 1)^2 + bx_1(x_1 - 2)} \\
&= \frac{(x_1 - 2)^2}{(a + b)x_1^2 - 2(a + b)x_1 + a}.
\end{aligned}$$

In order to factorize the denominator we proceed as follows: The solutions of the quadratic equation $(a + b)x_1^2 - 2(a + b)x_1 + a = 0$ in x_1 are

$$t_{1,2} = \frac{2(a + b) \pm \sqrt{D}}{2(a + b)} = 1 \pm \frac{\sqrt{b^2 + ab}}{a + b},$$

13.3. Fluidized Reactor-Heater System

where D is the discriminant $D = 4(a+b)^2 - 4a(a+b) = 4(b^2 + ab) > 0$.
With $d := \frac{\sqrt{b^2+ab}}{a+b}$, $t_1 := 1 + d$ and $t_2 := 1 - d$ we get

$$\psi(x) = \frac{(x_1 - 2)^2}{(x_1 - t_1)(x_1 - t_2)(a+b)}.$$

Formal integration of $h := (a+b)\psi$ with the computer algebra system "maple" [54] yields

$$\int_0^{x_1} h(x)\, dx = x_1 + \frac{4\ln(\frac{t_2 - x_1}{t_1 - x_1}) + (4t_1 - t_1^2)\ln(t_1 - x_1)}{t_2 - t_1}$$

$$+ \frac{(t_2^2 - 4t_2)\ln(t_2 - x_1) + (4 + t_1^2 - 4t_1)\ln t_1}{t_2 - t_1}$$

$$+ \frac{(4t_2 - 4 - t_2^2)\ln t_2}{t_2 - t_1}. \tag{13.19}$$

13.3.2 Numerical Results

With the standard double precision in the programming language C it seems to be difficult to evaluate the formula (13.19) correctly. Therefore it is necessary to transform the formula to a form more appropriate for floating point evaluation. The form

$$\int_0^{x_1} h(x)\, dx$$

$$= \frac{x_1 + (-d^2 - 2d + 1)\mathrm{lp}(d - x_1) + (d+1)^2(\mathrm{lp}(-d) - \mathrm{lp}(-d - x_1))}{2d}$$

$$\frac{-(d-1)^2 \mathrm{lp}(d)}{2d}$$

can be evaluated rather fast where $\mathrm{lp}(x) := \ln(1+x)$ can be computed better with the aid of an appropriate function in the programming language C. But in order to obtain a better accuracy one should use the following formula

$$\int_0^{x_1} h(x)\, dx = x_1$$

$$
\begin{aligned}
&+ \frac{d}{2}(\mathrm{lp}(-d) - \mathrm{lp}(d-x_1) - \mathrm{lp}(-d-x_1) - \mathrm{lp}(d)) \\
&+ \mathrm{lp}(-d) - \mathrm{lp}(d-x_1) - \mathrm{lp}(-d-x_1) + \mathrm{lp}(d) \\
&+ (\mathrm{lp}(-d) + \mathrm{lp}(d-x_1) - \mathrm{lp}(-d-x_1) - \mathrm{lp}(d))/(2d).
\end{aligned}
$$

So the bicriterial optimization problem which has to be solved reads as follows:

$$
\min \begin{pmatrix} p_{01} V(x_1, x_2)^\alpha x_2^\beta + p_{02} x_1^\gamma + p_{04} 1_{\{x_1 > \xi(x_2)\}} + p_{06,g}(x_1 - \xi(x_2))^\varsigma \\ p_{03}/x_1 - p_{05}(x_1 - \xi(x_2)) \end{pmatrix}
$$

subject to the constraints

$$
\begin{aligned}
(p_{31} V(x_1, x_2))^{2/3} &\leq \frac{x_2^2}{p_{11} x_1} \\
p_{11} p_{21}(1 + x_1) &\leq x_2 \\
x_2 &\leq \frac{1}{p_{41}} \\
(p_{51} x_1 + p_{52}) x_2 &\leq 1 + p_{53} x_1 \\
x_1 &\geq 0 \\
x_1 &\leq 0.99999
\end{aligned}
$$

where

$$
\begin{aligned}
V(x_1, x_2) &= \frac{p_{61}}{x_1(a+b)}(x_1 + \mathrm{lp}(-d) - \mathrm{lp}(d-x_1) - \mathrm{lp}(-d-x_1) + \mathrm{lp}(d)) \\
&+ \frac{d}{2}(\mathrm{lp}(-d) - \mathrm{lp}(d-x_1) - \mathrm{lp}(-d-x_1) - \mathrm{lp}(d)) \\
&+ (\mathrm{lp}(-d) + \mathrm{lp}(d-x_1) - \mathrm{lp}(-d-x_1) - \mathrm{lp}(d))/(2d))
\end{aligned}
$$

with $a := 55e^{-4770/x_2}$, $b := 0.000014 e^{-19270/x_2}$ and $d := \sqrt{\frac{b}{a+b}}$.

This multiobjective optimization problem is solved by the method of reference point approximation in the bicriterial case (Alg. 12.12) with the weights $t_1 = 0.02$ and $t_2 = 1$ (Step 5 of Alg. 12.12). Possible compromise solutions as best approximations from the image set of Edgeworth-Pareto optimal points are given in Table 13.3.

reference point	estimate as best approximation from the image set of EP optimal points	preimage of this estimate
(100000, 10000)	(152022, 11878.3)	(0.513385, 826.448)
(150000, 6000)	(211169, 5842.24)	(0.837227, 826.446)
(200000, 4000)	(231012, 5372.81)	(0.875274, 826.446)
(500000, 3000)	(489157, 4183.88)	(0.969510, 770.829)
(1000000, 3000)	(1084990, 3749.83)	(0.978737, 637.244)
(2000000, 3000)	(2038190, 3513.11)	(0.983297, 562.305)

Table 13.3: Compromise solutions

13.4 A Cross-Current Multistage Extraction Process

Finally we discuss another bicriteria optimization problem described by Kitagawa et al. [173], namely the design of a cross-current multistage extraction process. It is the aim to maximize the profit f_1 due to separation and to minimize the costs f_2 due to solvent consumption. The constraints consist mainly of a system of material balance equations.

In order to describe this problem mathematically, we consider for some fixed $n \in \mathbb{N}$ the design variables $x_1, \ldots, x_n$ and $u_1, \ldots, u_n$. Then the objective functions read as follows (see [173]):

$$f_1(x_1, \ldots, x_n, u_1, \ldots, u_n) := 0.2 - x_n$$

and

$$f_2(x_1, \ldots, x_n, u_1, \ldots, u_n) := \sum_{i=1}^n u_i.$$

The constraints are given as follows:

$$x_{i-1} = x_i + u_i \phi(x_i) \text{ for all } i \in \{1, \ldots, n\} \quad (13.20)$$

(where $x_0 := 0.2$),

$$0.2 \geq x_1 \geq \ldots \geq x_n > 0 \quad (13.21)$$

and
$$u_1, \ldots, u_n \geq 0 \qquad (13.22)$$

with

$$\phi(\alpha) := \begin{cases} 2.4\alpha & \text{for } 0 \leq \alpha \leq 0.05 \\ \begin{aligned} & 0.182 + 100(\alpha - 0.15)^3 \\ & -26175(\alpha - 0.15)^5 \\ & +3825000(\alpha - 0.15)^7 \\ & -158750000(\alpha - 0.15)^9 \end{aligned} & \text{for } 0.05 \leq \alpha \leq 0.15 \\ \begin{aligned} & 0.182 + 400(\alpha - 0.15)^3 \\ & -326400(\alpha - 0.15)^5 \\ & +140800000(\alpha - 0.15)^7 \\ & -20480000000(\alpha - 0.15)^9 \end{aligned} & \text{for } 0.15 \leq \alpha \leq 0.2 \end{cases}$$

(see Figure 13.5).

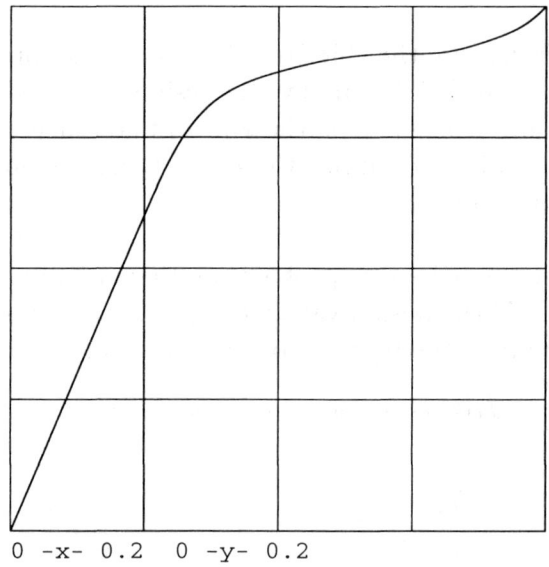

Figure 13.5: Graph of the function ϕ.

It is evident that the variables $u_1, \ldots, u_n$ can be eliminated from

13.4. A Cross-Current Multistage Extraction Process

the equations (13.20) with

$$u_i = \frac{x_{i-1} - x_i}{\phi(x_i)}.$$

Since $\phi(\alpha) > 0$ for all $\alpha \in (0, 0.2]$ and $x_{i-1} - x_i \geq 0$ for all $i \in \{1, \ldots, n\}$ by the inequalities (13.21), we get $u_i \geq 0$ for all $i \in \{1, \ldots, n\}$, i.e., the inequalities (13.22) are satisfied. So, this bicriterial optimization problem can be simplified to the problem

$$\min \begin{pmatrix} x_n \\ \sum_{i=1}^{n} \frac{x_{i-1} - x_i}{\phi(x_i)} \end{pmatrix} \tag{13.23}$$

subject to the constraints
$$0.2 \geq x_1 \geq \ldots \geq x_n > 0$$

where $x_0 := 0.2$.

The bicriterial optimization problem (13.23) can be solved numerically for different values of n. Fig. 13.6 shows the calculated approximations of the image set of Edgeworth Pareto optimal points for $n = 2$ and $n = 4$. Using the method of reference point approximation in the bicriterial case (Alg. 12.12) one gets compromise solutions for the weights $t_1 = 15$ and $t_2 = 1$ of the weighted Chebyshev norm given in Table 13.4.

Reference point		Estimate as best approximation from the image set of EP optimal points		Preimage of this estimate			
(0.001,	100.0)	(0.00010,	146.655)	(0.195,	0.075,	0.035,	0.0001)
(0.01,	10.0)	(0.00041,	6.1548)	(0.0426,	0.0090,	0.0019,	0.0004)
(0.05,	1.0)	(0.05007,	0.9073)	(0.1215,	0.0817,	0.0626,	0.0500)
(0.1,	0.5)	(0.10005,	0.5540)	(0.1409,	0.1239,	0.1112,	0.1000)
(0.15,	0.1)	(0.15002,	0.2744)	(0.1560,	0.1502,	0.1501,	0.1500)

Table 13.4: Compromise solutions for $n = 4$.

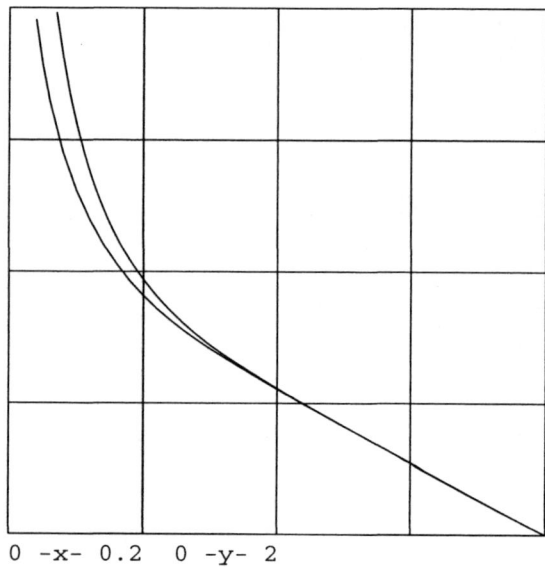

Figure 13.6: Approximation of the image set of Edgeworth Pareto optimal points for $n = 2$ (upper curve) and $n = 4$ (lower curve).

Notes

A classical source for bibliographical references to application problems of vector optimization is [287], a more recent one is [77].

The investigations of the first subsection are based on the papers [162] and [123]. For a standard reference text on antenna theory we refer to [62]. The description of several performance criteria used for the design of antennas is given in [3]. In [4], [5], [172] and [209] one finds examples of scalar optimization problems which are actually bicriterial problems where the second criterion is plugged in a constraint. The systematic application of multiobjective optimization theory to antenna problems is a relatively new area. We refer to [2], [6], [161] and [162] for some theoretical and numerical results.

The game theoretic approach to the optimization of FDDI computer networks is taken from [115]. Bicriterial problems for the optimization of timed polling systems are already considered by Klehmet

[174], [175]. The numerical results presented in Subsection 13.2.3 show that the proposed game theoretic approach is a useful tool for the improvement of the performance of the total FDDI ring. Further investigations indicate (see [114]) that only the use of the parameters TTRT and THT when the loads $\rho_1, \ldots, \rho_n$ are assumed to be constant do not lead to significant improvements of the mean waiting time at a specific station. Therefore, it is necessary to include more than these two types of parameters in the optimization process.

The multiobjective optimization problems in Sections 13.3 and 13.4 are given by Kitagawa et al. [173], and the presentation of these sections is based on [153].

Part V

Extensions to Set Optimization

Set optimization means optimization of sets or set-valued maps. It is an extension of vector optimization to the set-valued case. In the last decade there has been an increasing interest in set optimization. Although the notions "set-valued optimization" and "set optimization" are used in the literature, the second notion makes more sense - as an extension of vector optimization.

General optimization problems with set-valued constraints or a set-valued objective function are closely related to problems in stochastic programming, fuzzy programming and optimal control. If the values of a given function vary in a specified region, this fact could be described using a membership function in the theory of fuzzy sets or using information on the distribution of the function values. In this general setting probability distributions or membership functions are not needed because only sets are considered. Optimal control problems with differential inclusions belong to this class of set optimization problems as well. Set optimization seems to have the potential to become a bridge between different areas in optimization. And it is a substantial extension of standard optimization theory. Set-valued analysis is the most important tool for such an advancement in continuous optimization. And conversely, the development of set-valued analysis receives important impulses from optimization.

In this fifth part we investigate vector optimization problems with a set-valued objective map as special set optimization problems. We consider unconstrained as well as constrained problems of this type. The presented theory is an extension of the second part of this book. The main topics are basic concepts, differentiability notions, subdifferentials and the presentation of optimality conditions including the generalized Lagrange multiplier rule.

Chapter 14

Basic Concepts and Results of Set Optimization

In this chapter we consider vector optimization problems with a set-valued objective map which has to be minimized or maximized. For these set optimization problems we present basic concepts and first results.

For the investigation of a vector optimization problem with a set-valued objective map we need the following standard assumption.

Assumption 14.1. Let X and Y be real linear spaces, let S be a nonempty subset of X, let Y be partially ordered by a convex cone $C \subset Y$ (then $\leq_C$ denotes the corresponding partial ordering), and let $F : S \to 2^Y$ be a set-valued map.

Throughout this fifth part we generally assume that the domain of a set-valued map equals its effective domain, i.e. for every element of the domain the image is a nonempty set.

Under Assumption 14.1 we consider the set optimization problem

$$\min_{x \in S} F(x). \qquad (14.1)$$

A minimizer of this problem is introduced as follows.

Definition 14.2. Let Assumption 14.1 be satisfied, and let $F(S)$

$:= \bigcup_{x \in S} F(x)$ denote the image set of F. Then a pair $(\bar{x}, \bar{y})$ with $\bar{x} \in S$ and $\bar{y} \in F(\bar{x})$ is called a *minimizer* of the problem (14.1), if $\bar{y}$ is a minimal element of the set $F(S)$, i.e.

$$(\{\bar{y}\} - C) \cap F(S) \subset \{\bar{y}\} + C.$$

Example 14.3. Let Assumption 14.1 be satisfied.

(a) Assume that $f, g : S \to Y$ are given vector functions. Then $F : S \to 2^Y$ with

$$F(x) := \{y \in Y \mid f(x) \leq_C y \leq_C g(x)\}$$

is a possible set-valued map which may be used as an objective. If $f = g$ and C is pointed, then at every $x \in S$ a corresponding image y is uniquely determined, otherwise the values of y vary in the order interval $[f(x), g(x)]$ (see Fig. 14.1).

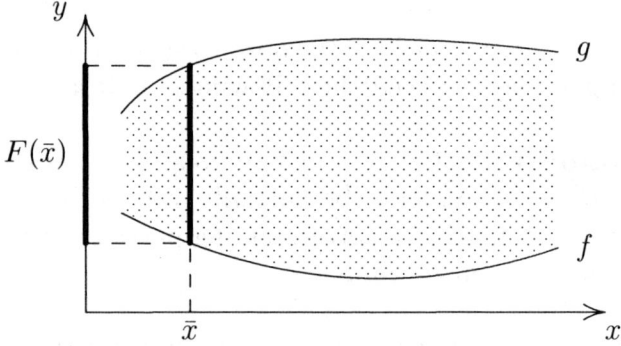

Figure 14.1: Illustration of the set-valued map F in Example 14.3,(a).

(b) One special case of the previous example is obtained if a vector function $\varphi : S \to Y$ is known and the y-values vary around $\varphi(x)$, i.e. we have

$$F(x) := \{y \in Y \mid \varphi(x) - \alpha \leq_C y \leq_C \varphi(x) + \beta\}$$

where $\alpha, \beta \in C$.

(c) Another special case appears if we admit relative errors around $\varphi(x)$. Again, we assume that a vector function $\varphi : S \to Y$ is known, and for an arbitrary $\varepsilon > 0$ we define

$$\begin{aligned} F(x) &:= \{y \in Y \mid \varphi(x) - \varepsilon\varphi(x) \leq_C y \leq_C \varphi(x) + \varepsilon\varphi(x)\} \\ &= \{y \in Y \mid (1-\varepsilon)\varphi(x) \leq_C y \leq_C (1+\varepsilon)\varphi(x)\}. \end{aligned}$$

The navigation of transportation robots leads to an industrial application of set optimization ([276]). The navigation and control of autonomous transportation robots is of particular importance. One uses ultrasonic sensors determining the smallest distance to an obstacle in the emission cone. Since the direction of the object cannot be identified in this cone, the location of the object is set-valued. Therefore, questions of navigation may lead to problems of set optimization.

It can be expected that the minimization of the set-valued map F in Example 14.3,(a) has something to do with the minimization of f. Therefore, under the assumptions given in Example 14.3,(a) we consider the single-valued vector optimization problem

$$\min_{x \in S} f(x). \tag{14.2}$$

Theorem 14.4. *Let Assumption 14.1 be satisfied, let C be pointed, let $f : S \to Y$ be a given function, and let $F : S \to 2^Y$ be defined as*

$$F(x) := \{y \in Y \mid f(x) \leq_C y\} \text{ for all } x \in S.$$

(a) *If $(\bar{x}, \bar{y})$ is a minimizer of the problem (14.1), then $\bar{y} = f(\bar{x})$ and $\bar{x}$ is a minimal solution of the problem (14.2).*

(b) *If $\bar{x}$ is a minimal solution of the problem (14.2), then $(\bar{x}, f(\bar{x}))$ is a minimizer of the problem (14.1).*

Proof.

(a) Since $(\bar{x}, \bar{y})$ is a minimizer of the problem (14.1) and C is pointed we have

$$(\{\bar{y}\} - C) \cap F(S) = \{\bar{y}\}. \tag{14.3}$$

Obviously it is

$$\bar{y} \in F(\bar{x}) \subset \bigcup_{x \in S} F(x) = F(S),$$

and, therefore, we conclude

$$(\{\bar{y}\} - C) \cap F(\bar{x}) = \{\bar{y}\}. \tag{14.4}$$

If we assume that $\bar{y} \neq f(\bar{x})$, we obtain because of $f(\bar{x}) \leq_C \bar{y}$ a contradiction to (14.4). Consequently, $\bar{y} = f(\bar{x})$, and by the equation (14.3) and $f(\bar{x}) \in f(S) \subset F(S)$ we get $(\{f(\bar{x})\} - C) \cap f(S) = \{f(\bar{x})\}$, i.e. $\bar{x}$ is a minimal solution of the problem (14.2).

(b) Assume that $(\bar{x}, f(\bar{x}))$ is not a minimizer of the problem (14.1). Then there is an $\tilde{x} \in S$ with

$$(\{f(\tilde{x})\} - C) \cap F(\tilde{x}) \neq \{f(\tilde{x})\}. \tag{14.5}$$

So we have for some $y \in F(\tilde{x})$

$$y \leq_C f(\tilde{x}), \ y \neq f(\tilde{x}) \ \text{(by (14.5))}$$

and

$$f(\tilde{x}) \leq_C y \ \text{(by the definition of } F(\tilde{x})).$$

Hence we get

$$f(\tilde{x}) \leq_C f(\bar{x}), \ f(\tilde{x}) \neq f(\bar{x}).$$

But then $\bar{x}$ is not a minimal solution of the problem (14.2).

□

The preceding theorem shows that in the special case discussed in Example 14.3,(a) the set optimization problem (14.1) is equivalent to the vector optimization problem (14.2) being simpler than the problem (14.1). Therefore, it is not necessary to work with such a general

Chapter 14. Basic Concepts and Results of Set Optimization   375

set-valued theory in this special case. Hence, a general set-valued theory makes only sense for set-valued maps whose lower boundary cannot be described by a function f as it is done in Example 14.3,(a).

Next, we mention another optimality notion for the set optimization problem (14.1). Whereas the concept of a minimizer considers only one point in the image $F(\bar{x})$, it seems to be more natural to use the whole image $F(\bar{x})$. Then, instead of a pair $(\bar{x}, \bar{y})$, one only considers the element $\bar{x}$ as it is known from standard optimization. The considered partial ordering has been independently introduced by Young [334], Nishnianidze [240] and presented by Kuroiwa [188] in a modified form. Therefore, this partial ordering is called KNY partial ordering.

Definition 14.5. Let Assumption 14.1 be satisfied. Then $\bar{x} \in S$ is called a *minimal solution* of the problem (14.1) if

$$F(x) \preccurlyeq F(\bar{x}), \quad x \in S \quad \implies \quad F(\bar{x}) \preccurlyeq F(x).$$

Here $\preccurlyeq$ denotes the KNY partial ordering for sets and is defined by

$$A \preccurlyeq B \quad :\iff \quad A \subset B - C \quad \text{and} \quad B \subset A + C$$

(A and B are arbitrary nonempty subsets of Y).

$A \preccurlyeq B$ means that for every $a \in A$ there is a $b \in B$ with $a \leq_C b$, and for every $b \in B$ there is an $a \in A$ with $a \leq_C b$. Since every element of both sets is considered, the concept of a minimal solution uses the whole set $F(\bar{x})$, and one does not consider only a special element $\bar{y}$ as in the definition of a minimizer. So this concept of a minimal solution seems to be more natural.

Fig. 14.2 illustrates the KNY partial ordering $\preccurlyeq$ introduced in Definition 14.5. The investigations in this book are based on the standard optimality concept presented in Definition 14.2.

Now we turn our attention to a C-convex set-valued map F.

Definition 14.6. Let Assumption 14.1 be satisfied, and, in addition, let S be convex. The set-valued map $F : S \to 2^Y$ is called C-*convex*, if for all $x_1, x_2 \in S$ and $\lambda \in [0, 1]$

$$\lambda F(x_1) + (1 - \lambda) F(x_2) \subset F(\lambda x_1 + (1 - \lambda) x_2) + C.$$

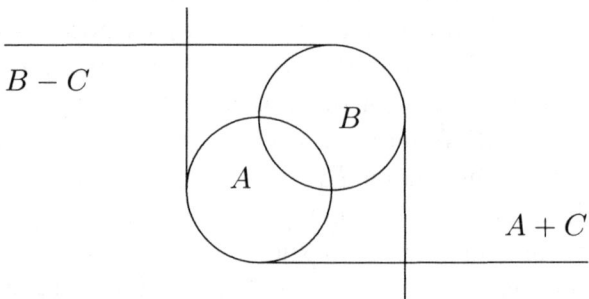

Figure 14.2: Illustration of the KNY partial ordering $\preccurlyeq$ (here we have $A \preccurlyeq B$).

A known result from convex analysis says that C-convexity of a map is characterized by the convexity of its epigraph (compare Thm. 2.6). We present its definition and then show this characterization.

Definition 14.7. Let Assumption 14.1 be satisfied, and, in addition, let S be convex. The set

$$\mathrm{epi}(F) := \{(x,y) \in X \times Y \mid x \in S,\ y \in F(x) + C\}$$

is called the *epigraph* of F.

Lemma 14.8. *Let Assumption 14.1 be satisfied, and, in addition, let S be convex. Then F is C-convex if and only if* $\mathrm{epi}(F)$ *is a convex set.*

Proof.

(a) Take arbitrary elements $(x_1, y_1), (x_2, y_2) \in \mathrm{epi}(F)$ and $\lambda \in [0,1]$. Because of the convexity of S we have

$$\lambda x_1 + (1-\lambda) x_2 \in S, \tag{14.6}$$

and since F is C-convex, we obtain

$$\lambda y_1 + (1-\lambda) y_2 \ \in\ \lambda (F(x_1) + C) + (1-\lambda)(F(x_2) + C)$$

$$\begin{aligned} &= \lambda F(x_1) + (1-\lambda)F(x_2) + C \\ &\subset F(\lambda x_1 + (1-\lambda)x_2) + C. \end{aligned} \qquad (14.7)$$

The conditions (14.6) and (14.7) imply

$$\lambda(x_1, y_1) + (1-\lambda)(x_2, y_2) \in \text{epi}(F).$$

Consequently, $\text{epi}(F)$ is a convex set.

(b) Now assume that $\text{epi}(F)$ is a convex set. Let $x_1, x_2 \in S$, $y_1 \in F(x_1)$, $y_2 \in F(x_2)$ and $\lambda \in [0,1]$ be arbitrarily given. Because of the convexity of $\text{epi}(F)$ we obtain

$$\lambda(x_1, y_1) + (1-\lambda)(x_2, y_2) \in \text{epi}(F)$$

implying

$$\lambda y_1 + (1-\lambda)y_2 \in F(\lambda_1 x_1 + (1-\lambda)x_2) + C.$$

Hence, F is C-convex.

$\square$

Notes

Set optimization problems have been investigated by many authors, for instance, there are papers on optimality conditions (e.g., [30], [242], [34], [37], [67], [214], [216], [155], [57], [207], [110]), duality theory (e.g., [254], [66], [215], [282]) and related topics (e.g., [340], [176], [188], [86], [113], [306]). For recent investigations we refer to the special issue [58]. For the definition of a minimizer we also refer to [213], [214], [155] and [150].

The KNY partial ordering has been originally defined by Young [334] and Nishnianidze [240]. Nishnianidze used it for the analysis of fixed points of set-valued maps. A modification independently given by Kuroiwa [188] has been used for the definition of minimal solutions. He has also presented different types of set partial orderings which may be used for the definition of minimal solutions. The KNY partial ordering opens a new and wide field of research. Here the investigation of the set space 2^Y plays an important role. Optimality notions and existence results using the KNY partial ordering can be found in [304], [189] and [190]. Duality investigations are carried out in [191]. The variational principle of Ekeland has also been investigated in this setting [305]. The KNY partial ordering turns out to be promising in set optimization.

Notice that the definition of C-convex set-valued maps and convex set-valued maps are different. For the definition of convex set-valued maps see, for instance, [11, p. 56–57].

Chapter 15

Contingent Epiderivatives

For the formulation of optimality conditions one needs an appropriate differentiability concept for set-valued maps. In this chapter we present the notion of contingent epiderivatives and generalized contingent epiderivatives. We show properties of these contingent epiderivatives and discuss the special case of real-valued functions.

15.1 Contingent Derivatives and Contingent Epiderivatives

The concept of contingent derivatives plays an important role in set-valued analysis. But it turns out that the contingent *epi*derivative is a better tool for the formulation of necessary and sufficient optimality conditions. Therefore, we investigate properties of this type of a derivative in detail.

Definition 15.1. Let $(X, \|\cdot\|_X)$ and $(Y, \|\cdot\|_Y)$ be real normed spaces, and let $F : X \to 2^Y$ be a set-valued map.

(a) The set
$$\operatorname{graph}(F) := \{(x, y) \in X \times Y \mid y \in F(x)\}$$
is called the *graph* of the map F.

(b) Let a pair $(\bar{x}, \bar{y}) \in \operatorname{graph}(F)$ be given. A set-valued map $D_c F(\bar{x}, \bar{y}) : X \to 2^Y$ whose graph equals the contingent cone to the graph of F at $(\bar{x}, \bar{y})$, i.e.

$$\operatorname{graph}(D_c F(\bar{x}, \bar{y})) = T(\operatorname{graph}(F), (\bar{x}, \bar{y})),$$

is called *contingent derivative* of F at $(\bar{x}, \bar{y})$.

The importance of this notion of a derivative is based on the fact that it extends the Fréchet differentiability concept very naturally to the set-valued case.

Remark 15.2. Let $(X, \|\cdot\|_X)$ and $(Y, \|\cdot\|_Y)$ be real normed spaces, let $f : X \to Y$ be a single-valued map assumed to be Fréchet differentiable at some $\bar{x} \in X$ with a surjective Fréchet derivative $f'(\bar{x})$. Then we conclude with Lyusternik's Theorem 3.49 which implies in essential that the contingent cone of an equality constraint equals the linearized cone:

$$\begin{aligned} &T(\operatorname{graph}(f), (\bar{x}, f(\bar{x}))) \\ &= T(\{(x,y) \in X \times Y \mid f(x) - y = 0\}, (\bar{x}, f(\bar{x}))\}) \\ &= \{(x,y) \in X \times Y \mid f'(\bar{x})(x) - y = 0\} \\ &= \operatorname{graph}(f'(\bar{x})). \end{aligned}$$

Hence, the Fréchet derivative $f'(\bar{x})$ coincides with the contingent derivative $D_c f(\bar{x}, f(\bar{x}))$ (see Fig. 15.1).

This remark shows that the concept of the contingent derivative is a quite natural extension of tangents. It is obvious that contingent derivatives have a rich structure and play a central role in set-valued analysis. And, therefore, this concept has also been used in set optimization. But it turns out that necessary optimality conditions and sufficient optimality conditions do not coincide under standard assumptions. This shows that contingent derivatives are not completely the right tool for the formulation of optimality conditions in set optimization.

15.1. Contingent Derivatives and Contingent Epiderivatives

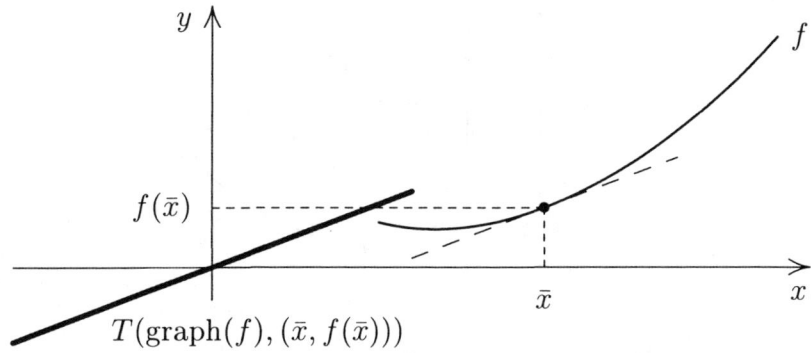

Figure 15.1: Illustration of the result in Remark 15.2.

In order to get optimality conditions generalizing the known classical conditions we come to another differentiability notion, the so-called contingent epiderivative.

Definition 15.3. Let $(X, \|\cdot\|_X)$ and $(Y, \|\cdot\|_Y)$ be real normed spaces, let Y be partially ordered by a convex cone $C \subset Y$, let S be a nonempty subset of X, and let $F : S \to 2^Y$ be a set-valued map. Let a pair $(\bar{x}, \bar{y}) \in X \times Y$ with $\bar{x} \in S$ and $\bar{y} \in F(\bar{x})$ be given. A single-valued map $DF(\bar{x}, \bar{y}) : X \to Y$ whose epigraph equals the contingent cone to the epigraph of F at $(\bar{x}, \bar{y})$, i.e.

$$\text{epi}(DF(\bar{x}, \bar{y})) = T(\text{epi}(F), (\bar{x}, \bar{y})),$$

is called *contingent epiderivative* of F at $(\bar{x}, \bar{y})$.

The essential differences between the definitions of the contingent derivative and the contingent epiderivative are that the graph is now replaced by the epigraph and the derivative is now single-valued. For an illustration of this notion see Fig. 15.2.

Next, we consider again the set-valued map in Example 3,(a). The contingent epiderivative of this map can be given with the aid of the contingent epiderivative of f.

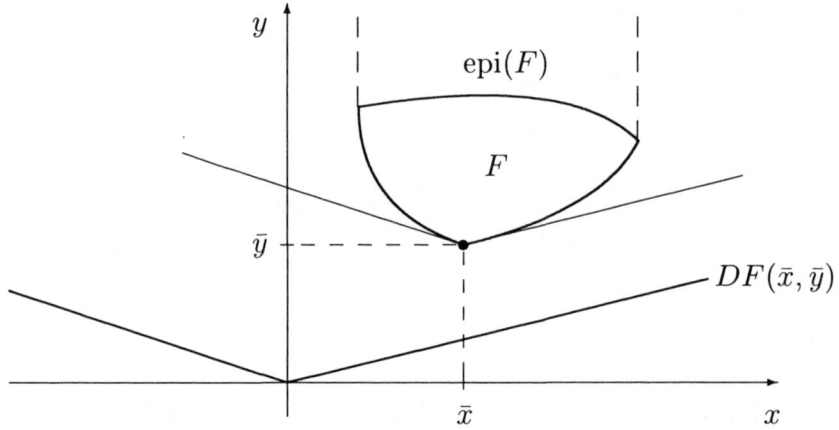

Figure 15.2: Illustration of the contingent epiderivative for $C = \mathbb{R}_+$.

Lemma 15.4. *Let Assumption 14.1 be satisfied, let $F : S \to 2^Y$ be given as*

$$F(x) := \{y \in Y \mid f(x) \leq_C y \leq_C g(x)\}$$

with $f, g : S \to Y$, and let $\bar{x} \in S$ be arbitrarily given. If the contingent epiderivative $DF(\bar{x}, f(\bar{x}))$ exists, then

$$DF(\bar{x}, f(\bar{x})) = Df(\bar{x}, f(\bar{x})).$$

Proof. Because of the definition of F we have

$$\text{epi}(F) = \{(x, y) \in X \times Y \mid x \in S, \ f(x) \leq_C y\} = \text{epi}(f),$$

and, therefore, we conclude

$$\begin{aligned}
\text{epi}(DF(\bar{x}, \bar{y})) &= T(\text{epi}(F), (\bar{x}, \bar{y})) \\
&= T(\text{epi}(f), (\bar{x}, \bar{y})) \\
&= \text{epi}(Df(\bar{x}, \bar{y})).
\end{aligned}$$

This leads to the assertion. □

15.2 Properties of Contingent Epiderivatives

For the presentation of various properties of contingent epiderivatives we use the following standard assumption in this section.

Assumption 15.5. Let $(X, \|\cdot\|_X)$ and $(Y, \|\cdot\|_Y)$ be a real normed spaces, let S be a nonempty subset of X, let Y be partially ordered by a convex cone $C \subset Y$, let $F : S \to 2^Y$ be a set-valued map, and let $\bar{x} \in S$ and $\bar{y} \in F(\bar{x})$ be given elements.

Our first result is an existence theorem for contingent epiderivatives in the special case $Y = \mathbb{R}$.

Theorem 15.6. *Let Assumption 15.5 be satisfied with $Y = \mathbb{R}$, and assume that there are functions $f, g : X \to \mathbb{R}$ with $\mathrm{epi}(f) \supset T(\mathrm{epi}(F), (\bar{x}, \bar{y})) \supset \mathrm{epi}(g)$. Then the contingent epiderivative $DF(\bar{x}, \bar{y})$ is given as*

$$DF(\bar{x}, \bar{y})(x) = \min\{y \in \mathbb{R} \mid (x, y) \in T(\mathrm{epi}(F), (\bar{x}, \bar{y}))\} \quad \textit{for all } x \in X. \tag{15.1}$$

Proof. We define the functional $DF(\bar{x}, \bar{y}) : X \to \mathbb{R} \cup \{-\infty\}$ by

$$DF(\bar{x}, \bar{y})(x) = \inf\{y \in \mathbb{R} \mid (x, y) \in T(\mathrm{epi}(F), (\bar{x}, \bar{y}))\} \quad \text{for all } x \in X.$$

Since $\mathrm{epi}(g) \subset T(\mathrm{epi}(F), (\bar{x}, \bar{y}))$, for every $x \in X$ there is at least one $y \in \mathbb{R}$ with $(x, y) \in T(\mathrm{epi}(F), (\bar{x}, \bar{y}))$. So, $DF(\bar{x}, \bar{y})$ is well-defined on X. Now we show that it is the contingent epiderivative. For this proof take an arbitrary $x \in X$. Then there is an infimal sequence $(y_n)_{n \in \mathbb{N}}$ converging to $DF(\bar{x}, \bar{y})$ with $(x, y_n) \in T(\mathrm{epi}(F), (\bar{x}, \bar{y}))$. Since the contingent cone is always closed in a normed space, we conclude

$$(x, DF(\bar{x}, \bar{y})(x)) \in T(\mathrm{epi}(F), (\bar{x}, \bar{y})).$$

By assumption, $-\infty < f(x) \le DF(\bar{x}, \bar{y})(x)$, and hence the equation (15.1) is satisfied. It follows from this equation that

$$\mathrm{epi}(DF(\bar{x}, \bar{y})) = T(\mathrm{epi}(F), (\bar{x}, \bar{y})).$$

Hence, $DF(\bar{x}, \bar{y})$ is the contingent epiderivative of F at $(\bar{x}, \bar{y})$. □

Corollary 15.7. *Let Assumption 15.5 be satisfied with $Y = \mathbb{R}$ and $S = X$, and, in addition, let $F : X \to \mathbb{R}$ be a single-valued and convex function being continuous at $\bar{x}$. Then the contingent epiderivative $DF(\bar{x}, \bar{y})$ is given by the equation (15.1).*

Proof. Since F is continuous at $\bar{x}$ and convex, its subdifferential $\partial F(\bar{x})$ (e.g., see [149]) is nonempty. Because of the convexity of F its epigraph is convex as well, and, therefore, the contingent cone $T(\text{epi}(F), (\bar{x}, \bar{y}))$ is convex (Theorem 3.47) and we obtain

$$\text{epi}(f) \supset T(\text{epi}(F), (\bar{x}, \bar{y})) + \{(\bar{x}, \bar{y})\} \supset \text{epi}(F)$$

with

$$f(x) := l(x - \bar{x}) + \bar{y} \text{ for all } x \in S$$

for a subgradient $l \in \partial F(\bar{x})$. Consequently, the assumption of Theorem 15.6 is fulfilled, and Theorem 15.6 leads to the assertion. □

The next result shows that contingent epiderivatives are unique, if they exist.

Theorem 15.8. *Let Assumption 15.5 be satisfied. If the contingent epiderivative $DF(\bar{x}, \bar{y})$ exists, then it is unique.*

Proof. Assume that $\bar{D}F(\bar{x}, \bar{y}) \neq DF(\bar{x}, \bar{y})$ is a contingent epiderivative as well. Then there is at least one $x \in S$ with

$$\bar{D}F(\bar{x}, \bar{y})(x) \neq DF(\bar{x}, \bar{y})(x)$$

and consequently

$$\text{epi}(\bar{D}F(\bar{x}, \bar{y})) \neq \text{epi}(DF(\bar{x}, \bar{y})) = T(\text{epi}(F), (\bar{x}, \bar{y})).$$

But this contradicts the assumption that $\bar{D}F(\bar{x}, \bar{y})$ is also a contingent epiderivative. □

15.2. Properties of Contingent Epiderivatives

If $(\bar{x}, \bar{y})$ belongs to the interior of $\mathrm{epi}(F)$, the contingent cone $T(\mathrm{epi}(F), (\bar{x}, \bar{y}))$ equals the product space $X \times Y$ and in this case the contingent epiderivative $DF(\bar{x}, \bar{y})$ does not exist.

The next theorem gives a relationship between the contingent derivative and the contingent epiderivative for C-convex maps.

Theorem 15.9. *Let Assumption 15.5 be satisfied, and, in addition, let $S = X$, let C be closed, and let F be C-convex. If the contingent derivative $D_c F(\bar{x}, \bar{y})$ and the contingent epiderivative $DF(\bar{x}, \bar{y})$ exist, then*
$$\mathrm{epi}(D_c F(\bar{x}, \bar{y})) \subset \mathrm{epi}(DF(\bar{x}, \bar{y})).$$

Proof. We have with Lemma 15.8

$$\begin{aligned}
\mathrm{epi}(DF(\bar{x}, \bar{y})) &= T(\mathrm{epi}(F), (\bar{x}, \bar{y})) \\
&= \mathrm{cl}(\mathrm{cone}(\mathrm{epi}(F) - \{(\bar{x}, \bar{y})\})) \\
&= \mathrm{cl}(\mathrm{cone}(\mathrm{graph}(F) - \{(\bar{x}, \bar{y})\} + (\{0_X\} \times C))) \\
&\supset \mathrm{cl}(\mathrm{cone}(\mathrm{graph}(F) - \{(\bar{x}, \bar{y})\})) + \mathrm{cl}(\{0_X\} \times C) \\
&= \mathrm{cl}(\mathrm{cone}(\mathrm{graph}(F) - \{(\bar{x}, \bar{y})\})) + (\{0_X\} \times C) \\
&\supset T(\mathrm{graph}(F), (\bar{x}, \bar{y})) + (\{0_X\} \times C) \\
&= \mathrm{epi}(D_c F(\bar{x}, \bar{y}))
\end{aligned}$$

(where "cone" denotes the cone generated by a set (Definition 1.15)). □

Now we are able to present a special property of contingent epiderivatives in the C-convex case: they are sublinear, if they exist. First, recall the definition of sublinearity in this abstract setting.

Definition 15.10. Let X be a real linear space, and let Y be a real linear space partially ordered by a convex cone $C \subset Y$. A map $f : X \to Y$ is called *sublinear* if

(a) $f(\alpha x) = \alpha f(x)$ for all $\alpha \geq 0$ and all $x \in X$ (positive homogenity),

(b) $f(x_1 + x_2) \leq_C f(x_1) + f(x_2)$ for all $x_1, x_2 \in X$ (subadditivity).

Theorem 15.11. *Let Assumption 15.5 be satisfied, and, in addition, let C be pointed, let S be convex, and let F be C-convex. If the contingent epiderivative $DF(\bar{x}, \bar{y})$ exists, then it is sublinear.*

Proof. Since F is C-convex, by Lemma 15.8 epi(F) is a convex set. Hence the contingent cone $T(\text{epi}(F), (\bar{x}, \bar{y}))$ is convex and, therefore, the epigraph of $DF(\bar{x}, \bar{y})$ is a convex cone. Now take any $\alpha > 0$ and any $x \in X$. Since epi($DF(\bar{x}, \bar{y})$) is a cone and $(x, DF(\bar{x}, \bar{y})(x)) \in$ epi($DF(\bar{x}, \bar{y})$), we get $(\alpha x, \alpha DF(\bar{x}, \bar{y})(x)) \in$ epi($DF(\bar{x}, \bar{y})$) implying

$$\alpha DF(\bar{x}, \bar{y})(x) \in \{DF(\bar{x}, \bar{y})(\alpha x)\} + C. \tag{15.2}$$

But with $(\alpha x, DF(\bar{x}, \bar{y})(\alpha x)) \in$ epi($DF(\bar{x}, \bar{y})$) we also obtain $(x, \frac{1}{\alpha} DF(\bar{x}, \bar{y})(\alpha x)) \in$ epi($DF(\bar{x}, \bar{y})$) resulting in

$$\frac{1}{\alpha} DF(\bar{x}, \bar{y})(\alpha x) \in \{DF(\bar{x}, \bar{y})(x)\} + C$$

or

$$\alpha DF(\bar{x}, \bar{y})(x) \in \{DF(\bar{x}, \bar{y})(\alpha x)\} - C. \tag{15.3}$$

Since C is pointed, we conclude from the conditions (15.2) and (15.3)

$$\alpha DF(\bar{x}, \bar{y})(x) = DF(\bar{x}, \bar{y})(\alpha x). \tag{15.4}$$

Moreover, from the condition (15.2) we obtain for $\alpha = 2$ and $x = 0_X$

$$2 DF(\bar{x}, \bar{y})(0_X) \in \{DF(\bar{x}, \bar{y})(0_X)\} + C$$

implying

$$DF(\bar{x}, \bar{y})(0_X) \in C \tag{15.5}$$

Since $(0_X, 0_Y) \in$ epi($DF(\bar{x}, \bar{y})$), we also have

$$DF(\bar{x}, \bar{y})(0_X) \in -C. \tag{15.6}$$

If we notice that C is pointed, we conclude $DF(\bar{x}, \bar{y})(0_X) = 0$, i.e., the equation (15.4) holds for $\alpha = 0$ as well. Hence, the contingent epiderivative is positively homogeneous. Next we show the subadditivity of $DF(\bar{x},\bar{y})$. Take arbitrary $x_1, x_2 \in X$. Since $(x_1, DF(\bar{x},\bar{y})(x_1)) \in \mathrm{epi}(DF(\bar{x},\bar{y}))$, $(x_2, DF(\bar{x},\bar{y})(x_2)) \in \mathrm{epi}(DF(\bar{x},\bar{y}))$ and $\mathrm{epi}(DF(\bar{x},\bar{y}))$ is convex, we have

$$(\frac{1}{2}x_1 + \frac{1}{2}x_2, \frac{1}{2}DF(\bar{x},\bar{y})(x_1) + \frac{1}{2}DF(\bar{x},\bar{y})(x_2)) \in \mathrm{epi}(DF(\bar{x},\bar{y}))$$

which implies

$$\frac{1}{2}(DF(\bar{x},\bar{y})(x_1) + DF(\bar{x},\bar{y})(x_2)) \in \underbrace{\{DF(\bar{x},\bar{y})(\frac{1}{2}(x_1+x_2))\}}_{=\frac{1}{2}DF(\bar{x},\bar{y})(x_1+x_2)} + C$$

or

$$DF(\bar{x},\bar{y})(x_1) + DF(\bar{x},\bar{y})(x_2) \in \{DF(\bar{x},\bar{y})(x_1+x_2)\} + C.$$

Hence, the contingent epiderivative is subadditive. □

Notice that the positive homogenity of $DF(\bar{x},\bar{y})$ can be proved without the additional convexity assumptions. For this proof we only need that C is pointed.

This theorem shows that contingent epiderivatives have a rich mathematical structure in the C-convex case. Using the generalized Hahn-Banach Theorem 3.13 one gets a linear map as lower bound of the sublinear contingent epiderivative, and, therefore, generalized subgradients can be introduced in the same way as it is done in convex analysis. These subdifferentials are investigated in the Sections 16.1 and 16.2.

15.3 Contingent Epiderivatives of Real-Valued Functions

In this section we investigate the relationship between the contingent epiderivative and the directional derivative in the special case that F is a single-valued function.

Our special assumption reads as follows:

Assumption 15.12. Let $(X, \|\cdot\|_X)$ be a real normed space, let $F : X \to \mathbb{R}$ be a single-valued function, and let $\bar{x} \in X$ be given.

Theorem 15.13. *Let Assumption 15.12 be satisfied. If the contingent epiderivative $DF(\bar{x}, F(\bar{x}))$ exists, then it is lower semicontinuous.*

Proof. Since the contingent cone is always closed in a normed space, the epigraph of the contingent epiderivative is closed as well, and we conclude with a standard result that $DF(\bar{x}, F(\bar{x}))$ is lower semicontinuous. $\square$

In order to give a relationship between the directional derivative and the contingent epiderivative we need the following lemma.

Lemma 15.14. *Let Assumption 15.12 be satisfied. If F is continuous at $\bar{x}$ and convex, then*

$$DF(\bar{x}, F(\bar{x}))(h) \geq F'(\bar{x})(h) \quad \text{for all } h \in X \qquad (15.7)$$

(where $F'(\bar{x})(h)$ denotes the directional derivative of F at $\bar{x}$ in the direction h).

Proof. Notice that $DF(\bar{x}, F(\bar{x}))$ exists by Corollary 15.7. Because of the convexity of F we obtain

$$\begin{aligned}
&\mathrm{epi}(DF(\bar{x}, F(\bar{x}))) \\
&= T(\mathrm{epi}(F), (\bar{x}, F(\bar{x}))) \\
&\subset T(\mathrm{epi}\{F(\bar{x}) + l(x - \bar{x}) | x \in X\}, (\bar{x}, F(\bar{x}))) \\
&= \mathrm{epi}\{l(h) | h \in X\} \text{ for all subgradients } l \in \partial F(\bar{x})
\end{aligned}$$

(notice that the subdifferential $\partial F(\bar{x})$ is nonempty (see [149])). So we conclude

$$DF(\bar{x}, F(\bar{x}))(h) \geq \max_{l \in \partial F(\bar{x})} l(h) = F'(\bar{x})(h) \quad \text{for all } h \in X.$$

$\square$

15.3. Contingent Epiderivatives of Real-Valued Functions

The next theorem shows that the inequality (15.7) is already an equality, or in other words, contingent epiderivative and directional derivative coincide in this special case.

Theorem 15.15. *Let Assumption 15.12 be satisfied. If F is continuous at $\bar{x}$ and convex, then the contingent epiderivative equals the directional derivative.*

Proof. By Corollary 15.7 $DF(\bar{x}, F(\bar{x}))$ exists. With Lemma 15.14 we have for a fixed $h \in X$

$$DF(\bar{x}, F(\bar{x}))(h) \geq l(h) \text{ for all } l \in \partial F(\bar{x}). \tag{15.8}$$

Next we define the set

$$\begin{aligned} T &:= \{(\bar{x} + \lambda h, F(\bar{x}) + DF(\bar{x}, F(\bar{x}))(\lambda h) \mid \lambda \geq 0\} \\ &= \{(\bar{x} + \lambda h, F(\bar{x}) + \lambda DF(\bar{x}, F(\bar{x}))(h) \mid \lambda \geq 0\} \end{aligned}$$
(by Theorem 15.11).

Since F is continuous at $\bar{x}$, epi(F) has a nonempty interior. Then we conclude $T \cap \text{int}(\text{epi}(F)) = \emptyset$. By Eidelheit's separation theorem (Theorem 3.16) there are a continuous linear functional $\bar{l}$ on X and real numbers β and γ with the property $(\bar{l}, \beta) \neq (0_{X^*}, 0)$ and

$$\bar{l}(x) + \beta \alpha \leq \gamma \leq \bar{l}(\bar{x} + \lambda h) + \beta(F(\bar{x}) + \lambda DF(\bar{x}, F(\bar{x}))(h))$$
$$\text{for all } (x, \alpha) \in \text{epi}(F) \text{ and } \lambda \geq 0. \tag{15.9}$$

With standard arguments (see [149, p. 57]) we obtain $-\frac{1}{\beta}\bar{l} \in \partial F(\bar{x})$, and with $x = \bar{x}$, $\alpha = F(\bar{x})$, $\lambda = 1$ we conclude from (15.9)

$$DF(\bar{x}, F(\bar{x}))(h) \leq -\frac{1}{\beta}\bar{l}(h). \tag{15.10}$$

The inequalities (15.8) and (15.10) imply

$$DF(\bar{x}, F(\bar{x}))(h) = \max_{l \in \partial F(\bar{x})} l(h) = F'(\bar{x})(h).$$

□

Example 15.16. Consider the set-valued map $F : X \to 2^{\mathbb{R}}$ (where $(X, \|\cdot\|_X)$ is a real normed space) with
$$F(x) = \{y \in \mathbb{R} \mid y \geq \|x\|\} \text{ for all } x \in X.$$
If we define $f : X \to \mathbb{R}$ with
$$f(x) = \|x\| \text{ for all } x \in X,$$
we have $\text{graph}(F) = \text{epi}(f)$. Consequently, we obtain with Theorem 15.15 for an arbitrary $\bar{x} \in X \setminus \{0_X\}$ and $\bar{y} := \|\bar{x}\|$
$$\begin{aligned} DF(\bar{x},\bar{y})(h) &= Df(\bar{x},f(\bar{x}))(h) \\ &= f'(\bar{x})(h) \\ &= \max\{l(h) \mid l \in \partial f(\bar{x})\} \\ &= \max\{l(h) \mid l \in X^*,\ l(\bar{x}) = \|\bar{x}\| \text{ and } \|l\|_{X^*} = 1\} \\ &\text{for all } h \in X \end{aligned}$$
(see Example 2.23).

Finally we consider Example 14.3, (a) for a set-valued map F in the special case of $Y = \mathbb{R}$

Corollary 15.17. *Let $(X, \|\cdot\|_X)$ be a real normed space, let $F : X \to 2^{\mathbb{R}}$ be given as*
$$F(x) := \{y \in \mathbb{R} \mid f(x) \leq y \leq g(x)\} \text{ for all } x \in X$$
with $f, g : X \to \mathbb{R}$, let $\bar{x} \in X$ be arbitrarily given, and let f be continuous at $\bar{x}$ and convex. Then the contingent epiderivative of F at $(\bar{x}, f(\bar{x}))$ exists and equals the directional derivative of f at $\bar{x}$.

Proof. It is obvious that $\text{epi}(F) = \text{epi}(f)$. Since f is a convex functional, its epigraph is convex and we get with Theorem 3.43
$$\text{epi}(f) = \text{epi}(F) \subset T(\text{epi}(F),(\bar{x},f(\bar{x}))) + \{(\bar{x},f(\bar{x}))\}.$$
f is continuous at $\bar{x}$ and convex and, therefore, there is a subgradient l of f at $\bar{x}$ (see [149]) with
$$\text{epi}(h) \supset T(\text{epi}(F),(\bar{x},f(\bar{x}))) + \{(\bar{x},f(\bar{x}))\}$$

for
$$h(x) := l(x - \bar{x}) + f(\bar{x}) \text{ for all } x \in X.$$
Hence, the assumptions of Theorem 15.6 are fulfilled and we conclude that the contingent epiderivative $DF(\bar{x}, f(\bar{x}))$ exists. By Lemma 15.4 $DF(\bar{x}, f(\bar{x}))$ equals $Df(\bar{x}, f(\bar{x}))$ which, by Theorem 15.15, equals the directional derivative of f at $\bar{x}$. □

15.4 Generalized Contingent Epiderivatives

In this section we use the following standard assumption:

Assumption 15.18. In addition to Assumption 15.5 let C be pointed.

The following concept extends a characterization of a contingent epiderivative given in Theorem 15.6 for a special case.

Definition 15.19. Let Assumption 15.18 be satisfied. A set-valued map $D_g F(\bar{x}, \bar{y}) : S - \{\bar{x}\} \to 2^Y$ is called *generalized contingent epiderivative* of F at $(\bar{x}, \bar{y})$ if

$$D_g F(\bar{x}, \bar{y})(x) = \text{Min}\,\{y \in Y \mid (x, y) \in T(\text{epi}(F), (\bar{x}, \bar{y}))\}$$
$$\text{for all } x \in S - \{\bar{x\}}$$

where Min $\{\ldots\}$ denotes the set of all minimal elements of the considered set.

Notice that for some $x \in S - \{\bar{x}\}$ the set $\{y \in Y \mid (x, y) \in T(\text{epi}(F), (\bar{x}, \bar{y}))\}$ may be empty. In this case we set $D_g F(\bar{x}, \bar{y})(x) = \emptyset$.

Next we show under appropriate assumptions that the generalized contingent epiderivative is a strictly positive homogeneous and subadditive map in the case of $S = X$.

Definition 15.20. Under the assumptions in Definition 15.10 a set-valued map $F : X \to 2^Y$ is called

(a) *strictly positive homogeneous* if
$$F(\alpha x) = \alpha F(x) \text{ for all } \alpha > 0 \text{ and all } x \in X,$$

(b) *subadditive* if
$$F(x_1) + F(x_2) \subset F(x_1 + x_2) + C \text{ for all } x_1, x_2 \in X.$$

If the properties under (a) with $\alpha \geq 0$ and (b) hold, then F is called *sublinear*.

Theorem 15.21. *Let Assumption 15.18 be satisfied, let $S = X$, and let for all $x \in X$, $D_g F(\bar{x}, \bar{y})(x) \neq \emptyset$. Then $D_g F(\bar{x}, \bar{y})$ is strictly positive homogeneous. Moreover, if F is C-convex and the set*
$$G(x) := \{y \in Y \mid (x, y) \in T(\text{epi}(F), (\bar{x}, \bar{y}))\} \tag{15.11}$$

fulfills the domination property for all $x \in X$ (i.e. $G(x) \subset \text{Min}\, G(x) + C$), then $D_g F(\bar{x}, \bar{y})$ is subadditive.

Proof. We take any $\alpha > 0$ and $x \in X$. Then we obtain

$$\begin{aligned}
\frac{1}{\alpha} D_g F(\bar{x}, \bar{y})(\alpha x) &= \text{Min}\left\{\frac{1}{\alpha} y \in Y \,\Big|\, (\alpha x, y) \in T(\text{epi}(F), (\bar{x}, \bar{y}))\right\} \\
&= \text{Min}\{u \in Y \mid (\alpha x, \alpha u) \in T(\text{epi}(F), (\bar{x}, \bar{y}))\} \\
&= \text{Min}\{u \in Y \mid (x, u) \in T(\text{epi}(F), (\bar{x}, \bar{y}))\} \\
&= D_g F(\bar{x}, \bar{y})(x).
\end{aligned}$$

Thus
$$D_g F(\bar{x}, \bar{y})(\alpha x) = \alpha D_g F(\bar{x}, \bar{y})(x),$$

and $D_g F(\bar{x}, \bar{y})$ is strictly positive homogeneous.

Next for $x_1, x_2 \in X$, $y_1 \in D_g F(\bar{x}, \bar{y})(x_1)$, $y_2 \in D_g F(\bar{x}, \bar{y})(x_2)$ we have $(x_1, y_1) \in T(\text{epi}(F), (\bar{x}, \bar{y}))$ and $(x_2, y_2) \in T(\text{epi}(F), (\bar{x}, \bar{y}))$. Since F

15.4. Generalized Contingent Epiderivatives

is C-convex, epi(F) is convex and then $T(\text{epi}(F),(\bar{x},\bar{y}))$ is a convex cone. Thus
$$(x_1 + x_2, y_1 + y_2) \in T(\text{epi}(F),(\bar{x},\bar{y}))$$
implying
$$D_g F(\bar{x},\bar{y})(x_1) + D_g F(\bar{x},\bar{y})(x_2) \subset G(x_1 + x_2)$$
with $G(x_1 + x_2)$ given by (15.11). By the domination property we have
$$G(x_1 + x_2) \subset \text{Min}\, G(x_1 + x_2) + C = D_g F(\bar{x},\bar{y})(x_1 + x_2) + C$$
resulting in
$$D_g F(\bar{x},\bar{y})(x_1) + D_g F(\bar{x},\bar{y})(x_2) \subset D_g F(\bar{x},\bar{y})(x_1 + x_2) + C.$$
□

Remark 15.22. Let Assumption 15.18 be satisfied, and let $F : X \to \mathbb{R}$ be a real convex functional. Then the generalized contingent epiderivative $D_g F$ is given by
$$D_g F(\bar{x},\bar{y})(x) = \text{Min}\, \{y \in \mathbb{R} \mid (x,y) \in T(\text{epi}(F),(\bar{x},\bar{y}))\} \text{ for all } x \in X$$
and $D_g F$ is single-valued. Under the assumptions of Theorem 15.21 $D_g F(\bar{x},\bar{y})$ is sublinear.

Now we give an existence theorem of $D_g F$.

Theorem 15.23. *Let Assumption 15.18 be satisfied, and let C be closed and Daniell. Let for every $x \in S$ the set $G(x)$ given by (15.11) have a lower bound. Then for all $x \in S$ $D_g F(\bar{x},\bar{y})(x)$ exists.*

Proof. Since the contingent cone is always closed in a normed space, then for every $x \in S$ $G(x)$ has a lower bound and is closed. From the existence theorem of minimal elements (see Theorem 6.3, (a)) Min $G(x)$ is nonempty, i.e. $D_g F(\bar{x},\bar{y})$ is well defined. □

Now we consider the relation between the generalized contingent epiderivative and the contingent epiderivative.

Theorem 15.24. *Let Assumption 15.18 be satisfied, let $S = X$, and let the domination property hold. If the contingent epiderivative $DF(\bar{x}, \bar{y})$ exists and the set $G(x)$ given by (15.11) fulfills the domination property for all $x \in X$, then*
$$\mathrm{epi}(DF(\bar{x}, \bar{y})) = \mathrm{epi}(D_g F(\bar{x}, \bar{y})).$$

Proof. By the definition of $D_g F$ we have
$$\begin{aligned} \mathrm{epi}(D_g F(\bar{x}, \bar{y})) &\subset T(\mathrm{epi}(F), (\bar{x}, \bar{y})) + \{0_X\} \times C \\ &= \mathrm{epi}(DF(\bar{x}, \bar{y})) + \{0_X\} \times C \\ &= \mathrm{epi}(DF(\bar{x}, \bar{y})) \end{aligned}$$

resulting in
$$\mathrm{epi}(D_g F(\bar{x}, \bar{y})) \subset \mathrm{epi}(DF(\bar{x}, \bar{y})).$$

Conversely, we suppose that $(x, \tilde{y}) \in \mathrm{epi}(DF(\bar{x}, \bar{y}))$ and $(x, \tilde{y}) \notin \mathrm{epi}(D_g F(\bar{x}, \bar{y}))$, i.e.
$$\tilde{y} \notin D_g F(\bar{x}, \bar{y})(x) + C$$

or
$$\tilde{y} \notin \mathrm{Min}\, \{y \in Y \mid (x, y) \in T(\mathrm{epi}(F), (\bar{x}, \bar{y}))\} + C. \tag{15.12}$$

Since $(x, \tilde{y}) \in \mathrm{epi}(DF(\bar{x}, \bar{y}))$, i.e. $(x, \tilde{y}) \in T(\mathrm{epi}(F), (\bar{x}, \bar{y}))$, then
$$\tilde{y} \in \{y \in Y \mid (x, y) \in T(\mathrm{epi}(F), (\bar{x}, \bar{y}))\}.$$

By the domination property there are $y_0 \in \mathrm{Min}\,\{y \in Y \mid (x, y) \in T(\mathrm{epi}(F), (\bar{x}, \bar{y}))\}$ and $c_0 \in C$ so that $\tilde{y} = y_0 + c_0$. Thus
$$\tilde{y} \in \mathrm{Min}\,\{y \in Y \mid (x, y) \in T(\mathrm{epi}(F), (\bar{x}, \bar{y}))\} + C,$$

contradicting the condition (15.12). Hence
$$\mathrm{epi}DF(\bar{x}, \bar{y}) = \mathrm{epi}D_g F(\bar{x}, \bar{y}).$$

□

Notes

The presentation of this chapter is based on the papers [155] and [57]. Already 1981 Aubin [9] has introduced the notion of contingent derivatives for set-valued maps. This concept is used in set-valued analysis (e.g., see [11]) and also in set optimization (e.g., see [67], [213] and [214]). Using this notion necessary optimality conditions ([67, Thm. 4.1]) and sufficient optimality conditions ([67, Thm. 4.2]) do not coincide under standard assumptions.

The concept of contingent epiderivative has been introduced by Aubin [9, p. 178] with the name "upper contingent derivative" for real-valued functions. Later the name "contingent epiderivative" is used in the context of extended real-valued functions (see [11]). Definition 15.3 can be found in [155].

In [11, p. 231] a result similar to that of Theorem 15.15 is mentioned for Fréchet differentiable functions. Generalized contingent epiderivatives have been introduced in [57] and [20].

Calculus rules for contingent epiderivatives can be found in [151]. Under special assumptions these derivatives can be determined on a computer ([89]).

Chapter 16

Subdifferential

There are different possibilities to introduce subgradients of set-valued maps. One possible approach is a generalization of the standard definition known from convex analysis (see also Definition 2.21). Another approach is based on a characterization of the subdifferential using directional derivatives (e.g., see [149, Lemma 3.25]). Instead of the directional derivative we now use the contingent epiderivative. Both approaches are presented in this chapter.

16.1 Concept of Subdifferential

In this section we present a possible generalization of the concept of the subdifferential of a convex functional to the case of a coneconvex set-valued map. For these investigations we have the following assumptions.

Assumption 16.1. Let Assumption 15.5 be satisfied, let S be convex, let $F : S \to 2^Y$ be C-convex, and let the contingent epiderivative $DF(\bar{x}, \bar{y})$ of F at $(\bar{x}, \bar{y})$ exist.

Definition 16.2. Let Assumption 16.1 be satisfied.

(a) A linear map $L : X \to Y$ with

$$L(x) \leq_C DF(\bar{x}, \bar{y})(x) \quad \text{for all } x \in X \qquad (16.1)$$

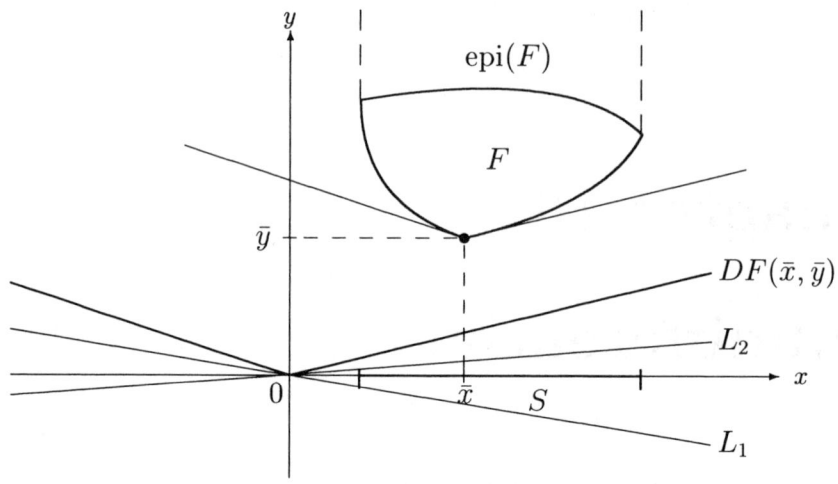

Figure 16.1: Subgradients of F at $(\bar{x}, \bar{y})$.

is called a *subgradient* of F at $(\bar{x}, \bar{y})$ (see Fig. 16.1).

(b) The set

$\partial F(\bar{x}, \bar{y})$
$:= \{L : X \to Y \text{ linear} \mid L(x) \leq_C DF(\bar{x}, \bar{y})(x) \text{ for all } x \in X\}$

of all subgradients L of F at $(\bar{x}, \bar{y})$ is called *subdifferential* of F at $(\bar{x}, \bar{y})$.

This definition is a natural extension of a known characterization of the subdifferential of a convex functional (e.g., see [149, Lemma 3.25]) Here the directional derivative is replaced by the contingent epiderivative and the usual $\leq$ ordering is replaced by the partial ordering $\leq_C$ induced by the convex cone C.

Obviously, the subdifferential is not defined, if the contingent epiderivative does not exist. Conditions ensuring the existence of the contingent epiderivative are given in Theorem 15.6. Notice also that the assumption of cone-convexity of F is actually not needed in Definition 16.2.

16.2 Properties of the Subdifferential

We now present basic properties of subdifferentials known from the convex single-valued case. First we remark under which assumptions the subdifferential is nonempty.

Theorem 16.3. *Let Assumption 16.1 be satisfied, and, in addition, let $S = X$, let C be pointed, and let Y have the least upper bound property. Then the subdifferential $\partial F(\bar{x}, \bar{y})$ is nonempty.*

Proof. By Theorem 15.11 the contingent epiderivative $DF(\bar{x}, \bar{y})$ is sublinear. Then, by the generalized Hahn-Banach Theorem 3.13, there is a linear map $L : X \to Y$ with

$$L(x) \leq_C DF(\bar{x}, \bar{y})(x) \text{ for all } x \in X.$$

Hence, the subdifferential $\partial F(\bar{x}, \bar{y})$ is nonempty. $\square$

Next, we show the convexity of the subdifferential.

Theorem 16.4. *Let Assumption 16.1 be satisfied. Then the subdifferential is convex.*

Proof. For an empty subdifferential the assertion is trivial. Take two arbitrary subgradients $L_1, L_2 \in \partial F(\bar{x}, \bar{y})$ and an arbitrary $\lambda \in [0, 1]$. Then we obtain

$$\begin{aligned} \lambda L_1(x) + (1-\lambda)L_2(x) &\leq_C \lambda DF(\bar{x}, \bar{y}) + (1-\lambda)DF(\bar{x}, \bar{y}) \\ &= DF(\bar{x}, \bar{y}) \text{ for all } x \in X. \end{aligned}$$

Hence

$$\lambda L_1 + (1-\lambda)L_2 \in \partial F(\bar{x}, \bar{y}).$$

$\square$

The next result shows that the subdifferential is closed under appropriate assumptions.

Theorem 16.5. *Let Assumption 16.1 be satisfied, and, in addition, let C be closed. If all subgradients are bounded, then the subdifferential is closed (in the linear space of all linear bounded maps).*

Proof. Choose an arbitrary sequence $(L_n)_{n \in \mathbb{N}}$ of subgradients converging to some linear bounded map L. Next, fix an arbitrary $x \in X$. Then we obtain

$$\|L_n(x) - L(x)\|_Y = \|(L_n - L)(x)\|_Y \leq \|\!|\!| L_n - L |\!|\!\|\, \|x\|_X \quad (16.2)$$

($|\!|\!| \cdot |\!|\!|$ denotes the operator norm). Since $\lim_{n \to \infty} L_n = L$, the inequality (16.2) implies

$$\lim_{n \to \infty} L_n(x) = L(x). \quad (16.3)$$

By the definition of the subgradients L_n we have

$$L_n(x) \leq_C DF(\bar{x}, \bar{y})(x)$$

or

$$L_n(x) \in \{DF(\bar{x}, \bar{y})(x)\} - C,$$

and with (16.3) and the assumption that C is closed we conclude

$$L(x) \in \{DF(\bar{x}, \bar{y})(x)\} - C.$$

Hence, L is a subgradient and, therefore, the subdifferential is closed. □

Notice that for $X = \mathbb{R}^n$ and $Y = \mathbb{R}^m$ linear maps are bounded, and in this special case the subdifferential is closed whenever C is closed.

The following result presents a condition under which the subdifferential is singleton.

Theorem 16.6. *Let Assumption 16.1 be satisfied, and, in addition, let C be pointed. If the contingent epiderivative $DF(\bar{x}, \bar{y})$ of F at $(\bar{x}, \bar{y})$ is linear, then $\partial F(\bar{x}, \bar{y}) = \{DF(\bar{x}, \bar{y})\}$.*

16.2. Properties of the Subdifferential

Proof. Since $DF(\bar{x}, \bar{y})$ is linear, $DF(\bar{x}, \bar{y})$ is a subgradient. Assume that there is another subgradient $L \neq DF(\bar{x}, \bar{y})$. Then we obtain
$$L(-x) \leq_C DF(\bar{x}, \bar{y})(-x) \quad \text{for all } x \in X$$
or
$$-L(x) \leq_C -DF(\bar{x}, \bar{y})(x) \quad \text{for all } x \in X.$$
This inequality implies (by addition of $L(x) + DF(\bar{x}, \bar{y})(x)$)
$$DF(\bar{x}, \bar{y})(x) \leq_C L(x) \quad \text{for all } x \in X.$$
Since C is pointed, we get with (16.1)
$$DF(\bar{x}, \bar{y}) = L,$$
a contradiction to our assumption. Hence, we conclude $\partial F(\bar{x}, \bar{y}) = \{DF(\bar{x}, \bar{y})\}$. □

Finally, we discuss the relationship of the presented definition of the subdifferential to the standard definition used in convex analysis. First, we need a special result for C-convex maps.

Lemma 16.7. *Let Assumption 16.1 be satisfied. Then*
$$F(x) - \{\bar{y}\} \subset \{DF(\bar{x}, \bar{y})(x - \bar{x})\} + C \text{ for all } x \in S.$$

Proof. Take arbitrary elements $x \in S$ and $y \in F(x)$. Then we define a sequence $(x_n, y_n)_{n \in \mathbb{N}}$ with
$$x_n := \bar{x} + \frac{1}{n}(x - \bar{x}) \text{ for all } n \in \mathbb{N}$$
and
$$y_n := \bar{y} + \frac{1}{n}(y - \bar{y}) \text{ for all } n \in \mathbb{N}.$$
Since S is a convex set and F is a C-convex map, it follows for all $n \in \mathbb{N}$
$$x_n = \left(1 - \frac{1}{n}\right)\bar{x} + \frac{1}{n}x \in S$$

and

$$y_n = \left(1 - \frac{1}{n}\right)\bar{y} + \frac{1}{n}y \in F\left(\left(1 - \frac{1}{n}\right)\bar{x} + \frac{1}{n}x\right) + C = F(x_n) + C.$$

So, $(x_n, y_n)_{n \in \mathbb{N}}$ is a sequence in the epigraph of F converging to $(\bar{x}, \bar{y})$. Moreover we obtain

$$\lim_{n \to \infty} n(x_n - \bar{x}, y_n - \bar{y}) = (x - \bar{x}, y - \bar{y}).$$

Consequently we get

$$(x - \bar{x}, y - \bar{y}) \in T(\text{epi}(F), (\bar{x}, \bar{y})) = \text{epi}(DF(\bar{x}, \bar{y}))$$

implying

$$y - \bar{y} \in \{DF(\bar{x}, \bar{y})(x - \bar{x})\} + C.$$

$\square$

Theorem 16.8. *Let Assumption 16.1 be satisfied. Then every subgradient L of F at $(\bar{x}, \bar{y})$ fulfills the inequality*

$$L(x - \bar{x}) \leq_C y - \bar{y} \quad \text{for all } x \in S \text{ and } y \in F(x).$$

Proof. By Lemma 16.7 we obtain

$$DF(\bar{x}, \bar{y})(x - \bar{x}) \leq_C y - \bar{y} \quad \text{for all } x \in S \text{ and } y \in F(x),$$

and with inequality (16.1) we conclude

$$L(x - \bar{x}) \leq_C y - \bar{y} \quad \text{for all } x \in S \text{ and } y \in F(x).$$

$\square$

16.3 Weak Subgradients

In this section we present the concept of weak subgradients. The existence of a weak subgradient for a general set-valued map is shown without the constraint that the considered map is a convex relation (i.e. a set-valued map with a convex graph).

Assumption 16.9. Let Assumption 15.5 be satisfied, and let C have a nonempty interior $\text{int}(C)$.

Definition 16.10. A continuous linear map $L \in L(X, Y)$ is called a *weak subgradient* of F at $\bar{x}$ if

$$F(x) - F(\bar{x}) - \{L(x - \bar{x})\} \subset Y \backslash (-\text{int}(C)) \quad \text{for all } x \in S. \quad (16.4)$$

Remark 16.11. If $F : X \to \mathbb{R}$ is a single-valued convex functional, then the condition (16.4) can be written as

$$F(x) - F(\bar{x}) - L(x - \bar{x}) \not< 0 \quad \text{for all } x \in X$$

or

$$F(x) \geq F(\bar{x}) + L(x - \bar{x}) \quad \text{for all } x \in X.$$

Hence, in this special case a weak subgradient is a subgradient known from convex analysis.

In order to prove the existence of a weak subgradient we need a technical lemma and the concept of upper semicontinuity.

Definition 16.12. Let Assumption 15.5 be satisfied. F is called *upper semicontinuous* at $\bar{x}$, if for any open set M in Y with $F(\bar{x}) \subset M$ there is a neighborhood N of the point $\bar{x}$ so that $F(N) \subset M$.

Lemma 16.13. *Let Assumption 16.9 be satisfied, let S have a nonempty interior $\text{int}(S)$, let S be a convex subset of X, let $F : S \to 2^Y$ be C-convex on S, let F be upper semicontinuous at $\bar{x} \in \text{int}\, S$, and* let $-F(\bar{x})$ have a strict lower bound. Then epi(F) is a convex subset of $X \times Y$ and int $(\text{epi}(F)) \neq \emptyset$.

Proof. By Lemma 14.8 epi(F) is convex. Now we prove that int $(\text{epi}(F)) \neq \emptyset$. Since $-F(\bar{x})$ has a strict lower bound, there is a $\tilde{y} \in Y$ with $F(\bar{x}) \subset \{\tilde{y}\} - \text{int}\, C$. Since $\bar{x} \in \text{int}\, S$ and F is upper semicontinuous at $\bar{x}$, there is some neighbourhood N of the zero in X so that $\{\bar{x}\} + N \subset S$ and
$$F(\{\bar{x}\} + N) \subset \{\tilde{y}\} - \text{int}\,(C).$$
For an arbitrarily chosen $\bar{y} \in \{\tilde{y}\} + \text{int}\,(C)$ there is an open neighborhood M of the zero in Y with
$$\{\bar{y}\} + M \subset \{\tilde{y}\} + \text{int}\,(C).$$
Thus we conclude
$$\begin{aligned}\{\bar{y}\} + M - F(\{\bar{x}\} + N) &\subset \{\tilde{y}\} + \text{int}\,(C) - (\{\tilde{y}\} - \text{int}\,(C)) \\ &\subset \text{int}\,(C) + \text{int}\,(C) \\ &\subset C.\end{aligned}$$
Hence we get
$$(\{\bar{x}\} + N, \{\bar{y}\} + M) \subset \text{epi}(F),$$
i.e. int $(\text{epi}(F)) \neq \emptyset$. $\square$

Remark 16.14. It is obvious from the proof of the preceding lemma that int $(K) \neq \emptyset$ for
$$K := \{(x, y) \in X \times Y \mid x \in S,\ y \in F(x) + \text{int}\,(C)\}.$$

Theorem 16.15. Let Assumption 16.9 be satisfied, let S be convex with a nonempty interior int (S), let $\bar{x} \in \text{int}\,(S)$, let $F : S \to 2^Y$ be C-convex and upper semicontinuous at $\bar{x}$, let $F(\bar{x}) - C$ be convex, let $F(\bar{x})$ and $-F(\bar{x})$ have a strict lower bound, and let the set equation
$$F(\bar{x}) \cap (F(\bar{x}) - \text{int}\,(C)) = \emptyset \tag{16.5}$$

16.3. Weak Subgradients

be fulfilled. Then there is a weak subgradient L of F at $\bar{x} \in \text{int}(S)$ satisfying for every $x \in S$ the property

$$L(x - \bar{x}) \notin -\text{int}(C) \iff L(x - \bar{x}) \in C.$$

Proof. We define the set $D := S - \{\bar{x}\}$ and the set-valued map $H : D \to 2^Y$ with

$$H(x) = F(x + \bar{x}) - F(\bar{x}) \quad \text{for all } x \in D.$$

Then $0_X \in \text{int}(D)$, D is convex, H is upper semicontinuous at 0_X, and $H(0_X)$ has a strict lower bound. In order to see that H is C-convex, take arbitrary $x_1, x_2 \in D$ and $\lambda \in (0, 1)$. Then it follows with the C-convexity of F and the convexity of $F(\bar{x}) - C$

$$\begin{aligned}
&\lambda H(x_1) + (1 - \lambda)H(x_2) \\
&= \lambda F(x_1 + \bar{x}) + (1 - \lambda)F(x_2 + \bar{x}) - \lambda F(\bar{x}) - (1 - \lambda)F(\bar{x}) \\
&\subset F(\lambda x_1 + (1 - \lambda)x_2 + \bar{x}) + C - F(\bar{x}) + C \\
&\subset H(\lambda x_1 + (1 - \lambda)x_2) + C.
\end{aligned}$$

Next we set

$$K := \{(x, y) \in X \times Y \mid x \in D, \ y \in H(x) + \text{int}(C)\}.$$

By Remark 16.14 we obtain $\text{int}(K) \neq \emptyset$. Now we show that $(0_X, 0_Y) \notin K$. Suppose that $(0_X, 0_Y) \in K$, then there is a $y \in H(0_X)$ so that $0 \in \{y\} + \text{int}(C)$ which implies $H(0_X) \cap (-\text{int}(C)) \neq \emptyset$, i.e.

$$(F(\bar{x}) - F(\bar{x})) \cap (-\text{int}(C)) \neq \emptyset$$

contradicting (16.5). By Eidelheit's separation theorem (Theorem 3.16) there is a nonzero $(-\rho, \sigma) \in X^* \times Y^*$ so that

$$-\rho(x) + \sigma(y) \geq 0 \quad \text{for all } (x, y) \in K. \tag{16.6}$$

If $\sigma = 0_{Y^*}$, then $-\rho(x) \geq 0$ for all $x \in D$. Because of $0_X \in \text{int}(D)$ we obtain $\rho = 0_{X^*}$ contradicting $(-\rho, \sigma) \neq (0_{X^*}, 0_{Y^*})$. Hence we get

$\sigma \neq 0_{Y^*}$. Moreover, observe from (16.6) that $\sigma \in C^*$. Then there is a $\bar{y} \in \text{int}(C)$ with $\sigma(\bar{y}) = 1$. We now define a map $L : X \to Y$ by

$$L(x) = \rho(x)\,\bar{y} \quad \text{for all } x \in X.$$

Obviously, L is linear and continuous. Next we assert for this map L

$$F(x) - F(\bar{x}) - \{L(x - \bar{x})\} \subset Y\backslash(-\text{int}(C)) \quad \text{for all } x \in S$$

or

$$y - L(x) \notin -\text{int}(C) \quad \text{for all } x \in D,\ y \in H(x). \tag{16.7}$$

Suppose that there are some $x \in D$ and some $y \in H(x)$ with

$$y - L(x) \in -\text{int}(C).$$

Because of $\sigma \in C^*\backslash\{0_{Y^*}\}$ we then get

$$0 > \sigma(y - L(x)) = \sigma(y) - \rho(x)\sigma(\bar{y}) = \sigma(y) - \rho(x).$$

This is a contradiction to the inequality (16.6). Hence, the condition (16.7) is fulfilled and, therefore, L is a weak subgradient. Finally, for every $x \in D$ we get

$$\begin{aligned} L(x) \notin -\text{int}(C) &\iff \rho(x)\bar{y} \notin -\text{int}(C) \\ &\iff \rho(x) \geq 0 \\ &\iff L(x) \in C. \end{aligned}$$

$\square$

Remark 16.16.

(a) The following implication shows that the assumption (16.5) is rather restrictive for the set $F(\bar{x})$:

$$\text{int}(F(\bar{x})) \neq \emptyset \implies F(\bar{x}) \cap (F(\bar{x}) - \text{int}(C)) \neq \emptyset.$$

Hence the assumption (16.5) can only be fulfilled for a set $F(\bar{x})$ with an empty interior.

Proof. If $\text{int}(F(\bar{x}))$ is nonempty, then there are a $\bar{y} \in F(\bar{x})$

and a neighborhood M of $\bar{y}$ so that $M \subset F(\bar{x})$. Consequently we obtain

$$F(\bar{x}) \cap (F(\bar{x}) - \text{int}(C)) \supset M \cap (M - \text{int}(C)) \neq \emptyset.$$

$\square$

(b) If $F : S \to Y$ is single-valued as a special case, then the assumption (16.5) is always fulfilled.

Notes

The theory of this chapter is based on the papers [18] and [57]. Lemma 16.7 can be found in [155]. The proof of this lemma makes use of the idea of proof of Corley [67, Thm. 3.1] (Thm. 3.1 in [67] is based on [10]).

Weak subgradients have been introduced by Yang [332].

Chapter 17
Optimality Conditions

Based on the concepts introduced in the preceding chapters we now present optimality conditions for set optimization problems. These conditions are discussed using contingent epiderivatives, subgradients and weak subgradients. The main section of this chapter is devoted to a generalization of the Lagrange multiplier rule. We present this multiplier rule as a necessary optimality condition. Assumptions ensuring that this multiplier rule is a sufficient optimality condition are also given.

17.1 Optimality Conditions with Contingent Epiderivatives

In this section we apply the concept of contingent epiderivatives in order to obtain optimality conditions for a set optimization problem . For these investigations we state the following assumption.

Assumption 17.1. Let $(X, \|\cdot\|_X)$ be a real normed space, let S be a nonempty subset of X, let $(Y, \|\cdot\|_Y)$ be a real normed space partially ordered by a convex cone $C \subset Y$ with nonempty interior int(C), and let $F : S \to 2^Y$ be a set-valued map.

Under this assumption we consider the set optimization problem

$$\min_{x \in S} F(x). \tag{17.1}$$

We know from vector optimization that the so-called weak minimality notion is the appropriate concept for the formulation of necessary and sufficient optimality conditions. This fact also holds for the set-valued case. Therefore, in addition to the concept of a minimizer (given in Definition 14.2) we now introduce the notion of a weak minimizer.

Definition 17.2. Let Assumption 17.1 be satisfied, and let $F(S) := \bigcup_{x \in S} F(x)$ denote the image set of F. Then a pair $(\bar{x}, \bar{y})$ with $\bar{x} \in S$ and $\bar{y} \in F(\bar{x})$ is called a *weak minimizer* of the problem (17.1), if $\bar{y}$ is a weakly minimal element of the set $F(S)$, i.e.

$$(\{\bar{y}\} - \text{int}\,(C)) \cap F(S) = \emptyset.$$

First we present a necessary optimality condition for the problem (17.1).

Theorem 17.3. *Let Assumption 17.1 be satisfied. If $(\bar{x}, \bar{y})$ is a weak minimizer of the problem (17.1) and the contingent epiderivative $DF(\bar{x}, \bar{y})$ exists, then*

$$DF(\bar{x}, \bar{y})(x - \bar{x}) \notin -int(C) \text{ for all } x \in S.$$

Proof. Let a pair $(\bar{x}, \bar{y})$ with $\bar{x} \in S$ and $\bar{y} \in F(\bar{x})$ be arbitrarily given. Assume that there is an $x \in S$ with

$$y := DF(\bar{x}, \bar{y})(x - \bar{x}) \in -\text{int}(C). \tag{17.2}$$

By the definition of the contingent epiderivative $(x - \bar{x}, y)$ belongs to the contingent cone of the epigraph of F at $(\bar{x}, \bar{y})$. Then there are a sequence $(x_n, y_n)_{n \in \mathbb{N}}$ in epi(F) and a sequence $(\lambda_n)_{n \in \mathbb{N}}$ of positive real numbers with $(\bar{x}, \bar{y}) = \lim_{n \to \infty} (x_n, y_n)$ and

$$(x - \bar{x}, y) = \lim_{n \to \infty} \lambda_n (x_n - \bar{x}, y_n - \bar{y}). \tag{17.3}$$

17.1. Optimality Conditions with Contingent Epiderivatives

Because of the condition (17.2) and the equation (17.3) there is an $N \in \mathbb{N}$ with
$$\lambda_n(y_n - \bar{y}) \in -\text{int}(C) \text{ for all } n \geq N$$
resulting in
$$y_n \in \{\bar{y}\} - \text{int}(C) \text{ for all } n \geq N \tag{17.4}$$
because C is a cone. Since $(x_n, y_n) \in \text{epi}(F)$ for an arbitrary $n \in \mathbb{N}$, there is a $\tilde{y}_n \in F(x_n)$ with
$$y_n \in \{\tilde{y}_n\} + C \text{ for all } n \in \mathbb{N}.$$
Consequently, we obtain with the condition (17.4) and the equality $\text{int}(C) + C = \text{int}(C)$ (compare Lemmas 1.12, (b) and 1.32, (a))
$$\tilde{y}_n \in \{y_n\} - C \subset \{\bar{y}\} - \text{int}(C) - C = \{\bar{y}\} - \text{int}(C) \text{ for all } n \geq N.$$
Hence, $(\bar{x}, \bar{y})$ is not a weak minimizer of the problem (17.1). □

This necessary condition generalizes a known necessary optimality condition in vector optimization (see Theorem 7.6). It is also sufficient under an appropriate convexity assumption.

Theorem 17.4. *Let Assumption 17.1 be satisfied, and, in addition, let S be a convex set and let F be C-convex. If the contingent epiderivative $DF(\bar{x}, \bar{y})$ exists at an $\bar{x} \in S$ and a $\bar{y} \in F(\bar{x})$ and*
$$DF(\bar{x}, \bar{y})(x - \bar{x}) \notin -\text{int}(C) \text{ for all } x \in S,$$
then $(\bar{x}, \bar{y})$ is a weak minimizer of the problem (17.1).

Proof. By assumption we have
$$\{DF(\bar{x}, \bar{y})(x - \bar{x})\} \cap (-\text{int}(C)) = \emptyset \text{ for all } x \in S$$
which implies
$$(\{DF(\bar{x}, \bar{y})(x - \bar{x})\} + C) \cap (-\text{int}(C)) = \emptyset \text{ for all } x \in S$$
(compare Lemma 4.13, (b)). Then we obtain with Lemma 16.7
$$\begin{aligned}(F(x) - \{\bar{y}\}) \cap (-\text{int}(C)) &\subset (\{DF(\bar{x}, \bar{y})(x - \bar{x})\} + C) \cap (-\text{int}(C)) \\ &= \emptyset \text{ for all } x \in S.\end{aligned}$$

This means that $\bar{y}$ is a weakly minimal element of the set $F(S)$ or, in other words, $(\bar{x}, \bar{y})$ is a weak minimizer of the problem (17.1). □

Theorem 17.3 and 17.4 immediately lead to a characterization of weak minimizers in convex set optimization.

Corollary 17.5. *Let Assumption 17.1 be satisfied, and, in addition, let S be a convex set and let F be C-convex. Let the contingent epiderivative $DF(\bar{x}, \bar{y})$ exist at an $\bar{x} \in S$ and a $\bar{y} \in F(\bar{x})$. The pair $(\bar{x}, \bar{y})$ is a weak minimizer of the problem (17.1) if and only if*

$$DF(\bar{x}, \bar{y})(x - \bar{x}) \notin -int(C) \text{ for all } x \in S.$$

This corollary shows the importance of the concept of the contingent *epi*derivative. With the aid of the contingent derivative it is not possible to give such a characterization of weak minimizers in the convex case.

Finally, we present a necessary and sufficient optimality condition for strong minimizers.

Definition 17.6. Let Assumption 17.1 be satisfied, and let $F(S) := \bigcup\limits_{x \in S} F(x)$ denote the image set of F. Then a pair $(\bar{x}, \bar{y})$ with $\bar{x} \in S$ and $\bar{y} \in F(\bar{x})$ is called a *strong minimizer* of the problem (17.1), if $\bar{y}$ is a strongly minimal element of the set $F(S)$, i.e.

$$F(S) \subset \{\bar{y}\} + C.$$

Theorem 17.7. *Let Assumption 17.1 be satisfied, and, in addition, let C be closed, let S be a convex set and let F be C-convex. Let the contingent epiderivative $DF(\bar{x}, \bar{y})$ exist at an $\bar{x} \in S$ and a $\bar{y} \in F(\bar{x})$. The pair $(\bar{x}, \bar{y})$ is a strong minimizer of the problem (17.1) if and only if*

$$DF(\bar{x}, \bar{y})(x - \bar{x}) \in C \text{ for all } x \in S. \qquad (17.5)$$

17.1. Optimality Conditions with Contingent Epiderivatives

Proof. (a) Assume that the condition (17.5) is not fulfilled, i.e. there is an $x \in S$ with

$$DF(\bar{x}, \bar{y})(x - \bar{x}) \notin C. \tag{17.6}$$

By the definition of the contingent epiderivative we have for $\hat{x} := x - \bar{x}$ and $\hat{y} := DF(\bar{x}, \bar{y})(x - \bar{x})$

$$(\hat{x}, \hat{y}) \in \text{epi}(DF(\bar{x}, \bar{y})) = T(\text{epi}(F), (\bar{x}, \bar{y})).$$

Consequently, there are a sequence $(x_n, y_n)_{n \in \mathbb{N}}$ of elements in $\text{epi}(F)$ and a sequence $(\lambda_n)_{n \in \mathbb{N}}$ of positive real numbers with $(\bar{x}, \bar{y}) = \lim_{n \to \infty} (x_n, y_n)$ and

$$(\hat{x}, \hat{y}) = \lim_{n \to \infty} \lambda_n (x_n - \bar{x}, y_n - \bar{y}).$$

Since C is closed, it follows from (17.6)

$$\lambda_n(y_n - \bar{y}) \notin C \text{ for sufficiently large } n \in \mathbb{N}$$

and

$$y_n \notin \{\bar{y}\} + C \text{ for sufficiently large } n \in \mathbb{N}. \tag{17.7}$$

Because of

$$(x_n, y_n) \in \text{epi}(F) \text{ for all } n \in \mathbb{N}$$

we write for every $n \in \mathbb{N}$

$$y_n = \tilde{y}_n + c_n \text{ with } \tilde{y}_n \in F(S) \text{ and } c_n \in C. \tag{17.8}$$

The conditions (17.7) and (17.8) imply

$$\tilde{y}_n + c_n \notin \{\bar{y}\} + C \text{ for sufficiently large } n \in \mathbb{N}$$

and

$$\tilde{y}_n \notin \{\bar{y}\} + C \text{ for sufficiently large } n \in \mathbb{N}.$$

Hence, $\bar{y}$ is no strongly minimal element of $F(S)$ and, therefore, $(\bar{x}, \bar{y})$ is no strong minimizer of the problem (17.1).

(b) Assume that the condition (17.5) is fulfilled. Then we conclude with Lemma 16.7

$$\begin{aligned} F(x) &\subset \{\bar{y} + DF(\bar{x},\bar{y})(x-\bar{x})\} + C \\ &\subset \{\bar{y}\} + C + C \\ &= \{\bar{y}\} + C \text{ for all } x \in S. \end{aligned}$$

Hence, $(\bar{x}, \bar{y})$ is a strong minimizer of the problem (17.1). □

Notice that the assumption of C-convexity is only needed for the proof of the sufficiency of the condition (17.5).

17.2 Optimality Conditions with Subgradients

For strong minimizers an optimality condition based on the subdifferential is now given. This result extends the well-known result of convex analysis that a point is a minimal point of a convex functional if and only if the null functional is a subgradient (e.g., see [149, Thm. 3.27]).

Theorem 17.8. *Let Assumption 16.1 be satisfied.*

(a) *If the null map is a subgradient of F at $(\bar{x}, \bar{y})$, then the pair $(\bar{x}, \bar{y})$ is a strong minimizer of the set optimization problem (17.1).*

(b) *In addition, let S equal X and let C be closed. If the pair $(\bar{x}, \bar{y})$ is a strong minimizer of the set optimization problem (17.1), then the null map is a subgradient of F at $(\bar{x}, \bar{y})$.*

Proof.

(a) By Theorem 16.8 we conclude

$$0_Y \leq_C y - \bar{y} \text{ for all } x \in S \text{ and } y \in F(x)$$

or

$$\bar{y} \leq_C y \text{ for all } y \in F(S),$$

i.e., $(\bar{x}, \bar{y})$ is a strong minimizer of the set optimization problem (17.1).

(b) By Theorem 17.7 we obtain for the strong minimizer $(\bar{x}, \bar{y})$

$$DF(\bar{x}, \bar{y})(x - \bar{x}) \in C \quad \text{for all } x \in S = X$$

or

$$0_Y \leq_C DF(\bar{x}, \bar{y})(x) \quad \text{for all } x \in X.$$

Hence, the null map is a subgradient of F at $(\bar{x}, \bar{y})$. □

The preceding theorem immediately implies the following corollary.

Corollary 17.9. *Let Assumption 16.1 be satisfied, and, in addition, let S equal X and let C be closed. The pair $(\bar{x}, \bar{y})$ is a strong minimizer of the set optimization problem (17.1) if and only if the null map is a subgradient of F at $(\bar{x}, \bar{y})$.*

17.3 Optimality Conditions with Weak Subgradients

In this section we derive a sufficient optimality condition for set optimization problems in terms of weak subgradients.

Theorem 17.10. *Let Assumption 16.9 be satisfied. If there is a weak subgradient L of F at $\bar{x} \in S$ so that*

$$L(x - \bar{x}) \in C \quad \text{for all } x \in S, \tag{17.9}$$

then for every $\bar{y} \in F(\bar{x})$ $(\bar{x}, \bar{y})$ is a weak minimizer of the set optimization problem (17.1), and we have the property

$$F(\bar{x}) \cap (F(\bar{x}) - \text{int}(C)) = \emptyset.$$

Proof. Since L is a weak subgradient of F at $\bar{x} \in S$, then

$$F(x) - F(\bar{x}) - \{L(x - \bar{x})\} \subset W \quad \text{for all } x \in S \qquad (17.10)$$

where $W := Y \setminus (-\text{int}(C))$. Thus for every $\bar{y} \in F(\bar{x})$ we have

$$F(x) - \{\bar{y}\} \subset \{L(x - \bar{x})\} + W \subset C + W = W \quad \text{for all } x \in S$$

resulting in

$$F(S) \cap (\{\bar{y}\} - \text{int}(C)) = \emptyset,$$

i.e. $\bar{y}$ is a weakly minimal element of $F(S)$. From (17.10) we have

$$F(\bar{x}) - F(\bar{x}) - \{L(0)\} \subset W$$

implying

$$F(\bar{x}) - F(\bar{x}) \subset W,$$

hence

$$F(\bar{x}) \cap (F(\bar{x}) - \text{int}(C)) = \emptyset.$$

$\square$

Remark 17.11. In the special case $S = X$ the assumption (17.9) reads

$$L(x - \bar{x}) \in C \quad \text{for all } x \in X.$$

If C is pointed, we then conclude

$$L(x) \in C \cap (-C) = \{0_Y\} \quad \text{for all } x \in X$$

which means that $L = 0_{L(X,Y)}$, or in other words: $0_{L(X,Y)}$ is a weak subgradient of F at $\bar{x} \in X$. Hence we obtain the standard assumption known from the theory of subgradients in convex analysis.

17.4 Generalized Lagrange Multiplier Rule

More than 200 years ago Lagrange presented his multiplier rule as an optimality condition for optimization problems with equality constraints (see [195]). In this section we extend the investigations in chapter 7 to general optimization problems with a set-valued objective map and a set-valued constraint, and we show that the Lagrange multiplier rule remains valid in such a general setting as well.

Throughout this section we use the following standard assumption.

Assumption 17.12. Let $(X, \|\cdot\|_X)$ be a real normed space, let $(Y, \|\cdot\|_Y)$ and $(Z, \|\cdot\|_Z)$ be real normed spaces partially ordered by convex pointed cones $C_Y \subset Y$ and $C_Z \subset Z$, respectively, let $\hat{S}$ be a nonempty subset of X, and let $F : \hat{S} \to 2^Y$ and $G : \hat{S} \to 2^Z$ be set-valued maps.

Under this assumption we consider the constrained set optimization problem

$$\left.\begin{aligned}&\min F(x)\\&\text{subject to the constraints}\\&G(x) \cap (-C_Z) \neq \emptyset\\&x \in \hat{S}.\end{aligned}\right\} \quad (17.11)$$

For simplicity let $S := \{x \in \hat{S} \mid G(x) \cap (-C_Z) \neq \emptyset\}$ denote the feasible set of this problem which is assumed to be nonempty. If G is single-valued, the constraint in (17.11) reduces to $G(x) \in -C_Z$ or $G(x) \leq_{C_Z} 0_Z$ generalizing equality and inequality constraints. If, in addition, F is single-valued, then the problem (17.11) is a general vector optimization problem.

On the basis of the concept of contingent epiderivatives we prove in Subsection 17.4.1 a multiplier rule as a necessary optimality condition of problem (17.11) and discuss a regularity assumption. In Subsection 17.4.2 assumptions are presented which guarantee that this multiplier rule is a sufficient optimality condition as well.

17.4.1 A Necessary Optimality Condition

We begin our investigations with a generalized Lagrange multiplier rule as a necessary optimality condition for set optimization problems.

Theorem 17.13. *Let Assumption 17.12 be satisfied. Let the cones C_Y and C_Z have a nonempty interior $int(C_Y)$ and $int(C_Z)$ respectively, let the set $\hat{S}$ be convex and let the maps F and G be C_Y-convex and C_Z-convex, respectively. Assume that $(\bar{x}, \bar{y}) \in X \times Y$ with $\bar{x} \in S$ and $\bar{y} \in F(\bar{x})$ is a weak minimizer of the problem (17.11). Let the contingent epiderivative of (F, G) at $(\bar{x}, (\bar{y}, \bar{z}))$ for an arbitrary $\bar{z} \in G(\bar{x}) \cap (-C_Z)$ exist. Then there are continuous linear functionals $t \in C_{Y^*}$ and $u \in C_{Z^*}$ with $(t, u) \neq (0_{Y^*}, 0_{Z^*})$ so that*

$$t(y) + u(z) \geq 0 \text{ for all } (y, z) = D(F, G)(\bar{x}, (\bar{y}, \bar{z}))(x - \bar{x}) \text{ with } x \in \hat{S}$$

and

$$u(\bar{z}) = 0.$$

If, in addition to the above assumptions, the regularity assumption

$$\{z \mid (y, z) \in D(F, G)(\bar{x}, (\bar{y}, \bar{z}))(cone(S - \{\bar{x}\}))\} + cone(C_Z + \{\bar{z}\}) = Z \quad (17.12)$$

is satisfied, then $t \neq 0_{Y^*}$.

Proof. In the product space $Y \times Z$ we define for an arbitrary $\bar{z} \in G(\bar{x}) \cap (-C_Z)$ the following set

$$M := \left[\bigcup_{x \in \hat{S}} D(F, G)(\bar{x}, (\bar{y}, \bar{z}))(x - \bar{x}) \right] + (C_Y \times (C_Z + \{\bar{z}\})).$$

The proof of this theorem consists of several steps. First, we prove two important properties of this set M and then we apply a separation theorem in order to obtain the multiplier rule. Finally, we show $t \neq 0_{Y^*}$ under the regularity assumption.

(a) We show that the nonempty set M is convex. We prove the convexity for the translated set $M' := M - \{(0_Y, \bar{z})\}$ and immediately get the desired result. For this proof we fix two arbitrary pairs

17.4. Generalized Lagrange Multiplier Rule

$(y_1, z_1), (y_2, z_2) \in M'$. Then there are elements $x_1, x_2 \in \hat{S}$ with

$$(y_i, z_i) \in D(F, G)(\bar{x}, (\bar{y}, \bar{z}))(x_i - \bar{x}) + (C_Y \times C_Z) \text{ for } i = 1, 2$$

resulting in

$$(x_i - \bar{x}, (y_i, z_i)) \in T(\text{epi}(F, G), (\bar{x}, (\bar{y}, \bar{z}))) \text{ for } i = 1, 2.$$

This contingent cone is convex because the map (F, G) is cone-convex and, therefore, by Lemma 14.8 the epigraph $\text{epi}(F, G)$ is a convex set. Then we obtain for all $\lambda \in [0, 1]$

$$\lambda(x_1 - \bar{x}, (y_1, z_1)) + (1 - \lambda)(x_2 - \bar{x}, (y_2, z_2)) \in T(\text{epi}(F, G), (\bar{x}, (\bar{y}, \bar{z})))$$

implying

$$(\lambda y_1 + (1-\lambda)y_2, \lambda z_1 + (1-\lambda)z_2)$$
$$\in D(F,G)(\bar{x}, (\bar{y}, \bar{z}))(\lambda x_1 + (1-\lambda)x_2 - \bar{x}) + (C_Y \times C_Z).$$

Consequently, the set M is convex.

(b) In the next step of the proof we show the equality

$$M \cap \Big[(-\text{int}(C_Y)) \times (-\text{int}(C_Z))\Big] = \emptyset. \tag{17.13}$$

Assume that this equality does not hold. Then there are elements $x \in \hat{S}$ and $(y, z) \in Y \times Z$ with

$$(y, z + \bar{z}) \in \Big[D(F, G)(\bar{x}, (\bar{y}, \bar{z}))(x - \bar{x}) + (C_Y \times (C_Z + \{\bar{z}\}))\Big]$$
$$\cap \Big[(-\text{int}(C_Y)) \times (-\text{int}(C_Z))\Big]. \tag{17.14}$$

implying
$$(x - \bar{x}, (y, z)) \in T(\text{epi}(F, G), (\bar{x}, (\bar{y}, \bar{z}))).$$

This means that there are sequences $(x_n, (y_n, z_n))_{n \in \mathbb{N}}$ of elements in $\text{epi}(F, G)$ and a sequence $(\lambda_n)_{n \in \mathbb{N}}$ of positive real numbers with

$$(\bar{x}, (\bar{y}, \bar{z})) = \lim_{n \to \infty} (x_n, (y_n, z_n))$$

and
$$(x - \bar{x}, (y, z)) = \lim_{n \to \infty} \lambda_n(x_n - \bar{x}, (y_n - \bar{y}, z_n - \bar{z})). \qquad (17.15)$$

Since $y \in -\text{int}(C_Y)$ by (17.14), we conclude $\lambda_n(y_n - \bar{y}) \in -\text{int}(C_Y)$ for sufficiently large $n \in \mathbb{N}$ resulting in

$$y_n \in \{\bar{y}\} - \text{int}(C_Y) \text{ for sufficiently large } n \in \mathbb{N}. \qquad (17.16)$$

Because of $(x_n, (y_n, z_n)) \in \text{epi}(F, G)$ for all $n \in \mathbb{N}$ there are elements $\hat{y}_n \in F(x_n)$ with

$$y_n \in \{\hat{y}_n\} + C_Y \text{ for all } n \in \mathbb{N}.$$

Together with (17.16) we obtain

$$\hat{y}_n \in \{\bar{y}\} - \text{int}(C_Y) - C_Y = \{\bar{y}\} - \text{int}(C_Y) \text{ for sufficiently large } n \in \mathbb{N}$$

or

$$(\{\bar{y}\} - \text{int}(C_Y)) \cap F(x_n) \neq \emptyset \text{ for sufficiently large } n \in \mathbb{N}. \qquad (17.17)$$

Moreover, from (17.14) we conclude $z + \bar{z} \in -\text{int}(C_Z)$ and with (17.15) we obtain

$$\lambda_n(z_n - \bar{z}) + \bar{z} \in -\text{int}(C_Z) \text{ for sufficiently large } n \in \mathbb{N}$$

or

$$\lambda_n \left(z_n - \left(1 - \frac{1}{\lambda_n}\right) \bar{z} \right) \in -\text{int}(C_Z) \text{ for sufficiently large } n \in \mathbb{N}$$

implying

$$z_n - \left(1 - \frac{1}{\lambda_n}\right) \bar{z} \in -\text{int}(C_Z) \text{ for sufficiently large } n \in \mathbb{N}. \qquad (17.18)$$

Since $y \neq 0_Y$ (by (17.14)), we conclude with (17.15) that

$$\lambda_n > 1 \text{ for sufficiently large } n \in \mathbb{N}.$$

17.4. Generalized Lagrange Multiplier Rule

By assumption we have $\bar{z} \in -C_Z$ and, therefore, we get from (17.18)

$$z_n \in -C_Z - \text{int}(C_Z) = -\text{int}(C_Z) \text{ for sufficiently large } n \in \mathbb{N}. \quad (17.19)$$

Because of $(x_n, (y_n, z_n)) \in \text{epi}(F, G)$ for all $n \in \mathbb{N}$ there are elements $\hat{z}_n \in G(x_n)$ with

$$z_n \in \{\hat{z}_n\} + C_Z \text{ for all } n \in \mathbb{N}.$$

Together with (17.19) we then get

$$\hat{z}_n \in \{z_n\} - C_Z \subset -\text{int}(C_Z) \text{ for sufficiently large } n \in \mathbb{N}$$

and

$$\hat{z}_n \in G(x_n) \cap (-C_Z) \text{ for sufficiently large } n \in \mathbb{N}. \quad (17.20)$$

Hence, for a sufficiently large $n \in \mathbb{N}$ we have $\hat{x}_n \in \hat{S}$, $(\{\bar{y}\} - \text{int}(C_Y)) \cap F(x_n) \neq \emptyset$ (by (17.17)) and $G(x_n) \cap (-C_Z) \neq \emptyset$ (by (17.20)) and, therefore, $(\bar{x}, \bar{y})$ is not a weak minimizer of the problem (17.12) which is a contradiction to the assumption of the theorem.

(c) In this step we now prove the first part of the theorem. By part (a) the set M is convex and by (b) the equality (17.13) holds. By Eidelheit's separation theorem (Theorem 3.16) there are continuous linear functionals $t \in Y^*$ and $u \in Z^*$ with $(t, u) \neq (0_{Y^*}, 0_{Z^*})$ and a real number $\gamma > 0$ so that

$$t(c_Y) + u(c_Z) < \gamma \leq t(y) + u(z) \quad (17.21)$$
$$\text{for all } c_Y \in -\text{int}(C_Y), c_Z \in -\text{int}(C_Z), (y, z) \in M.$$

Since $(0, \bar{z}) \in M$, we obtain from (17.21) for $c_Y = 0_Y$

$$u(c_Z) < u(\bar{z}) \text{ for all } c_Z \in -\text{int}(C_Z). \quad (17.22)$$

If we assume that $u(c_Z) > 0$ for a $c_Z \in -\text{int}(C_Z)$, we get a contradiction to (17.22) because C_Z is a cone. Therefore, we obtain

$$u(c_Z) \leq 0 \text{ for all } c_Z \in -\text{int}(C_Z)$$

resulting in $u \in C_{Z^*}$ because $C_Z \subset \mathrm{cl}(\mathrm{int}(C_Z))$. For $(0,\bar{z}) \in M$ and $c_Z = 0_Z$ we get from (17.21)

$$t(c_Y) < u(\bar{z}) \leq 0 \text{ for all } c_Y \in -\mathrm{int}(C_Y) \qquad (17.23)$$

(notice that $\bar{z} \in -C_Z$ and $u \in C_{Z^*}$). This inequality implies $t \in C_{Y^*}$. From (17.22) and (17.23) we immediately obtain $u(\bar{z}) = 0$. In order to prove the inequality of the multiplier rule we conclude from (17.21) with $c_Y = 0_Y$ and $c_Z = 0_Z$

$$t(y) + u(z) \geq 0 \text{ for all } (y,z) = D(F,G)(\bar{x},(\bar{y},\bar{z}))(x - \bar{x}) \text{ with } x \in \hat{S}.$$

Hence, the first part of the theorem is shown.

(d) Finally, we prove $t \neq 0_{Y^*}$ under the regularity assumption (17.12). For an arbitrary $\hat{z} \in Z$ there are elements $x \in \hat{S}$, $c_Z \in C_Z$ and nonnegative real numbers α and β with

$$\hat{z} = z + \beta(c_Z + \bar{z}) \text{ for } (y,z) = D(F,G)(\bar{x},(\bar{y},\bar{z}))(\alpha(x - \bar{x})).$$

Since $D(F,G)(\bar{x},(\bar{y},\bar{z}))$ is positively homogeneous by Theorem 15.11 (notice that we do not need convexity assumptions for this proof), we can write

$$(y,z) = \alpha D(F,G)(\bar{x},(\bar{y},\bar{z}))(x - \bar{x}) =: \alpha(\tilde{y},\tilde{z}).$$

Assume that $t = 0_{Y^*}$. Then we conclude from the multiplier rule

$$\begin{aligned} u(\hat{z}) &= u(z) + \beta u(c_Z + \bar{z}) \\ &= \underbrace{\alpha\, u(\tilde{z})}_{\geq 0} + \underbrace{\beta\, u(c_Z)}_{\geq 0} + \underbrace{\beta\, u(\bar{z})}_{= 0} \\ &\geq 0. \end{aligned}$$

Because $\hat{z}$ is arbitrarily chosen we have

$$u(\hat{z}) \geq 0 \text{ for all } z \in Z$$

implying $u = 0_{Z^*}$. But this is a contradiction to $(t,u) \neq (0_{Y^*}, 0_{Z^*})$. □

17.4. Generalized Lagrange Multiplier Rule

Theorem 17.13 extends the Lagrange multiplier rule as necessary optimality condition (compare Theorem 7.4) to set optimization. Since a minimizer of the problem (17.11) is also a weak minimizer (see Lemma 4.14), this multiplier rule is a necessary optimality condition for a minimizer as well.

The regularity condition in Theorem 17.13 extends the Kurcyusz-Robinson-Zowe regularity assumption (e.g., see [149]) to set optimization problems. It is weaker than a generalized Slater condition (compare Lemma 17.15). Although the regularity condition in Theorem 17.13 also includes the objective map F, one only uses the second component of the contingent epiderivative of (F, G).

It is important to note that the maps F and G are assumed to be cone-convex in Theorem 17.13 whereas convexity of the objective function and the constraint function is not needed in the single-valued scalar case (e.g., see [149, Thm. 5.3]). In fact, the cone-convexity is only needed in part (a) of the proof in order to obtain the convexity of the contingent cone. If we would modify the notion of the contingent epiderivative in such a way that we replace the contingent cone by Clarke's tangent cone which is always convex, we could drop the cone-convexity assumption in Theorem 17.13.

With the following example we illustrate the usefulness of the necessary condition in Theorem 17.13.

Example 17.14. Let $(X, \|\cdot\|_X)$ be a real normed space, and let $f, g, h : X \to \mathbb{R}$ be given functionals. Then we consider the set-valued map $F : X \to 2^{\mathbb{R}}$ with

$$F(x) := \{y \in \mathbb{R} \mid f(x) \leq y \leq g(x)\}$$

and the set-valued map $G : X \to 2^{\mathbb{R}}$ (which is actually single-valued) with

$$G(x) := \{h(x)\}.$$

Under these assumptions we investigate the optimization problem

$$\left.\begin{aligned}
&\min F(x) \\
&\text{subject to the constraints} \\
&G(x) \cap (-\mathbb{R}_+) \neq \emptyset \\
&x \in X.
\end{aligned}\right\} \qquad (17.24)$$

This is a special problem of the general type (17.11). Notice that the constraint is equivalent to the inequality $h(x) \leq 0$. If $f = g$ this problem reduces to a standard optimization problem. But if the data of the objective function of a standard problem are not exactly known, it makes sense to replace the objective by a set-valued objective representing fuzzy outcomes. In this example the values of the objective may vary between the values of two known functions.

Next, we assume that $(\bar{x}, f(\bar{x}))$ is a weak minimizer of problem (17.24), and that f and h are continuous at $\bar{x}$ and convex. Since

$$\mathrm{epi}(F, G) = \{(x, (y, z)) \in X \times \mathbb{R}^2 \mid x \in X, \ y \geq f(x), \ z \geq h(x)\},$$

we conclude

$$T(\mathrm{epi}(F, G), (\bar{x}, (f(\bar{x}), h(\bar{x})))) = \mathrm{epi}(f, g)'(\bar{x}),$$

i.e., the contingent epiderivative of (F, G) at $(\bar{x}, (f(\bar{x}), h(\bar{x})))$ exists and equals the directional derivative $(f, h)'(\bar{x}) = (f', h')(\bar{x})$ of (f, h) at $\bar{x}$ (see Corollary 15.17 for the case of one functional). Consequently, by the previous theorem there are nonnegative numbers t and u with $(t, u) \neq (0, 0)$ so that

$$tf'(\bar{x})(x - \bar{x}) + uh'(\bar{x})(x - \bar{x}) \geq 0 \text{ for all } x \in X$$

and

$$uh(\bar{x}) = 0.$$

If $f'(\bar{x})$ and $h'(\bar{x})$ are linear (e.g., in the case of Fréchet differentiability), we even conclude

$$tf'(\bar{x}) + uh'(\bar{x}) = 0_{X^*}$$

and

$$uh(\bar{x}) = 0.$$

Finally, we discuss the regularity condition of Theorem 17.13 for this problem. Assume that for every $z < 0$ there is an $x \in X$ with $z = h'(\bar{x})(x)$. Then $h'(\bar{x})(X) \supset -\mathbb{R}_+$ and because of $h(\bar{x}) \leq 0$ we obtain

$$h'(\bar{x})(\text{cone}(X - \{\bar{x}\})) + \text{cone}(\mathbb{R}_+ + \{h(\bar{x})\})$$
$$= \underbrace{h'(\bar{x})(X)}_{\supset -\mathbb{R}_+} + \begin{cases} \mathbb{R} & \text{if } h(\bar{x}) < 0 \\ \mathbb{R}_+ & \text{if } h(\bar{x}) = 0 \end{cases}$$
$$= \mathbb{R}.$$

Hence, the general regularity condition (17.12) is satisfied in this case.

The next lemma shows that a generalization of the well-known Slater condition implies the extended Kurcyusz-Robinson-Zowe constraint qualification.

Lemma 17.15. *Let Assumption 17.12 be satisfied, let $int(\hat{S}) \neq \emptyset$, let $\bar{x} \in S$, $\bar{y} \in F(\bar{x})$ and $\bar{z} \in G(\bar{x}) \cap (-C_Z)$ be arbitrarily given, and let the contingent epiderivative of (F, G) at $(\bar{x}, (\bar{y}, \bar{z}))$ exist. If there is an $\hat{x} \in int(\hat{S})$ with*

$$\bar{z} + z \in -int(C_Z) \text{ for } (y, z) = D(F, G)(\bar{x}, (\bar{y}, \bar{z}))(\hat{x} - \bar{x}),$$

then the regularity assumption (17.12) is fulfilled.

Proof. Take an arbitrary $\hat{z} \in Z$. Since $D(F, G)(\bar{x}, (\bar{y}, \bar{z}))$ is positive homogeneous by Theorem 15.11 (notice that we do not need convexity assumptions for this proof), we obtain for a sufficiently large $\lambda > 0$

$$\lambda(y, z) = D(F, G)(\bar{x}, (\bar{y}, \bar{z}))(\lambda(\hat{x} - \bar{x}))$$
$$\in D(F, G)(\bar{x}, (\bar{y}, \bar{z}))(\text{cone}(S - \{\bar{x}\}))$$

and

$$\hat{z} = \lambda z + \lambda \underbrace{\left[-\bar{z} - z + \frac{1}{\lambda}\hat{z} + \bar{z}\right]}_{\in C_Z}$$
$$\in \{\tilde{z} \mid (y, \tilde{z}) \in D(F, G)(\bar{x}, (\bar{y}, \bar{z}))(\text{cone}(S - \{\bar{x}\}))\}$$
$$+ \text{cone}(C_Z + \{\bar{z}\}).$$

Hence, we conclude

$$Z \subset \{\tilde{z} \mid (y, \tilde{z}) \in D(F, G)(\bar{x}, (\bar{y}, \bar{z}))(\text{cone}(S - \{\bar{x}\}))\} + \text{cone}(C_Z + \{\bar{z}\}).$$

Because the converse inclusion is trivial, the regularity assumption (17.12) is fulfilled. □

The following example shows that the regularity condition (17.12) can be satisfied although the regularity assumption in Lemma 17.15 is not fulfilled.

Example 17.16. We consider $X = Z = L_2[0, 1]$ with the natural ordering cone

$$C_Z := \{x \in L_2[0, 1] \mid x(t) \geq 0 \text{ almost everywhere on } [0, 1]\}$$

(notice that $\text{int}(C_Z) = \emptyset$). Take an arbitrary $a \in L_2[0, 1]$ and define the set-valued map $G : X \to 2^Z$ with

$$G(x) = \{-x + a\} + C_Z \text{ for all } x \in X.$$

Then we investigate the constraint of problem (17.11)

$$G(x) \cap (-C_Z) \neq \emptyset, \ x \in X$$

being equivalent to

$$-x + a \in -C_Z, \ x \in X.$$

For instance, choose the objective map $F : X \to 2^{\mathbb{R}}$ with

$$F(x) = \{\langle x, x \rangle\} \text{ for all } x \in X$$

($\langle \cdot, \cdot \rangle$ denotes the scalar product in X).

Since $\text{int}(C_Z) = \emptyset$, it is obvious that Lemma 17.15 is not applicable in this case. Therefore, we investigate the regularity assumption (17.12) in Theorem 17.13. For an arbitrary $\bar{x} \in X$ with $\bar{z} := -\bar{x} + a \in -C_Z$ we obtain with

$$\text{epi}(F, G) = \{(x, (y, z)) \in X \times \mathbb{R} \times Z \mid x \in X, \ y \geq \langle x, x \rangle, \ -x + a \leq_{C_Z} z\}$$

17.4. Generalized Lagrange Multiplier Rule

the equality
$$T(\text{epi}(F,G),(\bar{x},(\langle\bar{x},\bar{x}\rangle,\bar{z}))) = \text{epi}(2\langle x,\cdot\rangle,-\text{id})$$
implying
$$D(F,G)(\bar{x},(\bar{y},\bar{z})) = (2\langle x,\cdot\rangle,-\text{id})$$
(id denotes the identity). Then we get
$$-\text{id}(\text{cone}(X-\{\bar{x}\}))+\text{cone}(C_Z+\{\bar{z}\}) = X+\text{cone}(C_Z+\{\bar{z}\}) = X = Z,$$
i.e., the regularity condition (17.12) in Theorem 17.13 is fulfilled.

17.4.2 A Sufficient Optimality Condition

In this subsection we answer the question under which assumptions the multiplier rule in Theorem 17.13 is also a sufficient optimality condition. It is known from standard optimization theory (see Section 7.2) that convexity or generalized concepts like quasiconvexity play the essential role. Therefore, we begin with an extension of the quasiconvexity concept to set-valued maps.

Definition 17.17. Let $(X,\|\cdot\|_X)$ and $(Y,\|\cdot\|_Y)$ be real normed spaces, let $\hat{S}$ be a nonempty subset of X, let $\tilde{C}$ be a nonempty subset of Y and let $F: \hat{S} \to 2^Y$ be a set-valued map whose contingent epiderivative exists at $(\bar{x},\bar{y})$ with $\bar{x} \in \hat{S}$ and $\bar{y} \in F(\bar{x})$. The map F is called $\tilde{C}$-quasiconvex at $(\bar{x},\bar{y})$, if for all $x \in \hat{S}$
$$(F(x)-\{\bar{y}\})\cap\tilde{C} \neq \emptyset \implies (\{DF(\bar{x},\bar{y})(x-\bar{x})\}+C_Y)\cap\tilde{C} \neq \emptyset.$$

This notion extends a concept introduced in Definition 7.17 for problems in vector optimization. The following lemma shows that cone-convexity implies quasiconvexity in this set-valued setting.

Lemma 17.18. *Let $\hat{S}$ be a nonempty convex subset of a real normed space $(X,\|\cdot\|_X)$, let $\tilde{C}$ be a nonempty subset of the real normed space $(Y,\|\cdot\|_Y)$ partially ordered by a convex pointed cone $C_Y \subset Y$*

and let a set-valued map $F : \hat{S} \to 2^Y$ be given whose contingent epiderivative exists at $(\bar{x}, \bar{y})$ with $\bar{x} \in \hat{S}$ and $\bar{y} \in F(\bar{x})$. If F is C_Y-convex, then it is also $\tilde{C}$-quasiconvex at $(\bar{x}, \bar{y})$.

Proof. Choose an arbitrary $x \in \hat{S}$ with
$$(F(x) - \{\bar{y}\}) \cap \tilde{C} \neq \emptyset.$$
Since F is C_Y-convex, we conclude with Lemma 16.7
$$F(x) - \{\bar{y}\} \subset \{DF(\bar{x}, \bar{y})(x - \bar{x})\} + C_Y.$$
Consequently we obtain
$$(\{DF(\bar{x}, \bar{y})(x - \bar{x})\} + C_Y) \cap \tilde{C} \neq \emptyset.$$
$\square$

It is known from Theorem 7.20 that the quasiconvexity of a certain composite map completely characterizes the sufficiency of a multiplier rule. This idea is extended in the next theorem.

Theorem 17.19. *Let Assumption 17.12 be satisfied. Let the cone C_Y have a nonempty interior $int(C_Y)$, let the contingent derivative of (F, G) exist at $(\bar{x}, \bar{y})$ with $\bar{x} \in S$, $\bar{y} \in F(\bar{x})$ and $\bar{z} \in G(\bar{x})$. Moreover, assume that there are continuous linear functionals $t \in C_{Y^*} \setminus \{0_{Y^*}\}$ and $u \in C_{Z^*}$ with*
$$t(y) + u(z) \geq 0 \text{ for all } (y, z) = D(F, G)(\bar{x}, (\bar{y}, \bar{z}))(x - \bar{x}) \text{ with } x \in \hat{S} \tag{17.25}$$
and
$$u(\bar{z}) = 0. \tag{17.26}$$
Then $(\bar{x}, \bar{y})$ is a weak minimizer of F on
$$\tilde{S} := \{x \in \hat{S} \mid G(x) \cap (-C_Z + cone(\bar{z}) - cone(\bar{z})) \neq \emptyset\}$$
if and only if the map $(F, G) : \hat{S} \to 2^Y \times 2^Z$ is $\tilde{C}$-quasiconvex at $(\bar{x}, (\bar{y}, \bar{z}))$ with
$$\tilde{C} := (-int(C_Y)) \times (-C_Z + cone(\bar{z}) - cone(\bar{z})).$$

17.4. Generalized Lagrange Multiplier Rule

Proof. First we show under the given assumptions

$$\Big((\{y\} + C_Y) \times (\{z\} + C_Z)\Big) \cap \tilde{C} = \emptyset$$

for all $(y, z) = D(F, G)(\bar{x}, (\bar{y}, \bar{z}))(x - \bar{x})$ with $x \in \hat{S}$. (17.27)

For the proof of this assertion assume that there is an $x \in \hat{S}$ with

$$\Big((\{y\} + C_Y) \times (\{z\} + C_Z)\Big) \cap \tilde{C} \neq \emptyset$$

for $(y, z) = D(F, G)(\bar{x}, (\bar{y}, \bar{z}))(x - \bar{x})$,

i.e.

$$(\{y\} + C_Y) \cap (-\operatorname{int}(C_Y)) \neq \emptyset \qquad (17.28)$$

and

$$(\{z\} + C_Z) \cap (-C_Z + \operatorname{cone}(\bar{z}) - \operatorname{cone}(\bar{z})) \neq \emptyset. \qquad (17.29)$$

The condition (17.28) implies

$$y \in -C_Y - \operatorname{int}(C_Y) = -\operatorname{int}(C_Y),$$

and with the condition (17.29) we obtain

$$z \in -C_Z - C_Z + \operatorname{cone}(\bar{z}) - \operatorname{cone}(\bar{z}) = -C_Z + \operatorname{cone}(\bar{z}) - \operatorname{cone}(\bar{z}).$$

Consequently, we get with the equation (17.26)

$$t(y) + u(z) < 0$$

which contradicts the inequality (17.25). Hence, the set equation (17.27) is satisfied.

Now we come to the actual proof of this theorem. First, we assume that the map (F, G) is $\tilde{C}$-quasiconvex at $(\bar{x}, (\bar{y}, \bar{z}))$. Then we conclude with the equality (17.27)

$$\Big((F(x) - \{\bar{y}\}) \times (G(x) - \{\bar{z}\})\Big) \cap \tilde{C} = \emptyset \text{ for all } x \in \hat{S}.$$

Hence, there is no $x \in \hat{S}$ with

$$(F(x) - \{\bar{y}\}) \cap (-\operatorname{int}(C_Y)) \neq \emptyset$$

and
$$(G(x) - \{\bar{z}\}) \cap (-C_Z + \text{cone}(\bar{z}) - \text{cone}(\bar{z})) \neq \emptyset.$$
Consequently, there is no $x \in \hat{S}$ with
$$(F(x) - \{\bar{y}\}) \cap (-\text{int}(C_Y)) \neq \emptyset$$
and
$$G(x) \cap (-C_Z + \text{cone}(\bar{z}) - \text{cone}(\bar{z})) \neq \emptyset.$$
This means that $(\bar{x}, \bar{y})$ is a weak minimizer of F on $\tilde{S}$.

Finally, we assume that $(\bar{x}, \bar{y})$ is a weak minimizer of F on $\tilde{S}$. Then there is no $x \in \hat{S}$ with
$$(F(x) - \{\bar{y}\}) \cap (-\text{int}(C_Y)) \neq \emptyset$$
and
$$G(x) \cap (-C_Z + \text{cone}(\bar{z}) - \text{cone}(\bar{z})) \neq \emptyset$$
implying
$$(G(x) - \{\bar{z}\}) \cap (-C_Z + \text{cone}(\bar{z}) - \text{cone}(\bar{z})) \neq \emptyset.$$
Then we obtain
$$\Big((F(x) - \{\bar{y}\}) \times (G(x) - \{\bar{z}\})\Big) \cap \tilde{C} = \emptyset \text{ for all } x \in \hat{S}.$$
Together with the equality (17.27) we conclude that the map (F, G) is $\tilde{C}$-quasiconvex. $\square$

Notice that the set $\text{cone}(\bar{z}) - \text{cone}(\bar{z})$ in Theorem 17.19 equals the one dimensional linear subspace of Z generated by $\bar{z}$, i.e. $\{\lambda \bar{z} \in Z \mid \lambda \in \mathbb{R}\}$.

Based on the result of Theorem 17.19 we can now formulate the type of quasiconvexity which is needed for the multiplier rule to be a sufficient optimality condition.

Corollary 17.20. *Let Assumption 17.12 be satisfied. Let the cone C_Y have a nonempty interior $\text{int}(C_Y)$, and let the contingent derivative of (F, G) exist at $(\bar{x}, (\bar{y}, \bar{z}))$ with $\bar{x} \in S$, $\bar{y} \in F(\bar{x})$ and $\bar{z} \in G(\bar{x})$.*

17.4. Generalized Lagrange Multiplier Rule

If there are continuous linear functionals $t \in C_{Y^*}\setminus\{0_{Y^*}\}$ and $u \in C_{Z^*}$ with

$$t(y) + u(z) \geq 0 \text{ for all } (y,z) = D(F,G)(\bar{x},(\bar{y},\bar{z}))(x - \bar{x}) \text{ with } x \in \hat{S}$$

and

$$u(\bar{z}) = 0,$$

and if the map $(F,G) : \hat{S} \to 2^Y \times 2^Z$ is $\tilde{C}$-quasiconvex at $(\bar{x},(\bar{y},\bar{z}))$ with

$$\tilde{C} := (-\text{int}(C_Y)) \times (-C_Z + \text{cone}(\bar{z}) - \text{cone}(\bar{z})),$$

then $(\bar{x},\bar{y})$ is a weak minimizer of the problem (17.11).

Proof. By Theorem 17.19 $(\bar{x},\bar{y})$ is a weak minimizer of F on

$$\tilde{S} = \{x \in \hat{S} \mid G(x) \cap (-C_Z + \text{cone}(\bar{z}) - \text{cone}(\bar{z})) \neq \emptyset\}.$$

For every $x \in S$ we obtain

$$\emptyset \neq G(x) \cap (-C_Z) \subset G(x) \cap (-C_Z + \text{cone}(\bar{z}) - \text{cone}(\bar{z}))$$

implying $x \in \tilde{S}$. Hence, we have $S \subset \tilde{S}$ and $(\bar{x},\bar{y})$ is a weak minimizer of the problem (17.11). □

Example 17.21. We investigate the optimization problem in Example 17.14 again. Since F is $\mathbb{R}_+$-convex (notice that f is a convex functional) and G is $\mathbb{R}_+$-convex (notice that h is also a convex functional), the composite map $(F,G) : X \to 2^{\mathbb{R}} \times 2^{\mathbb{R}}$ has the required quasiconvexity property. Hence, if there are real numbers $t > 0$ and $u \geq 0$ with

$$tf'(\bar{x})(x - \bar{x}) + uh'(\bar{x})(x - \bar{x}) \geq 0 \text{ for all } x \in X$$

and

$$uh(\bar{x}) = 0,$$

then $(\bar{x}, f(\bar{x}))$ is a weak minimizer of the optimization problem (17.11) in Example 17.14.

If we combine Theorem 17.13 and Corollary 17.20 we obtain the main result of this section: a complete characterization of weak minimizers using the Lagrange multiplier rule.

Corollary 17.22. *Let the cones C_Y and C_Z have a nonempty interior $int(C_Y)$ and $int(C_Z)$ respectively, let the set $\hat{S}$ be convex and let the maps F and G be C_Y-convex and C_Z-convex, respectively. Assume that a pair $(\bar{x}, \bar{y}) \in X \times Y$ with $\bar{x} \in S$ and $\bar{y} \in F(\bar{x})$ is given. Let the contingent epiderivative of (F, G) at $(\bar{x}, (\bar{y}, \bar{z}))$ for an arbitrary $\bar{z} \in G(\bar{x}) \cap (-C_Z)$ exist. Moreover, let the regularity assumption (17.12) be satisfied. Then $(\bar{x}, \bar{y})$ is a weak minimizer of the problem (17.11) if and only if there are continuous linear functionals $t \in C_{Y^*} \setminus \{0_{Y^*}\}$ and $u \in C_{Z^*}$ with*

$$t(y) + u(z) \geq 0 \text{ for all } (y, z) = D(F, G)(\bar{x}, (\bar{y}, \bar{z}))(x - \bar{x}) \text{ with } x \in \hat{S}$$

and

$$u(\bar{z}) = 0.$$

Notes

The results of Section 17.1 are taken from [155]. Optimality conditions in set optimization were given by Corley [67] and Luc [213], [214] using contingent derivatives. Oettli [242] introduced a differentiability notion generalizing the Neustadt derivative known in the single-valued case. Here we present a necessary optimality condition for weak minimizers of the set optimization problem (17.1) using the concept of contingent epiderivatives. The proofs of Theorems 17.3 and 17.4 make use of the idea of proof of Corley [67, Thms. 4.1 and 4.2]. The optimality condition for strong minimizers in Theorem 17.7 is based on a result of Aubin and Ekeland [10].

Section 17.2 is based on [18], and the presentation in Section 17.3 follows [57].

Section 17.4 extends the investigations in Chapter 7 to the set-valued case. The results are taken from [110]. The basic idea for the first part of the proof of Theorem 17.13 has been given by Corley [67] using a different differentiability concept. This idea of proof has also been used by Luc and Malivert [216] (e.g., see Thm. 5.6) for the contingent derivative. They have already proved an optimality condition under a regularity assumption (a generalized Slater condition).

Optimality conditions for generalized contingent epiderivatives are published in [152].

Bibliography

[1] Achilles, A., Elster, K.-H., Nehse, R., "Bibliographie zur Vektoroptimierung (Theorie und Anwendungen)", *Math. Operationsforsch. Statist. Ser. Optim.* 10 (1979) 277–321.

[2] Angell, T.S., Kirsch, A., "Multicriteria optimization in antenna design", *Math. Methods Appl. Sci.* 15 (1992) 647–660.

[3] Angell, T.S., Kirsch, A., Kleinman, R.E., "Antenna control and optimization", *Proc. IEEE* 79 (1991) 1559-1568.

[4] Angell, T.S., Kleinman, R.E., "Generalized exterior boundary value problems and optimization for the Helmholtz equation", *J. Optim. Theory Appl.* 37 (1982) 469-497.

[5] Angell, T.S., Kleinman, R.E., "A Galerkin procedure for optimization in radiation problems", *SIAM J. Appl. Math.* 44 (1984) 1246-1257.

[6] Angell, T.S., Kleinman, R.E., Kirsch, A., "Multicriteria optimization in arrays", in: *Proc. J. Intern. de Nice sur les Antennes* (Nice, France, 1992).

[7] Arrow, K.J., Barankin, E.W., Blackwell, D., "Admissible points of convex sets", in: Kuhn, H.W., Tucker, A.W. (eds.), *Contributions to the theory of games* (Princeton University Press, 1953), pp. 87–91.

[8] Aubin, J.-P., *Mathematical methods of game and economic theory* (North-Holland, Amsterdam, 1979).

[9] Aubin, J.-P., "Contingent derivatives of set-valued maps and existence of solutions to nonlinear inclusions and differential inclusions", in: Nachbin, L. (ed.), *Mathematical analysis and applications, Part A* (Academic Press, New York, 1981), pp. 159–229.

[10] Aubin, J.-P., Ekeland, I., *Applied nonlinear analysis* (Wiley, New York, 1984).

[11] Aubin, J.-P., Frankowska, H., *Set-valued analysis* (Birkhäuser, Boston, 1990).

[12] Bacopoulos, A., "Nonlinear Chebychev approximation by vector-norms", *J. Approx. Theory* 2 (1969) 79–84.

[13] Bacopoulos, A., Godini, G., Singer, I., "On best approximation in vector-valued norms", *Colloq. Math. Soc. János Bolyai* 19 (1978) 89–100.

[14] Bacopoulos, A., Godini, G., Singer, I., "Infima of sets in the plane and applications to vectorial optimization", *Rev. Roumaine Math. Pures Appl.* 23 (1978) 343–360.

[15] Bacopoulos, A., Godini, G., Singer, I., "On infima of sets in the plane and best approximation, simultaneous and vectorial, in a linear space with two norms", in: Frehse, J., Pallaschke, D., Trottenberg, U. (eds.), *Special topics of applied mathematics* (North-Holland, Amsterdam, 1980), pp. 219–239.

[16] Bacopoulos, A., Singer, I., "On convex vectorial optimization in linear spaces", *J. Optim. Theory Appl.* 21 (1977) 175–188.

[17] Bacopoulos, A., Singer, I., "Errata corrige: On vectorial optimization in linear spaces", *J. Optim. Theory Appl.* 23 (1977) 473–476.

[18] Baier, J., Jahn, J., "On subdifferentials of set-valued maps", *J. Optim. Theory Appl.* 100 (1999) 233–240.

[19] Barbu, V., *Nonlinear semigroups and differential equations in Banach spaces* (Nordhoff, Leyden, 1976).

[20] Bednarczuk, E.M., Song, W., "Contingent epiderivative and its applications to set-valued optimization", *Control Cybernet.* 27 (1998) 375–386.

[21] Behringer, F.A., "Lexikographischer Ausgleich als Hypernorm-Bestapproximation und eigentliche Effizienz des linearen PARETO-Ausgleichs", *Z. Angew. Math. Mech.* 58 (1978) T461–T464.

[22] Benayoun, R., de Montgolfier, J., Tergny, J., Laritchev, O., "Linear programming with multiple objective functions: Step method (STEM)", *Math. Program.* 1 (1971) 366–375.

[23] Benson, H.P., "An improved definition of proper efficiency for vector maximization with respect to cones", *J. Math. Anal. Appl.* 71 (1979) 232–241.

[24] Benson, H.P., Morin, T.L., "The vector maximization problem: Proper efficiency and stability", *SIAM J. Appl. Math.* 32 (1977) 64–72.

[25] Bishop, E., Phelps, R.R., "The support functionals of a convex set", *Proc. Sympos. Pure Math.* 7 (1963) 27–35.

[26] Bitran, G.R., "Duality for nonlinear multiple-criteria optimization problems", *J. Optim. Theory Appl.* 35 (1981) 367–401

[27] Blaquière, A., *Topics in differential games* (North-Holland, Amsterdam, 1973).

[28] Blaquière, A., Juricek, L., Wiese, K.E., "Geometry of Pareto equilibria in N-person differential games", in: Blaquière [27], pp. 271–310.

[29] Borwein, J.M., "Proper efficient points for maximizations with respect to cones", *SIAM J. Control Optim.* 15 (1977) 57–63.

[30] Borwein, J.M., "Multivalued convexity and optimization: A unified approach to inequality and equality constraints", *Math. Program.* 13 (1977) 183–199.

[31] Borwein, J.M., "The geometry of Pareto efficiency over cones", *Math. Operationsforsch. Statist. Ser. Optim.* 11 (1980) 235–248.

[32] Borwein, J.M., "Convex relations in analysis and optimization", in: Schaible, S., Ziemba, W.T. (eds.), *Generalized concavity in optimization and economics* (Academic Press, New York, 1981), pp. 335–377.

[33] Borwein, J.M., "On the Hahn-Banach extension property", *Proc. Amer. Math. Soc.* 86 (1982) 42–46.

[34] Borwein, J.M., "A Lagrange multiplier theorem and a sandwich theorem for convex relations", *Math. Scand.* 48 (1981) 189–204.

[35] Borwein, J.M., "Continuity and differentiability properties of convex operators", *Proc. London Math. Soc. (3)* 44 (1982) 420–444.

[36] Borwein, J.M., "On the existence of Pareto efficient points", *Math. Oper. Res.* 8 (1983) 64–73.

[37] Borwein, J.M., "Adjoint process duality", *Math. Oper. Res.* 8 (1983) 403–434.

[38] Borwein, J.M., "Subgradients of convex operators", *Math. Operationsforsch. Statist. Ser. Optim.* 15 (1984) 179–191.

[39] Borwein, J.M., Nieuwenhuis, J.W., "Two kinds of normality in vector optimization", *Math. Program.* 28 (1984) 185–191.

[40] Borwein, J.M., Penot, J.P., Thera, M., "Conjugate convex operators", *J. Math. Anal. Appl.* 102 (1984) 399–414.

[41] Borwein, J.M., Zhuang, D.M., "Super efficiency in convex vector optimization", *ZOR Math. Methods Oper. Res* 35 (1991) 175–184.

[42] Boţ, R.I., "Duality and optimality in multiobjective optimization", Dissertation, Technical University of Chemnitz (Chemnitz, 2003).

[43] Bowman, V.J., "On the relationship of the Tchebycheff norm and the efficient frontier of multiple-criteria objectives", in: Thiriez, H., Zionts, S. (eds.), *Multiple criteria decision making, Jouy-en-Josas 1975* (Springer, Berlin, 1976).

[44] Breckner, W.W., "Dualität bei Optimierungsaufgaben in halbgeordneten topologischen Vektorräumen (I)", *Rev. Anal. Numér. Théor. Approx.* 1 (1972) 5–35.

[45] Brézis, H., Browder, F.E., "A general principle on ordered sets in nonlinear functional analysis", *Adv. Math.* 21 (1976) 355–364.

[46] Brockett, R.W., *Finite dimensional linear systems* (John Wiley, New York, 1970).

[47] Brucker, P., "Diskrete parametrische Optimierungsprobleme und wesentlich effiziente Punkte", *Z. Oper. Res.* 16 (1972) 189–197.

[48] Brumelle, S., "Duality for multiple objective convex programs", *Math. Oper. Res.* 6 (1981) 159–172.

[49] Burger, E., *Einführung in die Theorie der Spiele* (Walter de Gruyter, Berlin, Second Edition, 1966).

[50] Censor, Y., "Necessary conditions for Pareto optimality in simultaneous Chebyshev best approximation", *J. Approx. Theory* 27 (1979) 127–134.

[51] Cesari, L., Suryanarayana, M.B., "Existence theorems for Pareto problems of optimization", in: Russell, D.L. (ed.), *Calculus of variations and control theory* (Academic Press, New York, 1976), pp. 139–154.

[52] Cesari, L., Suryanarayana, M.B., "Existence theorems for Pareto optimization in Banach spaces", *Bull. Amer. Math. Soc.* 82 (1976) 306–308.

[53] Cesari, L., Suryanarayana, M.B., "Existence theorems for Pareto optimization; Multivalued and Banach space valued functionals", *Trans. Amer. Math. Soc.* 244 (1978) 37–65.

[54] Char, B.W., Geddes, K.O., Gentleman, W.M., Gonnet, G.H., "The design of Maple: A compact, portable and powerful computer algebra system", in: *Computer algebra* (Springer, Lecture Notes in Computer Science No. 162, Berlin, 1983).

[55] Charnes, A., Cooper, W., *Management models and industrial applications of linear programming, Vol. 1* (Wiley, New York, 1961).

[56] Chen G.-Y., "Necessary conditions of nondominated solutions in multicriteria decision making", *J. Math. Anal. Appl.* 104 (1984) 38–46.

[57] Chen, G.Y., Jahn, J., (1998) "Optimality conditions for set-valued optimization problems", *Math. Methods Oper. Res.* 48 (1998) 187–200.

[58] Chen, G.Y., Jahn, J., "Special issue on 'Set-valued optimization' ", *Math. Methods Oper. Res.* 48 (1998) Issue 2.

[59] Chew, K.L., "Maximal points with respect to cone dominance in Banach spaces and their existence", *J. Optim. Theory Appl.* 40 (1984) 1–53.

[60] Choo, E.U., Atkins, D.R., "Proper efficiency in nonconvex multicriteria programming", *Math. Oper. Res.* 8 (1983) 467–470.

[61] Collatz, L., Krabs, W., *Approximationstheorie* (Teubner, Stuttgart, 1973).

[62] Collin, R.E., Zucker, F.J., *Antenna theory, Part 1* (McGraw-Hill, New York, 1969).

[63] Corley, H.W., "An existence result for maximizations with respect to cones", *J. Optim. Theory Appl.* 31 (1980) 277–281.

[64] Corley, H.W., "Duality theory for maximizations with respect to cones", *J. Math. Anal. Appl.* 84 (1981) 560–568.

[65] Corley, H.W., "Duality theory for the matrix linear programming problem", *J. Math. Anal. Appl.* 104 (1984) 47–52.

[66] Corley, H.W., "Existence and Lagrangian duality for maximizations of set-valued functions", *J. Optim. Theory Appl.* 54 (1987) 489–501.

[67] Corley, H.W., "Optimality conditions for maximizations of set-valued functions", *J. Optim. Theory Appl.* 58 (1988) 1–10.

[68] Craven, B.D., "Strong vector minimization and duality", *Z. Angew. Math. Mech.* 60 (1980) 1–5.

[69] Craven, B.D., "Vector-valued optimization", in: Schaible, S., Ziemba, W.T. (eds.), *Generalized concavity in optimization and economics* (Academic Press, New York, 1981), pp. 661–687.

[70] Cristescu, R., *Topological vector spaces* (Noordhoff, Leyden, 1977).

[71] Cryer, C.W., Dempster, M.A.H., "Equivalence of linear complementarity problems and linear programs in vector lattice Hilbert spaces", *SIAM J. Control Optim.* 18 (1980) 76–90.

[72] Curtain, R.F., "The infinite-dimensional Riccati equation with applications to affine hereditary differential systems", *SIAM J. Control Optim.* 13 (1975) 1130–1143.

[73] Curtain, R.F., Pritchard, A.J., "The infinite-dimensional Riccati equation", *J. Math. Anal. Appl.* 47 (1974) 43–57.

[74] Curtain, R.F., Pritchard, A.J., "The infinite-dimensional Riccati equation for systems defined by evolution operators", *SIAM J. Control Optim.* 14 (1976) 951–983.

[75] Curtain, R.F., Pritchard, A.J., *Functional Analysis in Modern Applied Mathematics* (Academic Press, London, 1977).

[76] Curtain, R.F., Pritchard, A.J., *Infinite-dimensional linear systems theory* (Springer, Lecture Notes in Control and Information Sciences No. 8, Berlin, 1978).

[77] Das, I., "Nonlinear multicriteria optimization and robust optimality", Dissertation, Rice University (Houston, 1997).

[78] Day, M.M., *Normed linear spaces* (Springer, Berlin, Third Edition, 1973).

[79] di Guglielmo, F., "Nonconvex duality in multiobjective optimization", *Math. Oper. Res.* 2 (1977) 285–291.

[80] Dinkelbach, W., *Sensitivitätsanalysen und parametrische Programmierung* (Springer, Berlin, 1969).

[81] Dinkelbach, W., "Über einen Lösungsansatz zum Vektormaximumproblem", in: Beckmann, M. (ed.), *Unternehmensforschung Heute* (Springer, Lecture Notes in Operations Research and Mathematical Systems No. 50, Berlin, 1971) pp. 1–13.

[82] Dinkelbach, W., Dürr, W., "Effizienzaussagen bei Ersatzprogrammen zum Vektormaximumproblem", *Operations Research Verfahren XII* (1972) 69–77.

[83] Dinkelbach, W., *Entscheidungsmodelle* (de Gruyter, Berlin, 1982).

[84] Dunford, N., Schwartz, J.T., *Linear operators, Part I* (Interscience Publishers, New York, 1957).

[85] Dupré, R., Huckert, K., Jahn, J., "Lösung linearer Vektormaximumprobleme durch das STEM-Verfahren", in: Späth, H. (ed.), *Ausgewählte Operations Research Software in FORTRAN* (Oldenbourg, München, 1979).

[86] Dutta, J., Vetrivel, V., "Theorems of the alternative in set-valued optimization", Manuscript, (India, 1999).

[87] Edgeworth, F.Y., *Mathematical psychics* (Kegan Paul, London, 1881).

[88] Ehrgott, M., *Multicriteria optimization* (Springer, Lecture Notes in Economics and Mathematical Systems No. 491, Berlin, 2000).

[89] Eichfelder, G., "Tangentielle Epiableitung mengenwertiger Abbildungen", Diplomarbeit, University of Erlangen-Nürnberg (Erlangen, 2001).

[90] Eidelheit, M., "Zur Theorie der konvexen Mengen in linearen normierten Räumen", *Studia Math.* 6 (1936) 104–111.

[91] Ekeland, I., "On the variational principle", *J. Math. Anal. Appl.* 47 (1974) 324–353.

[92] Ekeland, I., Temam, R., *Convex analysis and variational problems* (North-Holland, Amsterdam, 1976).

[93] Elster, K.-H., Nehse, R., "Konjugierte Operatoren und Subdifferentiale", *Math. Operationsforsch. Statist. Ser. Optim.* 6 (1975) 641–657.

[94] Elster, K.-H., Nehse, R., "Necessary and sufficient conditions for the order-completeness of partially ordered vector spaces", *Math. Nachr.* 81 (1978) 301–311.

[95] Fandel, G., *Optimale Entscheidung bei mehrfacher Zielsetzung* (Springer, Lecture Notes in Economics and Mathematical Systems No. 76, Berlin, 1972).

[96] Fuchs, L., *Partially ordered algebraic systems* (Pergamon Press, Oxford, 1963).

[97] Fuchssteiner, B., Lusky, W., *Convex cones* (North-Holland, Amsterdam, 1981).

[98] Gale, D., *The theory of linear economic models* (McGraw-Hill, New York, 1960).

[99] Gale, D., Kuhn, H.W., Tucker, A.W., "Linear programming and the theory of games", in: Koopmans, T.C. (ed.), *Activity analysis of production and allocation* (John Wiley, New York, 1951), pp. 317–329.

[100] Gearhart, W.B., "On vectorial approximation", *J. Approx. Theory* 10 (1974) 49–63.

[101] Gearhart, W.B., "Compromise solutions and estimation of the noninferior set", *J. Optim. Theory Appl.* 28 (1979) 29–47.

[102] Gearhart, W.B., "Characterization of properly efficient solutions by generalized scalarization methods", *J. Optim. Theory Appl.* 41 (1983) 491–502.

[103] Geoffrion, A.M., "Proper efficiency and the theory of vector maximization", *J. Math. Anal. Appl.* 22 (1968) 618–630.

[104] Gerstewitz, C., Göpfert, A., Lampe, U., "Zur Dualität in der Vektoroptimierung", Abstract of a talk at the conference 'Mathematische Optimierung' (Vitte/Hiddensee, 1980).

[105] Gerth, C., Weidner, P., "Nonconvex separation theorems and some applications in vector optimization", *J. Optim. Theory Appl.* 67 (1990) 297–320.

[106] Girsanov, I.V., *Lectures on mathematical theory of extremum problems* (Springer, Lecture Notes in Economics and Mathematical Systems No. 67, Berlin, 1972).

[107] Goldberg, D.E., *Genetic algorithms in search, optimization, and machine learning* (Addison-Wesley, Reading, Massachusetts, 1989).

[108] Göpfert, A., Nehse, R., *Vektoroptimierung - Theorie, Verfahren und Anwendungen* (Teubner, Leipzig, 1990).

[109] Göpfert, A., Riahi, H., Tammer, C., Zalinescu, C., *Variational methods in partially ordered spaces* (Springer, New York, 2003).

[110] Götz, A., Jahn, J., "The Lagrange multiplier rule in set-valued optimization", *SIAM J. Optim.* 10 (1999) 331–344.

[111] Graef, F., private communication, University of Erlangen (Erlangen, 1992).

[112] Gros, C., "Generalization of Fenchel's duality theorem for convex vector optimization", *European J. Oper. Res.* 2 (1978) 368–376.

[113] Hamel, A., Löhne, A., "Minimal set theorems", Report No. 02–11, Department of Mathematics and Computer Science, University of Halle-Wittenberg (Halle, 2002).

[114] Häßler, S., "Wartezeitminimierung bei Token Bus- und FDDI-Kommunikationsnetzen", Diplomarbeit, University of Erlangen-Nürnberg (Erlangen, 1996).

[115] Häßler, S., Jahn, J., "Game-theoretic approach to the optimization of FDDI computer networks", *J. Optim. Theory Appl.* 106 (2000) 463–474.

[116] Hartley, R., "On cone-efficiency, cone-convexity and cone-compactness", *SIAM J. Appl. Math.* 34 (1978) 211–222.

[117] Hartwig, H., "Verallgemeinert konvexe Vektorfunktionen und ihre Anwendung in der Vektoroptimierung", *Math. Operationsforsch. Statist. Ser. Optim.* 10 (1979) 303–316.

[118] Henig, M.I., "Proper efficiency with respect to cones", *J. Optim. Theory Appl.* 36 (1982) 387–407.

[119] Henig, M.I., "A cone separation theorem", *J. Optim. Theory Appl.* 36 (1982) 451–455.

[120] Hestenes, M.R., *Calculus of variations and optimal control theory* (John Wiley, New York, 1966).

[121] Hille, E., Phillips, R.S., *Functional analysis and semi-groups* (American Mathematical Society Colloquium Publications Volume XXXI, Providence, 1957).

[122] Hillermeier, C.F., *Nonlinear multiobjective optimization: A generalized homotopy approach* (Birkhäuser, Basel, 2001).

[123] Hillermeier, C.F., Jahn, J., "Multiobjective optimization: Survey of methods, and industrial applications", *Surveys on Mathematics for Industry* (2003).

[124] Holmes, R.B., *A course on optimization and best approximation* (Springer, Lecture Notes in Mathematics No. 257, Berlin, 1972).

[125] Holmes, R.B., *Geometric functional analysis and its applications* (Springer, New York, 1975).

[126] Huang, S.C., "Note on the mean-square strategy of vector-valued objective functions", *J. Optim. Theory Appl.* 9 (1972) 364–366.

[127] Hurwicz, L., "Programming in linear spaces", in: Arrow, K.J., Hurwicz, L., Uzawa, H. (eds.), *Studies in linear and non-linear programming* (Stanford University Press, Stanford, 1958), pp. 38–102.

[128] Ijiri, Y., *Management goals and accounting for control* (North-Holland, Amsterdam, 1965).

[129] Ioffe, A.D., Tihomirov, V.M., *Theory of extremal problems* (North-Holland, Amsterdam, 1979).

[130] Isac, G., "Sur l'existence de l'optimum de Pareto", *Riv. Mat. Univ. Parma (4)* 9 (1983) 303–325.

[131] Isac, G., Bulavsky, V.A., Kalashnikov, V.V., *Complementarity, equilibrium, efficiency and economics* (Kluwer, Dordrecht, 2002).

[132] Isermann, H., "The enumeration of the set of all efficient solutions for a linear multiple objective program", *Oper. Res. Quart.* 28 (1977) 711–725.

[133] Isermann, H., "On some relations between a dual pair of multiple objective linear programs", *Z. Oper. Res. Ser. A* 22 (1978) 33–41.

[134] Isermann, H., "Duality in multiple objective linear programming", in: Zionts, S. (ed.), *Multiple criteria problem solving* (Springer, Lecture Notes in Economics and Mathematical Systems No. 155, Berlin, 1978), pp. 274–285.

[135] Isermann, H., "Users manual for the EFFACET computer package for solving multiple objective linear programming problems", Manuscript, University of Bielefeld (Bielefeld, 1984).

[136] Iwanow, E., Nehse, R., "On efficient and properly efficient points in vector optimization problems (in Russian)", *Wiss. Z. Tech. Hochsch. Ilmenau* 30 (1984) 55–60.

[137] Jacobson, D.H., Martin, D.H., Pachter, M., Geveci, T., *Extensions of linear-quadratic control theory* (Springer, Lecture Notes in Control and Information Sciences No. 27, Berlin, 1980).

[138] Jahn, J., "Duality in vector optimization", *Math. Program.* 25 (1983) 343–353.

[139] Jahn, J., "Zur vektoriellen linearen Tschebyscheff-Approximation", *Math. Operationsforsch. Statist. Ser. Optim.* 14 (1983) 577–591.

[140] Jahn, J., "Neuere Entwicklungen in der Vektoroptimierung", in: Steckhan, H., Bühler, W., Jäger, K.-E., Schneeweiß, Ch., Schwarze, J. (eds.), *Operations Research Proceedings 1983* (Springer, New York, 1984), pp. 511–519.

[141] Jahn, J., "Scalarization in vector optimization", *Math. Program.* 29 (1984) 203–218.

[142] Jahn, J., "A characterization of properly minimal elements of a set", *SIAM J. Control Optim.* 23 (1985) 649–656.

[143] Jahn, J., "Some characterizations of the optimal solutions of a vector optimization problem", *OR Spektrum* 7 (1985) 7-17.

[144] Jahn, J., "Existence theorems in vector optimization", *J. Optim. Theory Appl.* 50 (1986) 397–406.

[145] Jahn, J., *Mathematical vector optimization in partially ordered linear spaces* (Peter Lang, Frankfurt, 1986).

[146] Jahn, J., "Parametric approximation problems arising in vector optimization", *J. Optim. Theory Appl.* 54 (1987) 503–516.

[147] Jahn, J., "A method of reference point approximation in vector optimization", in: Schellhaas, H., van Beek, P., Isermann, H.,

Schmidt, R., Zijlstra, M., (eds.), *Operations Research Proceedings 1987* (Springer, Berlin, 1988), pp. 576–587.

[148] Jahn, J., "Vector optimization: Theory, methods, and application to design problems in engineering", in: Krabs, W., Zowe, J. (eds.), *Modern methods of optimization* (Springer, Berlin, 1992), pp. 127–150.

[149] Jahn, J., *Introduction to the theory of nonlinear optimization* (Springer, Berlin, Second Edition, 1996).

[150] Jahn, J., "Optimality conditions in set-valued vector optimization" in: Fandel, G., Gal, T., Hanne, T. (eds.), *Multiple criteria decision making, Hagen 1995* (Springer, Berlin, 1997), pp. 22–30.

[151] Jahn, J., Khan, A.A., "Some calculus rules for contingent epiderivatives", Manuscript, University of Erlangen-Nürnberg (Erlangen, 2002).

[152] Jahn, J., Khan, A.A., "Generalized contingent epiderivatives in set-valued optimization: Optimality conditions", *Numer. Funct. Anal. Optim.* 23 (2002) 807–831.

[153] Jahn, J., Klose, J., Merkel, A., "On the application of a method of reference point approximation to bicriterial optimization problems in chemical engineering", in: Oettli, W., Pallaschke, D. (eds.), *Advances in Optimization Proceedings 1991* (Springer, Lecture Notes in Economics and Mathematical Systems No. 382, Berlin, 1992), pp. 478–491.

[154] Jahn, J., Merkel, A., "Reference point approximation method for the solution of bicriterial nonlinear optimization problems", *J. Optim. Theory Appl.* 74 (1992) 87–103.

[155] Jahn, J., Rauh, R., "Contingent epiderivatives and set-valued optimization", *Math. Methods Oper. Res.* 46 (1997) 193–211.

[156] Jahn, J., Sachs, E., "Generalized convex mappings and vector optimization", Research report, North Carolina State University (Raleigh, 1983).

[157] Jahn, J., Sachs, E., "Generalized quasiconvex mappings and vector optimization", *SIAM J. Control Optim.* 24 (1986) 306–322.

[158] James, R., "Weak compactness and reflexivity", *Israel J. Math.* 2 (1964) 101–119.

[159] Jameson, G., *Ordered linear spaces* (Springer, Lecture Notes in Mathematics No. 141, Berlin, 1970).

[160] John, F., "Extremum problems with inequalities as subsidiary conditions", in: Friedrichs, K.O., Neugebauer, O.E., Stoker, J.J. (eds.), *Studies and essays: Courant anniversary volume* (Interscience Publishers, New York, 1948).

[161] Jüschke, A., "Ein bikriterielles Optimierungsproblem aus der Antennentheorie", Diplomarbeit, University of Erlangen-Nürnberg (Erlangen, 1994).

[162] Jüschke, A., Jahn, J., Kirsch, A., "A bicriterial optimization problem of antenna design", *Comput. Optim. Appl.* 7 (1997) 261-276.

[163] Juricek, L., "Games with coalitions", in: Blaquière [27], pp. 311-344.

[164] Kakutani, S., "Concrete representation of abstract M-spaces", *Ann. of Math. (2)* 42 (1941) 994-1024.

[165] Kaliszewski, I., *Quantitative Pareto analysis by cone separation technique* (Kluwer, Boston, 1994).

[166] Kantorovitch, L., "Lineare halbgeordnete Räume", *Recueil Mathématique* 44 (1937) 121-168.

[167] Kantorovitch, L., "The method of successive approximations for functional equations", *Acta Math.* 71 (1939) 63-97.

[168] Karlin, S., *Mathematical methods and theory in games, programming, and economics, Vol. I* (Addison-Wesley, 1959).

[169] Kawasaki, H., "A duality theorem in multiobjective nonlinear programming", *Math. Oper. Res.* 7 (1982) 95-110.

[170] Kelley, J.L., Namioka, I., *Linear topological spaces* (Van Nostrand, Princeton, 1963).

[171] Kirsch, A., Warth, W., Werner, J., *Notwendige Optimalitätsbedingungen und ihre Anwendung* (Springer, Lecture Notes in Economics and Mathematical Systems No. 152, Berlin, 1978).

[172] Kirsch, A., Wilde, P., "The optimization of directivity and signal-to-noise ratio of an arbitrary antenna array", *Math. Methods Appl. Sci.* 10 (1988) 153-164.

[173] Kitagawa, H., Watanabe, N., Nishimura, Y., Matsubara, M., "Some pathological configurations of noninferior set appearing in multicriteria optimization problems of chemical processes", *J. Optim. Theory Appl.* 38 (1982) 541-563.

[174] Klehmet, U., *Leistungsbewertung und Optimierung von zeitbegrenzten Polling-Systemen am Beispiel des Feldbusses PROFIBUS* (VDI Verlag, Reihe 10: Nr. 371, Düsseldorf, 1995).

[175] Klehmet, U., "Model supported parameter optimization of timed polling systems (PROFIBUS)", to appear.

[176] Klose, J., "Sensitivity analysis using the tangent derivative", *Numer. Funct. Anal. Optim.* 13 (1992) 143–153.

[177] Kolmogoroff, A.N., "A remark to the polynomials of P.L. Chebyshev differing from a given function at least (in Russian)", *Uspekhi Mat. Nauk* 3 (1948) 216–221.

[178] König, H., "Sublineare Funktionale", *Arch. Math. (Basel)* 23 (1972) 500–508.

[179] König, H., "Neue Methoden und Resultate aus Funktionalanalysis und Konvexer Analysis", *Operations Research Verfahren* 28 (1978) 6–16.

[180] König, H., "Der Hahn-Banach-Satz von Rodé für unendlichstellige Operationen", *Arch. Math. (Basel)* 35 (1980) 292–304.

[181] König, H., "On some basic theorems in convex analysis", in: Korte, B. (ed.), *Modern applied mathematics, optimization and operations research* (North-Holland, Amsterdam, 1982), pp. 107–144.

[182] Koopmans, T.C., "Analysis of production as an efficient combination of activities", in: Koopmans, T.C. (ed.), *Activity analysis of production and allocation* (Wiley, New York, 1951), pp. 33–97.

[183] Krabs, W., "Über differenzierbare asymptotisch konvexe Funktionenfamilien bei der nichtlinearen gleichmäßigen Approximation", *Arch. Ration. Mech. Anal.* 27 (1967) 275–288.

[184] Krabs, W., "Optimization and Approximation" (John Wiley, Chichester, 1979).

[185] Krabs, W., "Convex optimization and approximation", in: Korte, B. (ed.), *Modern applied mathematics* (North-Holland, Amsterdam, 1981), pp. 322–352.

[186] Krein, M.G., Rutman, M.A., "Linear operators leaving invariant a cone in a Banach space" (Uspekhi Mat. Nauk (N.S.) 3, 23 (1948) 3–95), *AMS Translation Series 1, Volume 10 (Functional Analysis and Measure Theory)* (Providence, 1962), pp. 199–325.

[187] Kuhn, H.W., Tucker, A.W., "Nonlinear programming", in: Neyman, J. (ed.), *Proceedings of the second Berkeley Symposium on mathematical Statistics and Probability* (University of California Press, Berkeley, 1951), pp. 481–492.

[188] Kuroiwa, D., "Natural criteria of set-valued optimization", Manuscript, Shimane University (Japan, 1998).

[189] Kuroiwa, D., "Existence theorems of set optimization with set-valued maps", Preprint, Shimane University (Japan, 1999).

[190] Kuroiwa, D., "On set-valued optimization", *Nonlinear Anal.* 47 (2001) 1395–1400.

[191] Kuroiwa, D., "Some duality theorems of set-valued optimization with natural criteria", in: Takahashi, W. (ed.), *Proceedings of the first International Conference on Nonlinear Analysis and Convex Analysis* (World Scientific, Singapore, 1999), pp. 221–228.

[192] Kusraev, A.G., "On necessary conditions for an extremum of nonsmooth vector-valued mappings", *Soviet Math. Dokl.* 19 (1978) 1057–1060.

[193] Kusraev, A.G., *Vector-valued duality and its applications (in Russian)* (Nauka, Moskow, 1985).

[194] Ladas, G.E., Lakshmikantham, V., *Differential equations in abstract spaces* (Academic Press, New York, 1972).

[195] Lagrange, J.L., *Théorie des fonctions analytiques* (Paris, 1797).

[196] Lampe, U., "Dualität und eigentliche Effizienz in der Vektoroptimierung", in: Guddat, J., Wendler, K. (eds.), *Bericht zur Arbeitstagung Vektoroptimierung* (Seminarbericht Nr. 37, Humboldt University Berlin, 1981), pp. 45–54.

[197] Landes, H., "Das Lemma von Bishop-Phelps – Weiterentwicklungen und Anwendungen", Diplomarbeit, University of Erlangen-Nürnberg (Erlangen, 1984).

[198] Lehmann, R., Oettli, W., "The theorem of the alternative, the key-theorem, and the vector-maximum problem", *Math. Program.* 8 (1975) 332–344.

[199] Leitmann, G., "Sufficiency theorems for optimal control", *J. Optim. Theory Appl.* 2 (1968) 285–292.

[200] Leitmann, G., "A note on a sufficiency theorem for optimal control", *J. Optim. Theory Appl.* 3 (1969) 76–78.

[201] Leitmann, G., *Cooperative and non-cooperative many players differential games* (Springer, CISM Courses and Lectures No. 190, Wien, 1974).

[202] Leitmann, G., *Multicriteria decision making and differential games* (Plenum Press, New York, 1976).

[203] Leitmann, G., *The calculus of variations and optimal control* (Plenum Press, New York, 1981).

[204] Leitmann, G., Liu, P.T., "A differential game model of labor-management negotiation during a strike", *J. Optim. Theory Appl.* 13 (1974) 427–435.

[205] Leitmann, G., Marzollo, A., *Multicriteria decision making* (Springer, CISM Courses and Lectures No. 211, Wien, 1975).

[206] Leitmann, G., Rocklin, S., Vincent, T.L., "A note on control space properties of cooperative games", *J. Optim. Theory Appl.* 9 (1972) 379–390.

[207] Li, Z., "A theorem of the alternative and its application to the optimization of set-valued maps", *J. Optim. Theory Appl.* 100 (1999) 365–375.

[208] Ljusternik, L.A., Sobolew, W.I., *Elemente der Funktionalanalysis* (Akademie-Verlag, Berlin, 1968).

[209] Lo, Y.T., Lee, S.W., Lee, Q.H., "Optimization of directivity and signal-to-noise ratio of an arbitrary antenna array", *Proc. IEEE* 54 (1966) 1033–1045.

[210] Lommatzsch, K., *Anwendungen der linearen parametrischen Optimierung* (Birkhäuser, Basel, 1979).

[211] Loridan, P., "ε-Solutions in vector minimization problems", *J. Optim. Theory Appl.* 43 (1984) 265–276.

[212] Luc, D.T., "On duality theory in multiobjective programming", *J. Optim. Theory Appl.* 43 (1984) 557–582.

[213] Luc, D.T., *Theory of vector optimization* (Springer, Berlin, 1989).

[214] Luc, D.T., "Contingent derivatives of set-valued maps and applications to vector optimization", *Math. Program.* 50 (1991) 99–111.

[215] Luc, D.T., Jahn, J., "Axiomatic approach to duality in optimization", *Numer. Funct. Anal. Optim.* 13 (1992) 305–326.

[216] Luc, D.T., Malivert, C., "Invex optimization problems", *Bull. Austral. Math. Soc.* 46 (1992) 47–66.

[217] Luenberger, D.G., *Optimization by vector space methods* (John Wiley, New York, 1969).

[218] Lyusternik, L.A., "Conditional extrema of functionals (in Russian)", *Mat. Sb.* 41 (1934) 390–401.

[219] Malivert, C., "Dualité en programmation linéaire multicritère", *Math. Operationsforsch. Statist. Ser. Optim.* 15 (1984) 555–572.

[220] Mangasarian, O.L., *Nonlinear programming* (McGraw-Hill, New York, 1969).

[221] Martin, R.H., Jr., *Nonlinear operators and differential equations in Banach spaces* (John Wiley, New York, 1976).

[222] Mazur, S., "Über konvexe Mengen in linearen normierten Räumen", *Studia Math.* 4 (1933) 70–84.

[223] Meinardus, G., *Approximation von Funktionen und ihre numerische Behandlung* (Springer, Berlin, 1964).

[224] Merkel, A., "Ein Verfahren der Referenzpunktapproximation für bikriterielle nichtlineare Optimierungsprobleme", Diplomarbeit, University of Erlangen-Nürnberg (Erlangen, 1989).

[225] Minami, M., "Weak Pareto optimality of multiobjective problems in a locally convex linear topological space", *J. Optim. Theory Appl.* 34 (1981) 469–484.

[226] Minami, W., "Weak Pareto optimality of multiobjective problems in a Banach space", *Bulletin of Mathematical Statistics* 19 (1981) 19–23.

[227] Minami, M., "Weak Pareto-optimal necessary conditions in a nondifferentiable multiobjective program on a Banach space", *J. Optim. Theory Appl.* 41 (1983) 451–461.

[228] Nachbin, L., *Topology and order* (Van Nostrand, Princeton, 1965).

[229] Nakano, H., *Semi-ordered linear spaces* (Maruzen, Tokyo, 1955).

[230] Nakayama, H., "Geometric consideration of duality in vector optimization", *J. Optim. Theory Appl.* 44 (1984) 625–655.

[231] Nakayama, H., "Duality theory in vector optimization: An overview", in: Serafini, P. (ed.), *Mathematics of multi objective optimization* (Springer, CISM Courses and Lectures No. 289, Wien, 1985), pp. 105–127.

[232] Nashed, M.Z., "Differentiability and related properties of nonlinear operators: Some aspects of the role of differentials in nonlinear functional analysis", in: Rall, L.B. (ed.), *Nonlinear functional analysis and applications* (Academic Press, New York, 1971) pp. 103–309.

[233] Nehse, R., "Strong pseudo-convex mappings in dual problems", *Math. Operationsforsch. Statist. Ser. Optim.* 12 (1981) 483–491.

[234] Nehse, R., "Duale Vektoroptimierungsprobleme vom Wolfe-Typ", in: Guddat, J., Wendler, K. (eds.), *Bericht zur Arbeitstagung Vektoroptimierung* (Seminarbericht Nr. 37, Humboldt-Universität Berlin, 1981), pp. 55–60.

[235] Nehse, R., "Bibliographie zur Vektoroptimierung - Theorie und Anwendungen (1. Fortsetzung)", *Math. Operationsforsch. Statist. Ser. Optim.* 13 (1982) 593–625.

[236] Nehse, R., "Zwei Fortsetzungssätze", *Wiss. Z. Tech. Hochsch. Ilmenau* 30 (1984) 49–57.

[237] Nieuwenhuis, J.W., "Supremal points and generalized duality", *Math. Operationsforsch. Statist. Ser. Optim.* 11 (1980) 41–59.

[238] Nieuwenhuis, J.W., "Properly efficient and efficient solutions for vector maximization problems in Euclidean space", *J. Math. Anal. Appl.* 84 (1981) 311–317.

[239] Nikaidô, H., "On von Neumann's minimax theorem", *Pacific J. Math.* 4 (1954) 65–72.

[240] Nishnianidze, Z.G., "Fixed points of monotonic multiple-valued operators (in Russian)", *Bull. Georgian Acad. Sci.* 114 (1984) 489–491.

[241] Oettli, W., "A duality theorem for the nonlinear vector-maximum problem", in: Prékopa, A. (ed.), *Colloquia Mathematica Societatis János Bolyai, 12. Progress in Operations Research, Eger (Hungary), 1974* (North-Holland, Amsterdam, 1976), pp. 697–703.

[242] Oettli, W., "Optimality conditions for programming problems involving multivalued mappings", in: Korte, B. (ed.), *Modern applied mathematics, optimization and operations research* (North-Holland, Amsterdam, 1980).

[243] Oettli, W., "Kolmogorov conditions for vectorial optimization problems" *OR Spektrum* 17 (1995).

[244] Pareto, V., *Cours d'economie politique* (F. Rouge, Lausanne, 1896).

[245] Pascoletti, A., Serafini, P., "Scalarizing vector optimization problems", *J. Optim. Theory Appl.* 42 (1984) 499–524.

[246] Peemöller, J., "Verallgemeinerte Quasikonvexitätsbegriffe", *Methods Oper. Res.* 40 (1981) 133–136.

[247] Penot, J.-P., "Calcul sous-differentiel et optimisation", *J. Funct. Anal.* 27 (1978) 248–276.

[248] Penot, J.-P., "L'optimisation à la Pareto: Deux ou trois choses que je sais d'elle", *Publications Mathématiques de Pau* (1978).

[249] Peressini, A.L., *Ordered Topological Vector Space* (Harper & Row, New York, 1967).

[250] Phelps, R.R., "Support cones in Banach spaces and their applications", *Adv. Math.* 13 (1974) 1–19.

[251] Podinovskij, V.V., Noghin, V.D., *Pareto optimization – Solution of multicriteria decision problems (in Russian)* (Nauka, Moskow, 1982).
[252] Polak, E., "On the approximation of solutions to multiple criteria decision making problems", in: Zeleny, M. (ed.), *Multiple criteria decision making, Kyoto 1975* (Springer, Berlin, 1976), pp. 271–281.
[253] Pontryagin, L.S., Boltyanskii, V.G., Gamkrelidze, R.V., Mishchenko, E.F., *The mathematical theory of optimal processes* (Interscience, New York, 1962).
[254] Postolică, V., "Vectorial optimization programs with multifunctions and duality", *Ann. Sci. Math. Québec* 10 (1986) 85–102.
[255] Reemtsen, R., "On level sets and an approximation problem for the numerical solution of a free boundary problem", *Computing* 27 (1981) 27–35.
[256] Reichenbach, G., " Der Begriff der eigentlichen Effizienz in der Vektoroptimierung", Diplomarbeit, University of Erlangen-Nürnberg (Erlangen, 1989).
[257] Reul, S., "Konzept für ein regelbasiertes Leistungsmanagement", Dissertation, Dresden University of Technology (Dresden, 1995).
[258] Roberts, R.W., Varberg, D.E., *Convex functions* (Academic Press, New York, 1973).
[259] Robertson, A.P., Robertson, W.J., *Topological vector spaces* (Cambridge University Press, Cambridge, 1966).
[260] Rockafellar, R.T., *Convex analysis* (Princeton University Press, Princeton, 1970).
[261] Rockafellar, R.T., *Conjugate duality and optimization* (SIAM, CBMS Lecture Note Series No. 16, Philadelphia, 1974).
[262] Rockafellar, R.T., *The theory of subgradients and its applications to problems of optimization – Convex and nonconvex functions* (Heldermann, Berlin, 1981).
[263] Rodé, G., "Eine abstrakte Version des Satzes von Hahn-Banach", *Arch. Math. (Basel)* 31 (1978) 474–481.
[264] Rodé, G., "Superkonvexität und schwache Kompaktheit", *Arch. Math. (Basel)* 36 (1981) 62–72.
[265] Rolewicz, S., "On a norm scalarization in infinite dimensional Banach spaces", *Control Cybernet.* 4 (1975) 85–89.
[266] Rosinger, E.E., "Duality and alternative in multiobjective optimization", *Proc. Amer. Math. Soc.* 64 (1977) 307–312.

[267] Rosinger, E.E., "Multiobjective duality without convexity", *J. Math. Anal. Appl.* 66 (1978) 442–450.

[268] Sachs, E., "Differenzierbarkeit in der Optimierungstheorie und Anwendung auf Kontrollprobleme", Dissertation, Darmstadt University of Technology (Darmstadt, 1975).

[269] Sachs, E., "Differentiability in optimization theory", *Math. Operationsforsch. Statist. Ser. Optim.* 9 (1978) 497–513.

[270] Salukvadze, M.E., "On the optimization of vector functionals (in Russian)", *Avtomat. i Telemekh.* 8 (1971) 5–15.

[271] Salukvadze, M.E., *Vector-valued optimization problems in control theory* (Academic Press, New York, 1979).

[272] Salz, W., "Die Untersuchung einer Vektoroptimierungsaufgabe, die zur Beschreibung kooperativer Differentialspiele dient", Dissertation, University of Bonn (Bonn, 1975).

[273] Schaefer, H.H., *Topological vector spaces* (Springer, New York, 1971).

[274] Schäffler, S., *Global optimization using stochastic integration* (S. Roderer Verlag, Regensburg, 1995).

[275] Schäffler, S., Schultz, R., Weinzierl, K., "Stochastic method for the solution of unconstrained vector optimization problems", *J. Optim. Theory Appl.* 114 (2002) 209–222.

[276] Schilling, K., Navigation of transportation robots (private communication, 2002).

[277] Schmitendorf, W.E., Moriarty, G., "A sufficiency condition for coalitive Pareto-optimal solutions", *J. Optim. Theory Appl.* 18 (1976) 93–102.

[278] Schniederjans, M.J., *Linear goal programming* (Petrocelli Books, Princeton, 1984).

[279] Schönfeld, P., "Some duality theorems for the non-linear vector maximum problem", *Unternehmensforschung* 14 (1970) 51–63.

[280] Schüler, M., Hafner, M., Isermann, M., "Model-based optimization of IC engines by means of fast neural networks, Part 2", *MTZ worldwide* 61 (2000) 28–31.

[281] Serafini, P., "A unified approach for scalar and vector optimization", in: Serafini, P. (ed.), *Mathematics of multi objective optimization* (Springer, CISM Courses and Lectures No. 289, Wien, 1985), pp. 89–104.

[282] Song, W., "A generalization of Fenchel duality in set-valued vector optimization", *Math. Methods Oper. Res.* 48 (1998) 259–272.

[283] Stadler, W., "A survey of multicriteria optimization or the vector maximum problem, Part I: 1776-1960", *J. Optim. Theory Appl.* 29 (1979) 1–52.

[284] Stadler, W., "A comprehensive bibliography on MCDM", in: Zeleny, M. (ed.), *A source book of multiple criteria decision making* (JAI-Press, Greenwich, 1984).

[285] Stadler, W., "Multicriteria optimization in mechanics (A survey)", *Appl. Mech. Rev.* 37 (1984) 277–286.

[286] Stadler, W., "Initiators of multicriteria optimization", in: Jahn, J., Krabs, W. (eds.), *Recent advances and historical development of vector optimization* (Springer, Berlin, 1987), pp. 3–47.

[287] Stadler, W. (ed.), *Multicriteria optimization in engineering and in the sciences* (Plenum Press, New York, 1988).

[288] Stalford, H., "Sufficient conditions for optimal control with state and control constraints", *J. Optim. Theory Appl.* 7 (1971) 118–135.

[289] Stalford, H.L., "Criteria for Pareto-optimality in cooperative differential games", *J. Optim. Theory Appl.* 9 (1972) 391–398.

[290] Starr, A.W., "Non-zero sum differential games: Concepts and models", Techn. Report 590, Division of Engineering and Applied Physics, Harvard University (Cambridge, 1969).

[291] Stern, R.J., Ben-Israel, A., "On linear optimal control problems with multiple quadratic criteria", in: Cochrane, J.L., Zeleny, M., (eds.), *Multiple criteria decision making* (University of South Carolina Press, Columbia, 1973).

[292] Steuer, R.E., "Repertoire of multiple objective linear programming test problems", Working paper in Business Administration, University of Kentucky (Lexington, 1978).

[293] Steuer, R.E., *Multiple criteria optimization: Theory, computation, and application* (John Wiley, New York, 1986).

[294] Steuer, R.E., Choo, E.-U., "An interactive weighted Tchebycheff procedure for multiple objective programming", *Math. Program.* 26 (1983) 326–344.

[295] Tammer, C., "Charakterisierung effizienter Elemente von Vektoroptimierungsaufgaben", Habilitationsschrift, Technical University of Merseburg (Halle, 1991).

[296] Tangemann, M., "Mean waiting time approximations for symmetric and asymmetric polling systems with time-limited service", in: Walke, B., Spaniol, O., (eds.), *Messung, Modellierung und Bewertung von Rechen- und Kommunikationssystemen* (Springer, Berlin, 1993) pp. 143–158.

[297] Tanino, T., "Conjugate maps and conjugate duality", in: Serafini, P. (ed.), *Mathematics of multi objective optimization* (Springer, CISM Courses and Lectures No. 289, Wien, 1985), pp. 129–155.

[298] Tanino, T., Sawaragi, Y., "Duality theory in multiobjective programming", *J. Optim. Theory Appl.* 27 (1979) 509–529.

[299] Tanino, T., Sawaragi, Y., "Conjugate maps and duality in multiobjective optimization", *J. Optim. Theory Appl.* 31 (1980) 473–499.

[300] Thibault, L., "Subdifferentials of nonconvex vector-valued functions", *J. Math. Anal. Appl.* 86 (1982) 319–344.

[301] Thibault, L., "On generalized differentials and subdifferentials of Lipschitz vector-valued functions", *Nonlinear Anal.* 6 (1982) 1037–1053.

[302] Tichomirov, V.M., *Grundprinzipien der Theorie der Extremalaufgaben* (Teubner-Texte zur Mathematik Bd. 30, Leipzig, 1982).

[303] Triebel, H., *Höhere Analysis* (VEB Deutscher Verlag der Wissenschaften, Berlin, 1972).

[304] Truong, X.D.H., "Optimal solution for set-valued optimization problems: The set optimization approach", Preprint No. 285, Institute of Applied Mathematics, University of Erlangen-Nürnberg (Erlangen, 2001).

[305] Truong, X.D.H., "Ekeland's variational principle for a set-valued map studied with the set optimization approach", Preprint No. 289, Institute of Applied Mathematics, University of Erlangen-Nürnberg (Erlangen, 2002).

[306] Truong, X.D.H., "Ekeland's variational principle for a set-valued map involving coderivatives", Preprint No. 295, Institute of Applied Mathematics, University of Erlangen-Nürnberg (Erlangen, 2002).

[307] Valadier, M., "Sous-différentiabilité de fonctions convexes à valeurs dans un espace vectoriel ordonné", *Math. Scand.* 30 (1972) 65–74.

[308] van Slyke, R.M., Wets, R.J.-B., "A duality theory for abstract mathematical programs with applications to optimal control theory", *J. Math. Anal. Appl.* 22 (1968) 679–706.

[309] van Tiel, J., *Convex analysis* (John Wiley, Chichester, 1984).

[310] Vincent, T.L., Grantham, W.J., *Optimality in parametric systems* (John Wiley, New York, 1981).

[311] Vincent, T.L., Leitmann, G., "Control-space properties of cooperative games", *J. Optim. Theory Appl.* 6 (1970) 91–113.

[312] Vogel, W., "Ein Maximum-Prinzip für Vektoroptimierungs-Aufgaben", *Operations Research Verfahren XIX* (1974) 161–184.

[313] Vogel, W., *Vektoroptimierung in Produkträumen* (Anton Hain, Meisenheim am Glan, 1977).

[314] Vogel, W., "Halbnormen und Vektoroptimierung", in: Albach, H., Helmstadter, E., Henn, R. (eds.), *Quantitative Wirtschaftsforschung, Wilhelm Krelle zum 60. Geburtstag* (Tübingen, 1977), pp. 703–714.

[315] Vogel, W., "Vectoroptimization without the use of functions", *Methods Oper. Res.* 40 (1981) 27–44.

[316] von Neumann, J., "Zur Theorie der Gesellschaftsspiele", *Math. Ann.* 100 (1928) 295–320.

[317] Vulikh, B.Z., *Introduction to the theory of partially ordered spaces* (Wolters-Noordhoff, Groningen, 1967).

[318] Wanka, G., "Kolmogorov-conditions for vectorial approximation problems", *OR Spektrum* 16 (1994) 53–58.

[319] Warga, J., *Optimal control of differential and functional equations* (Academic Press, New York, 1972).

[320] Weidner, P., "Ein Trennungskonzept und seine Anwendung auf Vektoroptimierungsverfahren", Dissertation B, University of Halle-Wittenberg (Halle, 1990).

[321] Wendell, R.E., Lee, D.N., "Efficiency in multiple objective optimization problems", *Math. Program.* 12 (1977) 406–414.

[322] Werner, J., "Der Satz von Ljusternik", Research Paper, University of Göttingen (Göttingen, 1983).

[323] Werner, J., *Optimization – Theory and applications* (Vieweg, Braunschweig, 1984).

[324] White, D.J., *Optimality and efficiency* (John Wiley, Chichester, 1982).

[325] White, D.J., "Vector maximization and Lagrange multipliers", *Math. Program.* 31 (1985) 192–205.

[326] Wierzbicki, A.P., "Penalty methods in solving optimization problems with vector performance criteria", Technical report of the Institute of Automatic Control, TU of Warsaw (Warsaw, 1974).

[327] Wierzbicki, A.P., "Basic properties of scalarizing functionals for multiobjective optimization", *Math. Operationsforsch. Statist. Ser. Optim.* 8 (1977) 55–60.

[328] Wierzbicki, A.P., "The use of reference objectives in multiobjective optimization", in: Fandel, G., Gal, T. (eds.), *Multiple criteria decision making – Theory and application* (Springer, Lecture Notes in Economics and Mathematical Systems No. 177, Berlin, 1980), pp. 468–486.

[329] Wierzbicki, A.P., "A mathematical basis for satisficing decision making", in: Morse, J.N. (ed.), *Organisations: Multiple agents with multiple criteria* (Springer, Lecture Notes in Economics and Mathematical Systems No. 190, Berlin, 1981), pp. 465–485.

[330] Wierzbicki, A.P., "A mathematical basis for satisficing decision making", *Math. Modelling* 3 (1982) 391–405.

[331] Winkler, K., "Aspekte Mehrkriterieller Optimierung $C(T)$-wertiger Abbildungen", Dissertation, University of Halle-Wittenberg (Halle, 2003).

[332] Yang, Q.X., "A Hahn-Banach theorem in ordered linear spaces and its applications", *Optimization* 25 (1992) 1–9.

[333] Younes, Y.M., "Studies on discrete vector optimization", Dissertation, University of Demiatta (Egypt, 1993).

[334] Young, R.C., "The algebra of many-valued quantities", *Math. Ann.* 104 (1931) 260–290.

[335] Yu, P.L., "A class of solutions for group decision problems", *Management Sci.* 19 (1973) 936–946.

[336] Yu, P.L., "Cone convexity, cone extreme points, and nondominated solutions in decision problems with multiobjectives", *J. Optim. Theory Appl.* 14 (1974) 319–377.

[337] Yu, P.L., Leitmann, G., "Compromise solutions, domination structures, and Salukvadze's solution", *J. Optim. Theory Appl.* 13 (1974) 362–378.

[338] Yu, P.L., Leitmann, G., "Nondominated decisions and cone convexity in dynamic multicriteria decision problems", *J. Optim. Theory Appl.* 14 (1974) 573–584.

[339] Yu, P.L., Zeleny, M., "The set of all nondominated solutions in linear cases and a multicriteria simplex method", *J. Math. Anal. Appl.* 49 (1975) 430–468.

[340] Zhuang, D., "Regularity and maximality properties of set-valued structures in optimization", Dissertation, Dalhousie University (Halifax, 1989).

[341] Zowe, J., "Subdifferentiability of convex functions with values in an ordered vector space", *Math. Scand.* 34 (1974) 69–83.

[342] Zowe, J., "Linear maps majorized by a sublinear map", *Arch. Math. (Basel)* 26 (1975) 637–645.

[343] Zowe, J., "A duality theorem for a convex programming problem in order complete vector lattices", *J. Math. Anal. Appl.* 50 (1975) 273–287.

[344] Zowe, J., "Konvexe Funktionen und konvexe Dualitätstheorie in geordneten Vektorräumen", Habilitationsschrift, University of Würzburg (Würzburg, 1976).

[345] Zowe, J., Kurcyusz, S., "Regularity and stability for the mathematical programming problem in Banach spaces", *Appl. Math. Optim.* 5 (1979) 49–62.

List of Symbols

X	real (topological) linear space	3, 24
0_X	zero element in X	4
X'	algebraic dual space of X	4
X^*	topological dual space of X	27
$S+T$	algebraic sum of two sets S and T	4
$S-T$	algebraic difference of two sets S and T	4
λS	$\lambda \in \mathbb{R}$, S nonempty set	4
$\mathrm{co}(S)$	convex hull of a set S	6
$\mathrm{cor}(S)$	algebraic interior (core) of a set S	6
$\mathrm{int}(S)$	interior of a set S	22
$\mathrm{lin}(S)$	algebraic closure of a set S	7
$\mathrm{cl}(S)$	closure of a set S	22
C	cone	8
$\mathrm{cone}(S)$	cone generated by a set S	12
$\leq$	partial ordering on a real linear space	13
$\leq_C$	partial ordering induced by a convex cone C	14
$[x,y]$	order interval between x and y	14
$C_{X'}$	dual cone for C_X	17
$C_{X'}^{\#}$	quasi-interior of the dual cone for C_X	17
$(x_i)_{i \in I}$	net	22
$\|\|\!\cdot\!\|\|$	vectorial norm	25
$\|\cdot\|$	norm	26
$(X, \|\cdot\|)$	normed space	26
$\langle .,. \rangle$	inner product	26
$(X, \langle .,. \rangle)$	Hilbert space	26
$\sigma(X, Y)$	weak topology on X generated by Y	27
$\sigma(X, X^*)$	weak topology	27
$\sigma(X^*, X)$	weak* topology	27

l_p	sequence space	32
l_∞	sequence space	33
$C(\Omega)$	space of continuous functions	33
$M(\Omega)$	space of bounded Radon measures	34
$L_p(\Omega)$	space of p-th power Lebesgue-integrable functions	34
$L_\infty(\Omega)$	space of essentially bounded functions	35
$\mathcal{D}$	space of functions with compact support in $\mathbb{R}^n$ having derivatives of all orders	35
$B(X,Y)$	space of bounded linear maps between X and Y	38
T^*	adjoint of a linear map T	38
$\text{epi}(f)$	epigraph of a map f	41
$f'(\bar{x})$	directional derivative, Gâteaux derivative, directional variation or Fréchet derivative of f at $\bar{x}$	46, 46, 47, 48
$\partial f(\bar{x})$	subdifferential of a map f at $\bar{x}$	52
$D_c F(\bar{x}, \bar{y})$	contingent derivative of F at $(\bar{x}, \bar{y})$	380
$DF(\bar{x}, \bar{y})$	contingent epiderivative of F at $(\bar{x}, \bar{y})$	381
$D_g F(\bar{x}, \bar{y})$	generalized contingent epiderivative of F at $(\bar{x}, \bar{y})$	391
$T(S, \bar{x})$	contingent cone to a set S at $\bar{x}$	91
$L(S, \bar{x})$	linearizing cone to a set S at $\bar{x}$	97
S_x	section of a set S	149
$W_{1,\infty}^m([t_0, t_1])$	special function space	246
$L_2([0, t_1], Z)$	special function space	270

Index

absolutely convex set 6
abstract complementary problem 105
abstract linear optimization problem 200, 206
abstract optimization problem 105, 162, 181, 192
adjoint equation 249, 258
adjoint map 38
affine linear map 43
algebraic boundary 7
algebraic closure 7
algebraic difference 4
algebraic dual space 4
algebraically bounded set 7
algebraically closed set 7
algebraically open set 7
alternation theorem 220, 235
antenna optimization 342
antisymmetric partial ordering 13
auxiliary problems 292
auxiliary programs 292

balanced set 5
Banach space 26
base 9, 12, 25, 62
best approximation 85
bicriterial optimization problem 314, 332
binary relation 13
bipolar theorem 82
bounded set 24

boundedly order complete topological linear space 31

Chebyshev vector approximation problem
    linear 227
    nonlinear 220
closed set 22
closure 22
cluster point of a net 23
compact set 23
complete set 24
compromise models 292
concave map 40
cone 8
cone-convex set-valued map 375, 385, 397
constraint set 162
contingent cone 91
contingent derivative 380
contingent epiderivative 381, 387
    existence theorem 383
    generalized 391
        existence theorem 393
continuous map 23
convergence of a net 22
convex hull 6
convex map 40
convex set 5
convex-like map 44
cooperative n player game 105, 243

core 6
C-quasiconvex map 175, 427

Daniell ordering cone 31
differentiably C_1-C_2-quasiconvex 178
differentiably C-quasiconvex 178
directed set 22
directional derivative 46, 387
directional variation 47
directionally differentiable 46
discrete optimization problem 336
dual cone 17
dual partial ordering 17
dual problem 189, 193, 229
dual set 190
duality theorem 191, 230
    converse 191
    strong converse 199, 231
    weak 193

Edgeworth-Pareto optimal point 284, 304
    essentially 290
    improperly 287
    necessary conditions 302
    properly 287
    sufficient conditions 293, 302
    strongly 289
    weakly 286, 305
Edgeworth-Pareto optimal solution 106
efficient solution 284
    essentially 290
    properly 287
    strongly 289
    weakly 286
Eidelheit's separation theorem 74
epigraph 41, 376

far field pattern 343
FDDI optimization 349
F.-John conditions 168
Fréchet derivative 48
Fréchet differentiable 48

Gâteaux derivative 46
Gâteaux differentiable 46
generated cone 12
goal programming 311
Graef-Younes method 337
graph 379

Hahn-Banach theorem
    basic version 68
    convex version 70
    extension version 69
    generalized basic version 71
    sandwich version 68
Hamiltonian map 258
Hamilton-Jacobi-Bellmann equations 266
Hausdorff space 23
Hilbert space 26

inductively ordered from above 62
inner product 26
interior 22
    element 22

James theorem 82, 83

Karush-Kuhn-Tucker conditions 168
KNY partial ordering 375
Kolmogorov condition,
    generalized 216
Krein-Rutman theorem 88, 89
Kurcyusz-Robinson-Zowe
    regularity assumption,
    generalized 423

Lagrange multiplier rule 432
    generalized 166, 168, 182, 418
Lagrangian map 168
least upper bound property 71
linear manifold 62
linear map 37
linearizing cone 97
linearly accessible element 7
locally convex space 25
locally convex toplogical
    linear space 25
lower bound 62
Lyusternik theorem 96

maximal element 62, 103
    properly 108
    strongly 107
    weakly 109
mean waiting time 352
metric 23
    space 23
metrizable topological space 23
mild solution 271
minimal element 62, 103
    almost properly 137
    properly 108
    strongly 107
    weakly 109
minimal solution 106, 162, 284, 375
    almost properly 228
    essentially 290
    local 177
    local weakly 177
    properly 108, 287
    strongly 289
    weakly 162, 286
minimizer 372
    strong 412
    weak 410, 432

Minkowski functional 29, 72, 118, 128
monotonically increasing
    functional 115
    strictly 116
    strongly 116
multiobjective optimization
    problem,
    general 172, 283
    linear 301, 319, 325
    nonconvex 304
    nonlinear 342

neighborhood 22, 23
net 22
nondominated point 284
norm 26
normal ordering cone 28
normal problem 195
normed space 26

objective map 162
one element 107
open set 22
optimal control 244
    weakly 245
optimization problems
    in chemical engineering 357, 363
order interval 14
order topology 28
ordering cone 14

partial ordering 13
partially ordered linear space 13
playable $(n+2)$-tuple 244
    optimal 244
    weakly optimal 244
pointed cone 8
Polak method,

modified 314
Pontryagin maximum principle 248, 260
   local 249, 258
primal problem 189, 193
primal set 190
proximinal set 85
pseudoconvex function 185
pseudoconvex map 179

quasi-complete topological linear space 24
quasiconvex function 185
quasiconvex map 174
quasi-interior 17

radiation efficiency 343
radiation pattern 343
real linear space 3
reference point 323
reference point approximation method,
   bicriterial case 332
   general case 324
reflexive normed space 27, 86, 153
regularity assumption 168
representation condition 222
reproducing cone 9, 11

second algebraic dual space 4
section 149
seminorm 26
separable topological space 22
separated topological space 23
separation theorem 75, 76, 81
   basic version 72
   for closed convex cones 79
sequence space 32
set optimization problem 371, 409
   constrained 417

   convex 412
set-valued map 371, 379
simultaneous approximation problem 213
single-valued optimization problem 373
Slater condition
   generalized 197, 423
space of
   bounded Radon measures 34
   continuous functions 33
   p-th power Lebesgue-integrable functions 34
stable problem 196
starshaped set 5
step method (STEM),
   modified 320
strictly positive homogeneous map 392
subadditive map 392
subdifferential 52, 398
subgradient 52, 398, 414
   weak 403, 415
sublinear functional 63
sublinear map 63, 385, 392

tangent 91
topological dual space 27
topological linear space 24
topological space 22
topology 21
   finer 22
   weak 27
   weak* 27
totally ordered set 61
transversality condition 249, 258
trivial cone 9
tunneling technique 316

Index 465

upper bound 62
upper semicontinuous 403

vector approximation problem
    105, 213
vectorial norm 25

weakly lower semicontinuous
    functional 85
Weierstraß theorem 82
weighted sum approach 292
weigthed Chebyshev approxima-
    tion problem 304
weighted Chebyshev norm 304,
    324

zero element 4, 107
Zorn's lemma 62

Druck: betz-druck GmbH, D-64291 Darmstadt
Verarbeitung: Buchbinderei Schäffer, D-67269 Grünstadt